Marine Bioactive Molecules for Biomedical and Pharmacotherapeutic Applications

Pallaval Veera Bramhachari •
Chanda Vikrant Berde
Editors

Marine Bioactive Molecules for Biomedical and Pharmacotherapeutic Applications

Springer

Editors
Pallaval Veera Bramhachari
Department of Biotechnology
Krishna University
Machilipatnam, Andhra Pradesh, India

Chanda Vikrant Berde
Ocean and Atmospheric Sciences
Goa University
Panaji, Goa, India

ISBN 978-981-99-6769-8 ISBN 978-981-99-6770-4 (eBook)
https://doi.org/10.1007/978-981-99-6770-4

This Springer imprint is published by the registered company Springer Nature Singapore Pte Ltd.
The registered company address is: 152 Beach Road, #21-01/04 Gateway East, Singapore 189721, Singapore

Paper in this product is recyclable.

Foreword

Marine natural products are important sources of biologically active agents, and various bioactive compounds have been extracted from marine organisms like tunicates, sponges, soft corals, and molluscs. The diversity of marine habitats and unique environmental conditions of the sea offer mostly untapped sources of potential drugs with superior chemical novelty. These biologically active compounds have been reported to modulate various biological activities and have anti-inflammatory, antifungal, and anticancer effects.

Natural marine products are increasingly targeted for studies of structure–activity relationships, as templates for molecular modeling and also for their synthesis and that of their analogs. With the emergence of new viruses and the growing resistance of infectious bacterial strains, it is important to keep looking to these alternative sources for new drugs. This can be done by studying the chemical profiles of the metabolites produced by these organisms, the mechanisms by which they are biosynthesized, and discovering the "whys" of their production. Very little work has been done using the chemical ecology of marine organisms in their environment. Therefore, recent research in marine biology has focused on searching for pharmacologically active compounds, but these efforts have been hindered by a lack of understanding of the ecology of these complex systems.

Marine Bioactive Molecules for Biomedical and Pharmacotherapeutic Applications, a volume published by Springer Nature, is important, and I strongly believe that it will attract readers working in the field. This volume has 14 chapters contributed by several competent academicians and scientists working on microbiome research throughout the world. I congratulate the editors of the book Dr. P. Veera Bramhachari and Dr. Chanda V. Berde for bringing out this volume with excellent contributions from scientists working on the microbiome and its applications in understanding microbiome interactions in agriculture and the environment.

Krishna University
Machilipatnam, Andhra Pradesh, India

Prof. G. Gnana Mani

Preface

Marine organisms produce biomolecules in response to the extreme physico-chemical environments in the marine ecosystems to which the marine organisms are exposed. These "biomolecules" are majorly secondary metabolites, chemicals that are not needed by the organism for basic or primary metabolic processes. These compounds help defend against predators, attract or inhibit other organisms from settling or growing on them, and provide chemical cues to synchronize reproduction among organisms that expel their eggs and sperm into the water. The mechanisms by which they prevent encroachment or predation interact with the same or similar enzymes and receptors that are involved in human disease processes. The biological diversity of the ocean offers great promise as a source of drugs for the future. The marine environment became a focus of natural products drug discovery research because of its relatively unexplored biodiversity compared to terrestrial environments.

Marine-derived bioactive compounds have been reported with different biological functions such as anticancer, antiviral, antiplasmodial, and antidiabetic properties along with other pharmaceutical activities like inhibition of the cell cycle, inhibition of kinase and phosphatase, antioxidant, neutritogenic, and anti-inflammatory effects. There is a continual need for new therapeutic agents, especially to treat a large variety of diseases for which there are no effective therapies. Many forms of cancer, viral and fungal infections, inflammatory diseases, and neurodegenerative diseases cannot be treated successfully. Development of resistance of pathogenic microorganisms to antibiotics and cancer cells to antitumor drugs requires the compensatory generation of new drugs. The pathogens responsible for dreadful infectious diseases have developed resistance to most available drugs. Thus, intense research and exploration of untouched resources should be the foresight. A unique source in the world's oceans, the deep ocean and geothermal vents, is now becoming the focus of considerable microbiological interest. Microbiological studies of the deep sea environment have shown the presence of obligate barophiles, which have been shown to have inhibitory activity toward colon cancer. This book is thus an effort to focus on the biomedical applications of marine-derived molecules in corroboration with the need of the day.

Compounds produced by marine microbes, macroalgae, sponges, and corals which grow in different global environments, including extreme environments, should be prioritized. Nonetheless a considerable increase in the number of marine drugs approved for clinical use has occurred in the past few decades, which may be attributed to increasing research on marine compounds in laboratories across the world. The large-scale collection, screening, and discovery of novel marine natural products have propelled new chemical entities into the clinic for the treatment of pain, cancer, and other disease states. Here, we review the history of drug discovery from marine natural products, and by describing selected examples, we examine the factors that contribute to new discoveries and the difficulties associated with translating marine-derived compounds into clinical trials. Providing an outlook into the future, we also examine the advances that may further expand the promise of drugs from the sea.

The objective of this book on *Marine Bioactive Molecules for Biomedical and Pharmacotherapeutic Applications* is to keep the readers informed about recent developments in marine microbial research and the challenges for researchers in the upcoming years. It is very important to exploit the resources from the vast marine ecosystems, especially the undiscovered and untouched econiches for the isolation of microorganisms as a tool for the production of biomolecules such as secondary bioactive compounds with medical applications. The volume consists of 14 chapters contributed by the author(s) with vast experience in the field of marine microbiology. The chapters cover the wide applications of marine biomolecules obtained from different sources and ecosystems. The main focus of the book is to bring to the forefront the enormous potential of the marine microbiome in medicine and pharmaceuticals that is waiting to be explored and exploited. This book will serve as a reference and guideline for students, researchers, and scientists working in marine microbiology. I express my sincere gratitude to all the contributors for their valuable contributions and support throughout.

Machilipatnam, Andhra Pradesh, India — Pallaval Veera Bramhachari
Panaji, Goa, India — Chanda Vikrant Berde

Acknowledgments

Our sincere thanks are extended to all the academicians and scientists who have contributed chapters and happily agreed to share their work on *Marine Bioactive Molecules for Biomedical and Pharmacotherapeutic Applications* in this volume.

This book is a stunning reflection of the seriousness with which several scientific minds are dedicated to the research community. We are extremely thankful to the contributors for paying continuous attention to my requests and bestowing faith and confidence in my competencies and capabilities. We shall always remain highly grateful to all contributors forever. These words cannot justify the worthiness of their quintessential efforts. We appreciate the excellent work of the authors and coauthors who were invited to contribute chapters to this book. The credit for making this book a reality goes to them. The editor and the review team of the book especially appreciate sharing expertise with the contributors. Each chapter is pretty informative and written as stand-alone contributions from several research institutes so that the reader can begin reading anywhere in the book depending upon his/her interests and needs.

At the same time, we also express our deepest gratitude to our family members for their kind support, which has prompted us to complete this assignment on time. We are also thankful to Krishna University and Goa University administrative officials, and my colleagues in the Department of Biotechnology, Krishna University, and School of Earth, Ocean and Atmospheric Sciences, Goa University, for their intellectual support. We are equally thankful to the Springer Nature Publishing Group for their full cooperation during the peer review and production of the volume.

We are thankful to our beloved teachers and mentors for their constant support and motivation at all stages of my progress.

Pallaval Veera Bramhachari
Chanda Vikrant Berde

About the Book

This book illustrates the importance and significance of marine ecosystem, which is still an untapped reservoir of numerous biologically active compounds and has considerable potential in biomedicine. With the goal of increasing the availability and chemical diversity of functional marine ingredients, much research has been developed using biotechnological tools to discover and produce new bioactive compounds. Over the last two decades, marine natural product research has observed a paradigm shift from terrestrial areas to oceans. Marine natural products provide a rich source of chemically diverse compounds that have significant potential to be developed as novel types of therapeutic agents. Strikingly, marine organisms produce a diverse group of secondary metabolites and have a broad spectrum of natural bioactivities for beneficial health effect, therefore gaining attention in biotechnology for the isolation of bioactive metabolites. These organisms are considered an excellent source of bioactive compounds that have a broad range of biological activities including antibacterial, antifungal, anticancer, and antiviral potential. The chapters provide current information pertaining to the myriad of structural motifs undiscovered from the marine ecosystem, and these molecules can serve as new potent antioxidant, immunomodulatory, anti-inflammatory, and anticancer agents. The chapters also describe recent advancements in the secondary metabolites from marine resources along with their biomedical applications. This underlines the pending potential for exploring new marine sources of bioactive compounds and the challenges related to their investigation. Finally, the book also elucidates a comprehensive yet representative description of a large number of challenges associated with the bioactivities and potentialities of marine bioactive metabolites for future nutraceutical, biomedical, and therapeutic applications.

Contents

Editors and Contributors

About the Editors

Pallaval Veera Bramhachari is currently an Associate Professor in the Department of Biosciences and Biotechnology at Krishna University, Machilipatnam, AP, India. He earlier served as a postdoctoral research scientist at the Bacterial Pathogenesis Laboratory, Queensland Institute of Medical Research (QIMR-Berghofer Research Center), Brisbane, Queensland, Australia, and a DBT postdoctoral fellow at the Department of Microbiology and Cell Biology, Indian Institute of Science (IISc), Bangalore. He completed his Ph.D. in Microbiology from Goa University, India.

Dr. Bramhachari's major research interests include the investigation of applied research in microbiology and cell biology, including the biochemical mechanism of bacterial biofilm formations, extracellular microbial polysaccharides, and the structure-function relationship of bacterial biosurfactants (glycolipids and rhamnolipids), as well as pathogenic and virulence characteristics of bacteria in basic and translational cancer research. He has published over 160 research articles in Scopus-indexed journals, authored and coauthored 8 international books, contributed to invited book chapters in prestigious peer-reviewed international publications, and presented 55 abstracts at various national and international conferences. He has been a reviewer and editorial board member of several international journals.

He is a member of numerous international scientific societies and organizations, such as the Indian Science

Congress, Society of Biotechnologists of India, the Society of Biological Chemists, India, Andhra Pradesh State Biodiversity Board, and Andhra Pradesh State Fisheries Board. He completed two major research projects for the SERB, DST NRDMS, and the Government of India. He was awarded a Travel Scholarship from QIMR in 2007, Australia, for attending the fourth Indo-Australian Biotechnology Conference in Brisbane, Australia. He was awarded a Young Scientist Travel Fellowship (2007) from the DST, Government of India, for attending the XVII Lancefield International Symposium at Porto Heli, Greece (2008). He was conferred with various prestigious awards: Science Education Research Board (SERB), Government of India-Young Scientist Award (2011) for a research project and nominated as Associate Fellow of Andhra Pradesh Academy of Sciences (APAS) (2016), Master Trainer: Andhra Pradesh English Communications Skills Project, British Council and APSCHE (2017), Andhra Pradesh State Best Scientist Award (2017), Dr. V. Ramalingaswamy Memorial Award for Biomedical Sciences (2019), KAPL Award (Industrial Biotechnology), SBTI Annual Award (2022), Dr. APJ Abdul Kalam Teacher Excellence Award (2022). He also obtained two Indian patents in 2017. He has more than 19 years of teaching and research experience in molecular microbiology, virology and immunology, Intellectual Property Rights, Patent Law and Entrepreneurship in Biotechnology, bioprocess engineering, and microbial and environmental biotechnology at the university level.

Chanda Vikrant Berde is currently Assistant Professor in Marine Microbiology program of School of Earth, Ocean and Atmospheric Sciences, Goa University, Goa, India. She has earlier served as Assistant Professor in the Postgraduate programs of the Departments of Microbiology and Biotechnology at Gogate Jogalekar College, Ratnagiri, Maharashtra, India (2010–2020); Scientist (Microbiology) at Vidya Pratishthan's School of Biotechnology, Baramati, Maharashtra, India (2005–2010) as well as Assistant Professor, Department of Biotechnology, Dhempe College of Arts and Science, Goa, India (2004–2005). She has completed her doctoral and postdoctoral studies from the Department of Microbiology, Goa University, Goa. She has more than 18 years of teaching experience in General Microbiology, Genetic Engineering, Biochemistry, Bioprocess and Biotechnology, IPR, Archaea, Extremophilic Microbiology, Estuarine Microbiology, Nanobiotechnology, etc. Her research experience spans over 20 years, and the areas of her research interests include bioactive biomolecules, endophytic microorganisms, PGPR, microbial metabolites of industrial importance, microbial remediation, and media formulations. She has guided more than 80 master's research projects in subjects of Microbiology, Biotechnology, and Marine Microbiology. She is a recognized guide for Ph.D. in Marine Microbiology at Goa University with more than 50 research articles published in peer-reviewed national and international journals and authorship or co-authorship of numerous book chapters. She is the editor of three books published by reputed publishers. She has completed two research projects funded by DBT, Govt. of India, and University of Mumbai, to her credit. She is a member of many scientific societies and organizations, importantly Mangrove Society of India; Microbiologists Society, India; and Society for Environment and Biodiversity Conservation, Ratnagiri. She has actively participated in social awareness programs like mangrove conservation and plantation, COVID-19 awareness, AIDS awareness, and tree plantations.

Contributors

Pragati Abhyankar Department of Microbiology, Haribhai V. Desai College of Arts, Science and Commerce, Pune, India

Abullais Ansari Camp Education Society's Dr. Arvind Telang Senior College of Arts, Science and Commerce, Pune, Maharashtra, India

V. Anu Prasanna Department of Zoology, Yogi Vemana University, Kadapa, Andhra Pradesh, India

Ashok Bankar MES Abasaheb Garware College, Affiliated to Savitribai Phule Pune University, Pune, Maharashtra, India

Chanda Vikrant Berde Ocean and Atmospheric Sciences, Goa University, Panaji, Goa, India

Vikrant B. Berde Department of Zoology, Arts, Commerce & Science College, Lanja, Ratnagiri, Maharashtra, India

Prajakta P. Bhende Department of Biological Sciences, BITS Pilani, Sancoale, Goa, India

Judith M. Braganca Department of Biological Sciences, Birla Institute of Technology and Science, Pilani, K K Birla Goa Campus, Pilani, Goa, India

Thummala Chandrasekhar Department of Environmental Science, Yogi Vemana University, Kadapa, Andhra Pradesh, India

Anasuya Ganguly Department of Biological Sciences, BITS Pilani, Sancoale, Goa, India

Trupti Gokhale Department of Biotechnology, BITS Pilani, Dubai Campus, Dubai International Academic City, Dubai, United Arab Emirates

Saqib Hassan Department of Biotechnology, School of Bio and Chemical Engineering, Sathyabhama Institute of Science and Technology, Chennai, Tamil Nadu, India
American Society for Microbiology, Washington, DC, USA

Laxmi Jadhav MES Abasaheb Garware College, Affiliated to Savitribai Phule Pune University, Pune, Maharashtra, India

C. Jayanthy Department of Bioinformatics, School of Bio and Chemical Engineering, Sathyabama Institute of Science and Technology, Chennai, Tamil Nadu, India

Yogini Kanade Department of Microbiology, Haribhai V. Desai College of Arts, Science and Commerce, Pune, India

Hasita Karri Department of Biotechnology, Jawaharlal Nehru Technological University (JNTU), Kakinada, Andhra Pradesh, India

George Seghal Kiran Department of Food Science and Technology, School of Life Sciences, Pondicherry University, Puducherry, India

Harish Babu Kolla Department of Biotechnology, Vignan's Foundation for Science, Technology and Research (Deemed to be University), Guntur, Andhra Pradesh, India

Rohini Krishna Kota Centre for Innovation in Molecular and Pharmaceutical Science (CIMPS), Dr. Reddy's Institute of Life Sciences, University of Hyderabad Campus, Hyderabad, India

Madhubalaji Chegu Krishnamurthi Plant Cell Biotechnology Department, CSIR-Central Food Technological Research Institute, Mysore, Karnataka, India

Y. Santhosh Kumar Faculty of Allied Health Sciences, Chettinad Hospital and Research Institute, Chettinad Academy of Research and Education (Deemed to be University), Chennai, Tamil Nadu, India

Chintha Lalitha Microbiology Department, Dr. V.S. Krishna Government Degree College, Visakhapatnam, Andhra Pradesh, India

Devanabanda Mallaiah Department of Biotechnology and Bioinformatics, Yogi Vemana University, Kadapa, India

Kabilan Mani Center for Molecular Medicine & Therapeutics, PSG Institute of Medical Sciences and Research, Coimbatore, India

Vaibhav A. Mantri Applied Phycology & Biotechnology Division, CSIR-Central Salt and Marine Chemicals Research Institute, Bhavnagar, India
Academy of Scientific and Innovative Research (AcSIR), Ghaziabad, India

Doniya Elze Mathew Applied Phycology & Biotechnology Division, CSIR-Central Salt and Marine Chemicals Research Institute, Bhavnagar, India
Academy of Scientific and Innovative Research (AcSIR), Ghaziabad, India

Ramu Meenatchi Department of Biotechnology, College of Science and Humanities, SRM Institute of Science and Technology, Chennai, Tamil Nadu, India

Rajalakshmy Menon Department of Biotechnology, BITS Pilani, Dubai Campus, Dubai International Academic City, Dubai, United Arab Emirates

Devika N. Nagar Department of Biological Sciences, Birla Institute of Technology and Science, Pilani, K K Birla Goa Campus, Pilani, Goa, India

Smita Patil Camp Education Society's Dr. Arvind Telang Senior College of Arts, Science and Commerce, Pune, Maharashtra, India

Rajashree Patwardhan Department of Microbiology, Haribhai V. Desai College of Arts, Science and Commerce, Pune, India

G. B. Priyadharshini Department of Bioinformatics, School of Bio and Chemical Engineering, Sathyabama Institute of Science and Technology, Chennai, Tamil Nadu, India

Prakash Narayana Reddy Department of Microbiology, Dr. V.S. Krishna Government Degree College, Visakhapatnam, Andhra Pradesh, India

K. Riazunnisa Department of Biotechnology and Bioinformatics, Yogi Vemana University, Kadapa, Andhra Pradesh, India

Ashwini Sarje Camp Education Society's Dr. Arvind Telang Senior College of Arts, Science and Commerce, Pune, Maharashtra, India

Joseph Selvin Department of Microbiology, School of Life Sciences, Pondicherry University, Puducherry, India

Aesha Sharma Department of Biological Sciences, BITS Pilani, Sancoale, Goa, India

Pramod B. Shinde Natural Products & Green Chemistry Division, CSIR-Central Salt and Marine Chemicals Research Institute, Bhavnagar, India
Academy of Scientific and Innovative Research (AcSIR), Ghaziabad, India

Aastha Soni Applied Phycology & Biotechnology Division, CSIR-Central Salt and Marine Chemicals Research Institute, Bhavnagar, India

Neeru Sood Department of Biotechnology, BITS Pilani, Dubai Campus, Dubai International Academic City, Dubai, United Arab Emirates

G. V. Swarnalatha Department of Biochemistry, Rayalaseema University, Kurnool, Andhra Pradesh, India

Rincy Thomas Department of Biotechnology, BITS Pilani, Dubai Campus, Dubai International Academic City, Dubai, United Arab Emirates

Sonal Tiwari Plant Cell Biotechnology Department, CSIR-Central Food Technological Research Institute, Mysore, Karnataka, India

Pallaval Veera Bramhachari Department of Biotechnology, Krishna University, Machilipatnam, Andhra Pradesh, India

D. Vijaya Lakshmi Department of Microbiology, Yogi Vemana University, Kadapa, Andhra Pradesh, India

Abbreviations

(AChE)	Acetylcholinesterase activity
(APP)	Amyloid precursor protein
(BACE1)	Beta-amyloid cleaving enzyme-1
(CNS)	Central nervous system
(GABA)	γ-amino butyric acid
(NO)	Nitric oxide
5-FU	5-Fluorouracil
AATP	Adenine arabinoside triphosphate
ACE	Angiotensin-converting enzyme
AGE	Advanced glycation end products
AgNPs	Silver nanoparticles
AHL	Acyl-homoserine lactone
AIDS	Acquired immunodeficiency syndrome
ALCL	Anaplastic large cell lymphoma
AMP	Adenosine mono phosphate
antiSMASHSHell	Antibiotics and secondary metabolite analysis
AP-1	Activator protein 1
ART	Antiretroviral therapy
ASCT	Autologous stem cell transplantation
ATCC	American Type Culture Collection
AuNPs	Gold nanoparticles
AuNWA	Au nanowire arrays
Avian RT	Avian retrovirus
BAC	Bacterial artificial chromosome
Balb/3T3	Murine fibroblasts
BHA	Beta-hydroxy acid
BJ Cells	Normal human skin fibroblast
BLAST	Basic local alignment search tool
bp	Base pair
C/N	Carbon/nitrogen
C2 position	Position of second carbon
CaO	Calcium oxide
CCBH	Culture collection of hospital acquired bacteria

CD	Circular dichroism
CD	Cluster of differentiation
CDK	Cyclin-dependent kinases
cDNA	Complementary DNA
CdS	Cadmium sulfide
CdTe	Cadmium telluride
CDW	Cell dry weight
CEA	Carcinoembryonic antigen
CIFE	Central Institute of Fisheries Education
CMC	Carboxymethyl cellulose
CNC	Cellulose nanocrystals
Cos-7	Fibroblast cell line
C-PCY	C-Phycocyanin
CREB	cAMP response element-binding protein
CRG	Carrageenan
CS	Chondroitin sulfate
CU-CPT	*N*-[(3-chloro-6-fluorobenzo[*b*]thien-2-yl)carbonyl]-D-phenylalanine
CuNPs	Copper nanoparticles
CuO	Copper oxide
CYP3A4	Cytochrome P450 3A4
DAPK	Death associated protein kinase
DDS	Drug delivery systems
DGC	Diffusion growth chamber
DHA	Docosahexaenoic acid
DMSO	Dimethyl sulfoxide
DNA	Deoxyribonucleic acid
DPPH	1,1-Diphenyl-2-picryl-hydrazyl
ECM	Extracellular matrix
EDC	1-Ethyl-3-(3-dimethylaminopropyl) carbodiimide
eDNA	Environmental deoxynucleic acid
ELISA	Enzyme-linked immunosorbent assay
ELP	Eukaryotic-like protein
EMBL	European Molecular Biology Laboratory
EPA	Eicosapentaenoic acid
EPR	Electron paramagnetic resonance spectroscopy
EPS	Exopolysaccharides
FACS	Fluorescent-activated cell sorting
FAD	Flavin adenine dinucleotide
FADD	Fas associated death domain
FADH2	Reduced flavin adenine dinucleotide
FASTA	Fast alignment
FISH	Fluorescence in situ hybridization
FTase	Farnesyl transferase
GAGs	Glycosaminoglycans

GFP	Green fluorescent protein
GG	Arrangement of α-L-guluronic acid monomer
GGF	Glial growth factor
GM	Heteropolymeric arrangement of α-L-guluronic acid and β-D-mannuronic acid monomer
Gp120	Glycoprotein 120
Gp160	Glycoprotein 160
Gp41	Glycoprotein 41
GTase	Geranyl transferase
HA	Hyaluronic acid
HAB	Harmful algal bloom
HaCaT	Human keratinocyte cell line
HEK001	Normal human epidermal keratinocytes transformed with HPV-16
HeLa	Human cervical cancer cell line
HHL	*N*-hexanoyl-homoserine lactone
HIV	Human immunodeficiency virus
HMA	High microbial abundance
hMSCs	Human mesenchymal stromal cells
HPLC	High-pressure liquid chromatography
HSL	Homoserine lactone
HSP90	Heat shock protein 90
HSV	Herpes simplex virus
HUVECs	Human umbilical vein endothelial cells
i.e.	That is
IAV	Influenza A viruses
IC50	Half maximal inhibitory concentration
ILs	Interleukins
IPA	Indole-3-pyruvic acid
IκBα	NFκB inhibitor α
kb	Kilobase
kDa	Kilodaltons
KS	Ketoacyl synthase
L	Liter
L929	Mouse fibroblast cell line
LB	Luria-Bertani broth
M2G	Mycosporine-2-glycine
Ma	Million years ago
MBC	Minimum bactericidal concentration
MC3T3-E1	Clonal murine cell line of immature osteoblast
MDR	Multidrug resistance
METREX	Metabolite regulated expression
MG-63	Human osteoblast-like cells
MgO	Magnesium oxide
MIC	Minimum inhibitory concentration

ML	Milliliters
MLH	MutL Homolog
MM	Arrangement of β-D-mannuronic acid monomer
MMA	Mycosporine-like amino acid
MMAE	Monomethyl auristatin
MMP	Matrix metalloproteinase
MP	Marine polysaccharides
MPF	Mitosis promoting factor
mRNA	Messenger ribonucleic acid
MRSA	Methicillin-resistant *Staphylococcus aureus*
MSCs	Mesenchymal stromal cells
MSH	Melanocyte-stimulating hormone
MSSA	Methicillin-sensitive *Staphylococcus aureus*
MTCC	Microbial type culture collection
MTT	3-(4,5-Dimethylthiazolyl-2)-2,5-diphenyltetrazolium bromide
NADP	Nicotinamide adenine dinucleotide phosphate
NADPH	Reduced nicotinamide adenine dinucleotide phosphate
NB	Nutrient broth
NB-G	Nutrient broth with glucose
NB-GY	Nutrient broth with glycerol
NB-X	Nutrient broth with xylose
NCBI	National Center for Biotechnology Information
NER	Nucleotide excision repair
NF-kB	Nuclear k-light chain enhancer of activated B cells
NGS	Next-generation sequencing
NHDF	Human dermal fibroblasts
NIH-3T3	Mouse embryonic fibroblasts
NiNPs	Nickel nanoparticles
nm	Nanometer
NMF	Natural moisturizing factor
NMR	Nuclear magnetic resonance
NP	Nanoparticles
NPC	Nasopharyngeal carcinoma
NRPS	Non-ribosomal peptide synthetases
NRU	Neutral red uptake
NS	Nanostructures
NSCLC	Non-small cell lung cancer
ORF	Open reading frame
OSMAC	One strain many compounds
P(3HB/4HB)	Poly 3-hydroxybutyrate-4-hydroxybutyrate
P(3HB-*co*-3HV)	Poly 3-hydroxybutyrate-*co*-3-hydroxy valerate
P(3HB-*co*-4HB)	Poly 3-hydroxybutyrate-*co*-4-hydroxybutyrate
P(3HHx-*co*-3HO)	Poly (3 hydroxyhexanoate-*co*-3 hydroxy octanoate)
P3HB	Poly 3-hydroxybutyric acid

P4HB	Poly 4-hydroxybutyric acid
PBMCs	Peripheral blood mononuclear cells
PCD	Programmed cell death
PCL	Poly-ε-caprolactone
PCR	Polymerase chain reaction
PCR-DGGE	Polymerase chain reaction—denaturing gradient gel electrophoresis
PCY	Phycocyanin
PDLLA	poly-DL-lactide
PDX	Polydioxanone
PEA	Phenylethylamine alkaloid
PHA	Polyhydroxyalkanoates
PHBV	Poly 3-hydroxybutyrate-*co*-hydroxyvalerate
PKA	Protein kinase
PKC	Protein kinase C
PKS	Polyketide synthase
PLLA	Poly-L-lactic acid
PRISM	PRediction Informatics for Secondary Metabolomes
PSI-Blast	Position Specific Iterative Basic Local Alignment Search Tool
PSII	Photosystem II
PUA	Polyunsaturated aldehyde
PUFA	Polyunsaturated fatty acid
PVA	Protodeoxy violaceinic acid
QDs	Quantum dot
RB	Retinoblastoma
RFLP	Restriction length fragment polymorphism
RGD	Arginyl-glycyl-aspartic acid
RNA	Ribonucleic acid
ROS	Reactive oxygen species
R-PCY II	R-Phycocyanin II
R-PCY	R-Phycocyanin
rRNA	Ribosomal ribonucleic acid
S.C	Subcutaneously
Saos-2	Human osteosarcoma cell line
SARS	Severe acute respiratory syndrome
SCG	Single-cell genome
SeNPs	Selenium nanoparticles
SH-SY5T	Human neuronal cells
SIGEX	Substrate-induced gene expression
SOLiD	Sequencing by oligonucleotide ligation and detection
spp	Species
T/C-28a2	Osteochondral cell lines
TEM	Transmission electron microscopy
THP-1	Human acute monocytic leukemia cells
TLR	Toll-like receptor

TNFα	Tumor necrosis factor α
UV	Ultraviolet
UV-A	Ultraviolet-A
UV-B	Ultraviolet-B
VEGF	Vascular endothelial growth factor
WHO	World Health Organization
ZnO	Zinc oxide
μg	Microgram

New Vistas and Frontiers of Marine Bioactive Molecules in Biomedical and Pharmacotherapeutic Applications

1

Pallaval Veera Bramhachari and Chanda Vikrant Berde

Abstract

There is an abundant supply of bioactive molecules in the marine environment's wide variety of creatures. Although these natural bioactive metabolites have significant biological activities, there are certain restrictions on how they can be used in biomedical research. Adaptation mechanisms of marine microbiomes must be significantly distinct from terrestrial microbiomes due to the unique challenges of marine life. An essential strategy for marine organisms to adapt to their marine environment is the creation of specialized secondary metabolites. The creation of physiologically active secondary metabolites is a key mechanism of adaptation. The biological actions of these metabolites make them promising candidates for future human therapeutics. Several investigations have shown many natural compounds isolated from marine microbiota, and other higher marine species have microbial origins. However, many obstacles are in the way of fully tapping the therapeutic potential of marine species. This study aims to highlight the unique properties of biomolecules found in marine environments and offer context for recent research papers on their potential medical uses.

Keywords

Marine environment · Marine organisms · Bioactive metabolites · Adaption mechanism · Biomedical applications

P. Veera Bramhachari (✉)
Department of Biotechnology, Krishna University, Machilipatnam, Andhra Pradesh, India

C. V. Berde
School of Earth, Ocean and Atmospheric Sciences, Goa University, Taleigao Plateau, Goa, India
e-mail: chanda.berde@unigoa.ac.in

P. Veera Bramhachari, C. V. Berde (eds.), *Marine Bioactive Molecules for Biomedical and Pharmacotherapeutic Applications*,
https://doi.org/10.1007/978-981-99-6770-4_1

1.1 Introduction

Marine environments excite scientists due to their unique platforms, which have sparked a renewed interest in the hunt for novel biochemicals. Marine species' function in generating a new therapeutic benefit is evolving as an arable field, encouraged by the vastness and diversity of the marine environment. Numerous species have adapted to marine life because of the extreme conditions. The ability of marine creatures to produce chemicals with potential biotechnological applications has piqued the scientific community's curiosity. Their resilience to temperature, pH, salt, and pollutants makes them potential biocatalysts for novel and sustainable manufacturing methods that can be implemented via biotechnology.

The discovery of bioactive molecules from marine species as possible therapeutic agents against various diseases, advancements in aquatic food product technology, and the creation of high-value goods for health management all point to marine research as a promising new frontier in medicine. Bioactive components and functional food ingredients obtained from the sea are gaining attention as a possible source of health benefits. An undiscovered pool of bioactive chemicals with great pharmacological and biological application lies in the Indian subcontinent's tremendous marine and coastal ecosystem biodiversity. Exploring bioactive chemicals and developing biologically usable goods has benefits as health products due to the underutilization of certain groupings of marine species. Researchers have an unquenchable thirst for finding novel pharmacological leads in aquatic environments. Due to their unique chemical properties and broad spectrum of biological activity, marine natural products have been widely identified as a valuable resource for drug discovery and pharmacology.

Marine microbiota and fauna are valuable sources of bioactive compounds for the food and pharmaceutical, and biomedicine industries. Bioactive compounds can be isolated from macroalgae (seaweeds), microalgae, echinoderms, crustaceans (for example, crab, shrimp, crayfish and lobster), cephalopods (such as cuttlefish, octopus and squid), mollusks (including scallop, mussel, clam, snail, oyster, abalone, and conch) and fish. Marine resources can also be a rich source of biological agents like tunicates, sponges, algae, sea whips, and more.

Since marine natural products can exhibit significant biological activity tenfold higher than terrestrial animals, they have progressively attracted major scientific interest. Marine natural products have repeatedly demonstrated their biomedical potential through studies of their biological functions, such as their antithrombotic, anticoagulant, antiinflammatory, anti-hypertension, anti-diabetic, heart-protecting, and neuro-protecting effects (Karthikeyan et al. 2022). There is a growing body of research suggesting that the chemical compounds found in marine natural products, such as polysaccharides, polyphenols, sterols, carotenoids, diterpenoids, alkaloids, amino sulfonate, marine toxins, and macrolide-type compounds, have great potential against neurodegenerative diseases. More chances for marine compounds are created due to the discovery of additional pharmacological targets, encouraging more research groups to study new medication candidates. As a result, marine resources provide us with a vast pool of potential medications for further study. Research into the molecular mechanisms, toxicity, and side reactions of active ingredients and the advancement of in vivo and in vitro studies is crucial to developing novel drugs; we must describe their health implications (da Silva 2021). Metagenomic research has shown that marine extremophile prokaryotes are a rich resource for novel genes and, by extension, novel bioproducts like enzymes and other bioactive metabolites (Dalmaso et al. 2015).

Over 36,000 chemicals have been extracted from marine micro- and macro-organisms. Carbohydrates, pigments, polyphenols, peptides, proteins, essential fatty acids, vitamins, and minerals are some of the most often isolated and studied bioactive substances in marine environments. Therefore, marine compounds can be used as therapeutic components in functional food, nutraceuticals, dietary supplements, prebiotics, and pharmaceuticals (Suleria et al. 2016; Barreca et al. 2020) due to their antioxidant, antithrombotic, antihypertensive, anticoagulant, antiinflammatory, antiproliferative, and cardio-protective activity. Synergistic or improved therapeutic effects have been demonstrated for several combinations of marine-based bioactive chemicals or extracts (Ebrahimi et al. 2023), including antioxidant, antiinflammatory, anti-obesity, and chemo-preventative actions.

Bifunctional secondary metabolites, which evolved in marine species due to harsh environmental conditions, are abundant. They differ greatly from terrestrial species in structure and function. This is why there has been a recent surge in interest in finding ways to utilize bioactive lead compounds derived from marine species in the service of human health preservation. Therefore novel methods for isolating and testing marine-based compounds have significantly widened the scope of medication development.

Bioactive compounds isolated from natural sources that can effectively act on molecular targets implicated in various disorders are currently the subject of intense research. The majority of sponges, mollusks, tunicates, coelenterates, and crustaceans, as well as seawater, fish, and algae, are cultured for the study of bacteria and fungi. It has been found that marine bacteria, fungi, and macro-organisms, including sponges, algae, and corals, develop unique marine secondary metabolites (MSMs) with specific and intricate chemical structures. Chemical structures with various health benefits have been abundant in marine environments. Antioxidant, anti-neuroinflammatory, cholinesterase inhibitory, and neuronal death-inhibiting properties are only some of how marine algae are currently being studied for their biological activities and neuroprotective effects. The potential therapeutic benefits of marine-derived bioactive chemicals have increased their interest in the application. Antibiotic, antiviral, neurodegenerative, anticancer, and antiinflammatory effects are pharmaceutically relevant bioactivities in marine natural products (MNPs). This review particularly emphasizes the potential benefits of marine bioactive chemicals and how they may help treat various ailments.

Marine microorganisms release important metabolites that control both internal and external physiological processes. The clinical relevance of these microbial compounds has attracted a lot of attention. Hopefully, these microbial metabolites can be used to treat various ailments, including cancer, inflammation, and neurological disorders (Khotimchenko et al. 2022). Proteasome inhibitors, therapeutic enzymes, bacteriocins, polyamines, and flavonoids are microbial metabolites with significant therapeutic potential. This review particularly emphasizes future research with new biologically active natural marine microbial compounds and highlights their mechanism of action with a focus on chemical potential.

1.2 Marine Bioactive Natural Compounds

Based on structural and functional differences, the bioactive substances can be broadly classified into the following classes: (1) amino acids; (2) phenolic compounds; (3) sterols; (4) carotenoids; (5) vitamins; (6) polysaccharides; and seven fatty acids. Non-flavonoid phenolic compounds include things like astragalus, hydrolyzable tannins, and phenolic acids, while flavonoid phenolic compounds include things like flavonoids, flavonols, flavanols, isoflavones, condensed tannins, and anthocyanins.

Some marine natural products have shown substantial neuroprotection functions, making them the most promising candidates for these substances. The ultimate goal of this work was to identify novel and effective drug compounds that are effective against Alzheimer's disease and other neurodegenerative diseases. So we set out to detail the potential impact of bioactive compounds derived from marine organisms, such as polysaccharides, carotenoids, polyphenols, sterols, and alkaloids, as drug candidates (Hu et al. 2023).

Polysaccharides, carotenoids, polyphenols, sterols, and alkaloids, all of which have biological and pharmacological properties, have been found by numerous researchers to be abundant in the marine environment in recent years. Half of all species on Earth are found in marine environments, which cover more than 70% of the planet (Hosseini et al. 2020). The chemical characteristics of marine bioactive substances distinguish them from terrestrial goods in terms of their biological activity (Bălaşa et al. 2020). Drugs that can treat human ailments are desperately needed; thus, researchers are examining these sources more than ever (Altmann 2017; Khotimchenko et al. 2022).

1.3 Bioprospection of Marine-Derived Bioactive Metabolites

Over the past few decades, there has been a dramatic surge in marine bioprospecting. Marine invertebrates have been the primary contributor thus far (Hu et al. 2015). Bioprospecting will help unravel the enigma of the bioactive metabolites from marine microbes (Sekurova et al. 2019). Bioprospecting focuses on cytostatic and cytotoxic chemicals from a pharmacological and biomedical standpoint. Because of their protective function in the natural environment, the medical industry's insatiable demand for new anticancer medications, the widespread availability of convenient in vitro test tools, and the financial backing from government organizations, these compounds have seen rapid development in recent years. More than half of the novel marine natural products discovered between 1985 and 2012 were chemicals with anticancer properties, according to research published in 2015. Next up are antibacterial chemicals (Hu et al. 2015). Antiinflammatory, anti-fungal, antiviral, and analgesic compounds are gaining ground. Marine natural products have been classified into chemical classes based on their molecular structures. These classes include polyketides, terpenes, steroidal or triterpene saponins, carbohydrates, aliphatic chemicals, amino acids, alkaloids, peptides, lipopeptides, and proteins. According to a 2015 study by Hu et al., peptides comprise the bulk of all bioactive substances.

Interestingly, many marine organisms are inaccessible for precise taxonomic and physiological characterizations since they cannot be cultivated under artificial circumstances. However, with modern, cutting-edge molecular tools, we can now detect and characterize microbiome communities without needing cultivation. The concept of genome mining greatly enhances the likelihood of discovering novel bioactive structures. Screening of functional gene clusters from uncultured microorganisms and communities of diverse species is made possible by metagenomic techniques, which rely on cloning and transforming these sequences into a suitable host-vector system and its production in the host cells. Understanding the production and physiological significance of the intriguing metabolites would require additional omics technologies such as proteome, transcriptome, and metabolome study (Lindequist 2016). Therefore, the pharmaceutical industry anticipates sourcing for the forthcoming generation of antimicrobial drugs from the marine environment.

1.4 Sustainable Exploitation of the Marine Environment

Creating a reliable method for finding new biomolecules in the ocean is a significant problem in marine biotechnology. Marine creatures are highly varied, making them an attractive bio-prospecting target. These species create unique biomolecules, including bioactive compounds and enzymes of medical and industrial significance. A recent investigation of proteins in the ocean discovered thousands of novel families with unknown functions, making microbes a particularly under-sampled and promising field. The distance between these components and their prospective industrial collaborators must be closed. Understanding how model organisms, such as bacteria, adjust their metabolite synthesis in response to environmental perturbations will be a major task over the coming decade. Understanding the function of bioactive chemicals in communities, in the struggle for limited resources, and the fight against predators and pathogens is essential if we are to reach our aim. The connection between biodiversity and chemodiversity can be better understood, as this will encourage parallel investigations in taxonomy, phylogeny, phylogeography, and chemistry (Heip et al. 2009).

The study of these bioactive molecules is a major focus of the rapidly expanding fields of marine microbiology and the chemistry of marine natural compounds. According to several studies, natural products recovered from higher marine species like marine invertebrates are often of microbial origin, particularly bacteria or cyanobacteria. However, microorganisms like those in a symbiotic relationship with higher species cannot be grown in isolation in a pure culture. Their development is directly related to the behavior of their hosts. Due to their special qualities, they have risen to the top of the list for cutting-edge marine biotechnology. Large sets of genes in marine bacteria and other microbes regulate metabolite production. Genome sequencing analyses have revealed that several genes have a role in producing secondary metabolites. This subset of the microbes' biosynthetic genes is likely transcribed during culture in axenic conditions. Many genes involved in biosynthesis are repressed and not expressed in vitro. The whole biochemical potential of these microbes has not yet been exploited, which is a significant issue (Martins et al. 2014). In this setting, the so-called metagenome analysis, which exploits the genomic data of entire communities, gains increasing relevance. The genomes of microbes and microbial communities that cannot be cultured are now within reach.

Macroalgae are a major source of bioactive chemicals alongside invertebrates and microbes (Hu et al. 2015). However, the development of metagenomic methodologies, genome mining, and heterologous biosynthesis, in addition to other genetic technologies and bioinformatics methods, has sped up the identification and made any remaining undiscovered MNPs with unique structures and intriguing marine microbe bioactivities available. Notably, implementing diverse methodologies and exploratory methods may facilitate the efficient exploitation of innovative MNPs with different systems. The widespread availability of secondary metabolites from MNPs suggests their possible application in pharmaceuticals, foods, cosmetics, and medicine. For these reasons, its potential industrial and academic applications are extensive, expanding its worth further. Understanding marine secondary metabolites is essential for developing future treatments. Because they include a variety of bioactive compounds that can work together to treat a wide range of ailments, they serve as a vital resource for drug discovery and development.

1.5 Marine Environment in Pharmacological and Biomedical Applications

1.5.1 Marine Bacteria

Structurally varied and effective bioactive chemicals are considered particularly abundant among marine microorganisms, particularly bacteria. Antimicrobial substances have been found to originate most frequently from the six phyla of marine bacteria (Actinobacteria, Bacteroidetes, Cyanobacteria, Firmicutes, Planctomycetes, and Proteobacteria). More than 93% of the research studies in the previous 25 years focus on Actinobacteria, Firmicutes, and Proteobacteria as the key taxa related to antimicrobial action. Also, marine microbes are showing promise as a source of new medicinal compounds.

Natural products are abundant in marine bacteria, with a surprisingly small number of species accounting for the vast bulk of compounds observed (Subramani and Aalbersberg 2013). Proteases, lipases, and glycoside hydrolases are all examples of hydrolase enzymes, which are widely used in industrial processes due to their novel, distinctive traits and properties of tolerance to extreme industrial circumstances. Because of their metabolic, ecological, evolutionary, and industrial potential, it is important to research and understand these diverse bacteria (Trincone 2010).

Actinobacteria are currently the most important source of secondary metabolites for antibacterial medicines for medical therapy originating from natural sources. Marine bioactive substances such as carbohydrates, pigments, polyphenols, peptides, proteins, and essential fatty acids have been the subject of extensive research for various purposes. Antioxidant, antithrombotic, anticoagulant, antiinflammatory, antiproliferative, antihypertensive, anti-diabetic, and cardio-protective actions are just some of the biological benefits associated with these chemicals in addition to their rheological effects, which are useful in the food sector.

1.5.2 Marine Fungi

Isolating secondary metabolites from marine fungi has greatly advanced the study of natural products. Many of these compounds are of interest in the medicinal, agricultural, and bioeconomy sectors because of their structural and functional distinctiveness. Maritime fungi have evolved metabolic pathways to manufacture various chemical scaffolds to survive the harsh biotic and abiotic conditions in the maritime habitat. Marine sponges and algae are just two of many places to find various fungi. In addition to Penicillin, *Aspergillus, Cladosporium, Fusarium, Trichoderma, Acremonium, Phoma*, and *Chaetomium* are often reported taxa. Polyketides are the most commonly reported secondary metabolite after alkaloids, terpenoids, peptides, and lipids. These chemicals have various pharmacological effects, such as those against microorganisms, cancer, viruses, oxidation, inflammation, and more. The enormous potential of these micro factories can be broadened by using next-generation sequencing methods to learn about the biology of marine fungi (Kamat et al. 2023).

1.5.3 Macroalgae

Primary producers in the food web, microalgae, alter the microbial environment by releasing specialized compounds and/or establishing symbiotic partnerships with microbes. Many important human nutrients can be found in microalgae, including polyunsaturated fatty acids like eicosapentaenoic acid (EPA) and docosahexaenoic acid (DHA), polysaccharides, vitamins (including vitamins A, B6, B9, B12, C, D, E, and K), and carotenoids (astaxanthin, -carotene). Microalgal bioactive substances are more secure because of the controlled conditions of their artificial cultivation. While fish oil has long been considered a reliable source of omega-3 DHA, the presence of antibiotics, heavy metals, and other contaminants in fish has restricted its usage. However, microalgae include pre-formed omega-3 DHA, which can aid in the alleviation of mental diseases (such as post-traumatic stress disorder and depression) while also reducing or preventing health hazards (Charles et al. 2019). Microalgae have been used to produce antigens for viruses such as influenza, HIV, hepatitis B, human papillomavirus, Zika virus, and Ebola Zaire virus (Ramos-Vega et al. 2021). Microalgae have undergone preclinical assessment for several vaccines containing microalgal antigens, showing promise as a rapid, inexpensive vaccine source (Yang et al. 2022).

In addition, seaweed has long been appreciated for its bioactive chemicals, which have promising applications in nutraceuticals, medicines, and functional foods. Despite some variation between different types of seaweed, most contain sufficient protein to support human health (Pangestuti and Kim 2015). Proteolysis of seaweed yields polysaccharides and peptides with a wide range of biological activities, including antibacterial, anticancer, antiproliferative, and antioxidant effects (Shannon and Abu-Ghannam 2016). In addition to lowering the risk of cardiovascular disease and cerebrovascular disease, the various monounsaturated, polyunsaturated, and long-chain polyunsaturated fatty acids (>18 carbon atoms) found in seaweed have been shown to boost brain cell activity and facilitate proper nervous system development (Mišurcová et al. 2011)

1.5.4 Marine Collagen

Marine collagen's many biological activities and therapeutic properties, including antioxidant and antiinflammatory activity, neuroprotective and anti-aging effects, and healing, are drawing increasing attention in the modern medical community. Marine collagen is also biodegradable and has a greater ability to penetrate lipid-free interfaces. Bioactivities such as antibacterial, antioxidant, antiinflammatory, immunomodulatory, and ACE-inhibitory effects were discovered in low-molecular-weight collagen peptides (di- and tripeptides), especially those having Pro or Hyp C-terminal residues. Applying tilapia collagen to a wound stimulated healing by increasing keratinocyte proliferation, fibroblast and myofibroblast differentiation, and extracellular matrix (ECM) production (Ge et al. 2020; Rigogliuso et al. 2023), which is yet another important property of collagen peptides. As a result, it is clear that marine collagen peptides exhibit interesting biological activities and can be efficiently used to treat various pathologies, including aging skin, bone defects, sarcopenia, wound healing, periodontal disease, gastroesophageal reflux, osteoarthritis, rheumatoid arthritis (Wang 2021). Marine collagen has been hypothesized to be an interesting source of bioactive peptides, with exciting pharmacological applications as functional foods and in drug delivery due to its abundance in marine organisms (fish, starfish, sponges, jellyfish, etc.). This is because some collagen-derived peptides resist gastrointestinal digestion and can reach the bloodstream intact. Dental tissue engineering, regeneration of the oral mucosa, repair of spinal cord injuries and nerve regeneration, regeneration of the skin and wound healing, regeneration of the cornea, and many other uses have been documented using marine collagen. Oslan et al. (2022) and Rigogliuso et al. (2023) suggest that marine-derived collagen can be used as a medication delivery system for treating cancer and other disorders.

1.5.5 Marine Peptides

In addition to their dietary roles, proteins and peptides play crucial roles in the body by regulating cell signaling pathways and engaging in other vital biological processes. Antioxidant, antibacterial, skin-protecting, apatite-regulating, bone-protecting, cardioprotective, anticancer, pain-relieving, anti-diabetic, and sexual and reproductive health advantages are only some of the purported benefits of peptides derived from marine species. Peptides have a wide range of applications, including in the production of medications, cosmetics, and dietary supplements. More research is mandated for these protein hydrolysates or peptides to be validated as therapeutic and for the structure-activity link to be understood. Constructing computational prediction models to comprehend the structure-activity association of peptides better and create peptide-based cosmetics and pharmaceuticals is necessary. As far as the editors can tell, peptides from marine creatures that have been approved for health claims are either uncommon or unreported in functional foods, dietary supplements, pharmaceuticals, and cosmetics. Bioactive peptides extracted from edible marine species present many opportunities for creating innovative goods (Fernando et al. 2023).

Furthermore, proteins and peptides found in the ocean have been shown to slow the development of neurodegenerative conditions like Parkinson's and Alzheimer's disease (Ryu and Kim 2013; Ghoran & Kijjoa 2021). Despite its potential for advancing industrial methods for synthesizing functional peptides, there has been a lack of research into this regime for marine peptides. It is possible to break down the proteins of seaweed, seafood, and by-products of seafood processing using proteolytic enzymes derived from plants, animals, or microbes, yielding a wide array of peptides with potential bioactivities. Several high-quality studies (Chai et al. 2017; Cermeño et al. 2020; Tacias-Pascacio et al. 2021) address recent advances in the discovery and possible applications of peptides produced from edible seaweed, fish, and marine invertebrates.

Functional foods, cosmetics, and pet food use peptides because of their high edibility and bioactivities (Saito 2012; Chai et al. 2017). These peptides range in size from 2 to 20 amino acids and have molecular weights of less than 2500 Da. The content and amino acid sequence of these peptides determines their bioactivity. Based on their structural properties, amino acid content, and sequencing, these peptides may have antihypertensive, antioxidative, hypocholesterolemic, antibacterial, immunomodulatory, opiate-like, mineral binding, and antithrombotic effects (Sarmadi and Ismail 2010). Strikingly, molecules doliculide, aplyronine A, sphinxolide, hectochlorin, lamellarin D, and spisulosine are only a few of the bioactive chemicals found in mollusks that have been shown to have antibacterial and anticancer activities. Drugs derived from natural products have been proposed as a potential solution to the problem of cancer patients developing multidrug resistance due to repeated courses of cytotoxic chemotherapy and the selective pressure exerted by the tumor microenvironment (Wyer et al. 2022).

1.6 Limitations on Bioactive Metabolite Research

The review authors approached the information offered here with great caution because most of the studies reviewed were conducted in mouse and in vitro models, despite the fact that research into natural compounds of marine origin presents a tremendous possibility in therapeutics. Therefore, although some studies have drawn attention to the therapeutic potential of specific molecules of marine origin concerning various disorders, it should be noted that an improved understanding of the implications of marine compounds in humans will only be achievable with more preclinical research and the implementation of clinical studies, so that in this way, it is possible to investigate and comprehend the potential of these substances of marine origin, thus bringing confidence. Nevertheless, because there aren't enough isolated and purified chemicals available, determining bioactivity has become challenging (Guimarães et al. 2023).

1.7 The Way Forward...

The automation of spectroscopy is crucial to combinatorial and technological advances in marine research, given that it paves the way for investigating the properties of novel compounds isolated from the intended marine species. Secondly, a more mechanistic approach using AI for drug development has allowed the discovery of new lead compounds targeting specific molecular flaws. Bioactivity study against various life-threatening diseases, in conjunction with high-throughput screening of numerous potential drug targets, will effectively disclose the potentially helpful biological features of marine natural materials. Identifying novel bioactive chemicals among marine metabolites provides a potential source for future therapeutic candidates. This means that the novel compounds will provide a wealth of material for bioactivity and drug discovery studies in the years to come. With their wide range of health benefits, medicinal and biomedical goods derived from marine flora and fauna are appealing to the food and pharmaceutical industries. Bioactive from marine organisms have the potential to maintain and improve human health and well-being, as evidenced by the increasing demand for marine-based functional food ingredients and nutraceutical formulations over the past decade and the accompanying rise in patent applications and approvals.

1.8 Future Perspective and Conclusions

Marine biotechnology involves researching marine environments to create new therapeutics, chemicals, enzymes, and manufacturing techniques. Biomaterials, medical diagnostics, aquaculture, seafood safety, bioremediation, and biofouling are all fields in which it has made significant contributions. Careful study of bioactive chemicals produced by marine microorganisms has the potential to have far-reaching effects on the marine economy and biomedicine. While there is no doubt that our knowledge of microbial metabolites has expanded greatly over the past few years, there is still a long way to go before we fully comprehend the intriguing possibilities of marine microbial metabolites. The moment is now to discover the potential of marine microbes in the field of marine natural products, and this promising scientific era appears to be the ideal time. The study of secondary metabolites from marine bacteria and fungi as a source of new bioactive metabolites has been continuously rising, as seen by the plethora of publications on bioactive metabolite research and the dramatic growth in quantities of these molecules.

However, many challenges remain to be overcome, such as the guided transformation of synthetic gene clusters, the design of simulated screening libraries for natural products, the activation of specific biosynthetic gene clusters, and the solution to the problem. Chemical and bioactivity diversity of secondary metabolites of marine microorganisms have received strong technical support thanks to the rapid development of marine exploration and bioinformatics. Recent successes in marine medication discovery, approval, and therapeutic application highlight the vast scope of marine natural compounds. Currently, seven medications on the market are derived from marine creatures, most of which are anticancer treatments. However, many obstacles are in the way of fully tapping the therapeutic potential of marine species. Marine pharmaceutical and neutraceutical development depend primarily on close multidisciplinary cooperation among biologists, chemists, biotechnologists, chemists, physicians, etc. The ongoing fight against antibiotic-resistant bacteria and numerous awful illnesses may, therefore can, win with the help of chemicals produced by marine microorganisms.

References

Altmann K-H (2017) Drugs from the oceans: marine natural products as leads for drug discovery. Chim Int J Chem 71:646–652

Bălaşa AF, Chircov C, Grumezescu AM (2020) Marine biocompounds for neuroprotection—a review. Mar Drugs 18:290

Barreca M, Spano V, Montalbano A, Cueto M, Diaz Marrero AR, Deniz I et al (2020) Marine anticancer agents: an overview with a particular focus on their chemical classes. Mar Drugs 18: 619. https://doi.org/10.3390/md18120619

Cermeño M, Kleekayai T, Amigo-Benavent M, Harnedy-Rothwell P, FitzGerald RJ (2020) Current knowledge on the extraction, purification, identification, and validation of bioactive peptides from seaweed. Electrophoresis 41:1694–1717

Chai TT, Law YC, Wong FC, Kim SK (2017) Enzyme-assisted discovery of antioxidant peptides from edible marine invertebrates: a review. Mar Drugs 15:42

Charles C, Msagati T, Swai H, Chacha M (2019) Microalgae: an alternative natural source of bioavailable omega-3 DHA for promotion of mental health in East Africa. Scientific African 6: e00187. https://doi.org/10.1016/j.sciaf.2019.e00187
da Silva APG (2021) Fighting coronaviruses with natural polyphenols. Biocatal Agric Biotechnol 37:102179
Dalmaso GZ, Ferreira D, Vermelho AB (2015) Marine extremophiles: a source of hydrolases for biotechnological applications. Mar Drugs 13:1925–1965
Ebrahimi B, Baroutian S, Li J, Zhang B, Ying T, Lu J (2023) Combination of marine bioactive compounds and extracts for the prevention and treatment of chronic diseases. Front Nutr 9: 1047026
Fernando IPS, Jayawardena TU, Wu J (2023) Marine proteins and peptides: production, biological activities, and potential applications. Food Innov Adv 2(2):69–84
Ge B, Wang H, Li J, Liu H, Yin Y, Zhang N, Qin S (2020) Comprehensive assessment of Nile tilapia skin (Oreochromis Niloticus) collagen hydrogels for wound dressings. Mar Drugs 18:178
Ghoran SH, Kijjoa A (2021) Marine-derived compounds with anti-Alzheimer's disease activities. Mar Drugs 19:410
Guimarães PL, Tavares DQ, Carrião GS, Oliveira MEH, Oliveira CR (2023) Potential of marine compounds in the treatment of neurodegenerative diseases: a review. Braz J Biol 83:e266795
Heip C, Hummel H, van Avesaath P, Appeltans W, Arvanitidis C, Aspden R et al (2009) Marine biodiversity and ecosystem functioning 2009. Printbase, Dublin, p 100
Hosseini SF, Rezaei M, McClements DJ (2020) Bioactive functional ingredients from aquatic origin: a review of recent progress in marine-derived nutraceuticals. Crit Rev Food Sci Nutr 62:1242–1269
Hu Y, Chen J, Hu G, Yu J, Zhu X, Lin Y, Chen S, Yuan J (2015) Statistical research on the bioactivity of new marine natural products discovered during the 28 years from 1985 to 2012. Mar Drugs 13:202–221
Hu D et al (2023) Application of marine natural products against Alzheimer's disease: past, present and future. Mar Drugs 21(1):43
Kamat S, Kumar S, Philip S, Kumari M (2023) Secondary metabolites from marine fungi: current status and application. In: Microbial biomolecules. Academic Press, New York, pp 181–209
Karthikeyan A, Joseph A, Nair BG (2022) Promising bioactive compounds from the marine environment and their potential effects on various diseases. J Genet Eng Biotechnol 20:14
Khotimchenko YS, Silachev DN, Katanaev VL (2022) Marine natural products from the Russian Pacific as sources of drugs for neurodegenerative diseases. Mar Drugs 20:708
Lindequist U (2016) Marine-derived pharmaceuticals—challenges and opportunities. Biomol Ther 24(6):561
Martins A, Vieira H, Gaspar H, Santos S (2014) Marketed marine natural products in the pharmaceutical and cosmeceutical industries: tips for success. Mar Drugs 12:1066–1101
Mišurcová L, Machů L, Orsavová J (2011) Seaweed minerals as nutraceuticals. In: Advances in food and nutrition research. Elsevier, Amsterdam, pp 371–390
Oslan SNH, Li CX, Shapawi R, Mokhtar RAM, Noordin WNM, Huda N (2022) Extraction and characterization of bioactive fish by-product collagen as promising for potential wound healing agent in pharmaceutical applications: current trend and future perspective. Int J Food Sci 2022: 1–10
Pangestuti R, Kim S-K (2015) Chapter 6—Seaweed proteins, peptides, and amino acids. In: Tiwari BK, Troy DJ (eds) Seaweed sustainability. Academic Press, San Diego, CA, pp 125–140. ISBN 978-0-12-418697-2
Ramos-Vega A, Monreal-Escalante E, Dumonteil E, Bañuelos-Hernández B, Angulo C (2021) Plantmade vaccines against parasites: bioinspired perspectives to fight against Chagas disease. Expert Rev Vaccines 5:1–16. https://doi.org/10.1080/14760584.2021.1893170
Rigogliuso S, Campora S, Notarbartolo M, Ghersi G (2023) Recovery of bioactive compounds from marine organisms: focus on the future perspectives for pharmacological, biomedical and regenerative medicine applications of marine collagen. Molecules 28(3):1152

Ryu B, Kim SK (2013) Potential beneficial effects of marine peptide on human neuron health. Curr Protein Pept Sci 14:173–176
Saito T (2012) Antihypertensive peptides. In: Tokyo MK (ed) Functional amino acids & functional peptides—technology & marketing. CMC Press, Tokyo, pp 60–71
Sarmadi BH, Ismail A (2010) Antioxidative peptides from food proteins: a review. Peptides 31: 1949–1956
Sekurova ON, Schneider O, Zotchev SB (2019) Novel bioactive natural products from bacteria via bioprospecting, genome mining and metabolic engineering. Microb Biotechnol 12:828–844
Shannon E, Abu-Ghannam N (2016) Antibacterial derivates of marine algae: an overview of pharmacological mechanisms and applications. Mar Drugs 14:81
Subramani R, Aalbersberg W (2013) Culturable rare actinomycetes: diversity, isolation and marine natural product discovery. Appl Microbiol Biotechnol 97:9291–9321
Suleria HAR, Gobe G, Masci P, Osborne SA (2016) Marine bioactive compounds and health promoting perspectives; innovation pathways for drug discovery. Trends Food Sci Technol 50: 44–55
Tacias-Pascacio VG, Castañeda-Valbuena D, Morellon-Sterling R, Tavano O, Berenguer-Murcia Á et al (2021) Bioactive peptides from fisheries residues: a review of use of papain in proteolysis reactions. Int J Biol Macromol 184:415–428
Trincone A (2010) Potential biocatalysts originating from sea environments. J Mol Catal B Enzym 66:241–256
Wang H (2021) A review of the effects of collagen treatment in clinical studies. Polymer 13:3868
Wyer S, Townsend DM, Ye Z, Kourtidis A, Choo YM, de Barros ALB et al (2022) Recent advances and limitations in the application of kahalalides for the control of cancer. Biomed Pharmacother 148:112676
Yang Y, Hassan SH, Awasthi MK, Gajendran B, Sharma M, Ji MK, Salama ES (2022) The recent progress on the bioactive compounds from algal biomass for human health applications. Food Biosci 51:102267

2 Bioprospection of Marine Sponge Microbiome for Bioactive Metabolites Employing Advanced Metagenomics Tools

Chanda Vikrant Berde, Vikrant Balkrishna Berde, Prakash Narayana Reddy, and Pallaval Veera Bramhachari

Abstract

The marine environment is an invaluable natural goods and biomaterials resource. Sponges, abundant members of marine benthic ecosystems, are a significant source of bioactive secondary metabolites, including chitin, chitosan, ceramic, biosilica, and collagen. The sponge microbiome includes several different species of symbiotic bacteria. These symbiotic bacteria may produce the bioactive compounds found in marine sponges. In the laboratory, most prokaryotes are unculturable, and the metabolite productivity of the cultured bacteria is inadequate, quantity-wise. Consequently, metagenomics can be employed to analyze and produce bioactive substances. Using metagenome mining, discovering new natural product biosynthetic routes from different bacterial consortia is feasible. The basic technologies utilized are identifying the biosynthetic genes in the metagenome of sponge microbiomes and creating a metagenomic library. Separation of biosynthetic gene clusters that might eventually be utilized to generate sustainable natural product sources through heterologous expression. The expression of these genes in readily culturable bacteria

C. V. Berde (✉)
School of Earth, Ocean and Atmospheric Sciences, Goa University, Panaji, Goa, India
e-mail: chanda.berde@unigoa.ac.in

V. B. Berde
Department of Zoology, Arts, Commerce & Science College, Lanja, Ratnagiri, Maharashtra, India

P. N. Reddy
Microbiology Department, Dr. V.S. Krishna Government Degree College (Autonomous), Visakhapatnam, Andhra Pradesh, India

P. Veera Bramhachari
Department of Biotechnology, Krishna University, Machilipatnam, Andhra Pradesh, India

P. Veera Bramhachari, C. V. Berde (eds.), *Marine Bioactive Molecules for Biomedical and Pharmacotherapeutic Applications*,
https://doi.org/10.1007/978-981-99-6770-4_2

results in higher yields of bioactive natural compounds. This review primarily emphasizes metagenomic technologies for using the metabolic pathways of symbiotic bacteria linked with marine sponges.

Keywords

Sponges · Microbiome · Metagenomics · Biosynthetic gene clusters · Metabolites

2.1 Introduction

Developing novel treatment strategies for infectious re-emerging diseases and emerging infections caused by multidrug-resistant pathogens, the main cause of increased mortalities is of utmost necessity (Abdelmohsen et al. 2010). These cannot be tackled with the existing antibiotics and treatment methods. Hence there is a need to discover bioactive compounds with different modes of action (Zhang et al. 2005).

The pharmaceutical industries are dependent upon natural products and their analogs, as leads for novel drug discovery as well as development (Singh and MacDonald 2010). Secondary metabolites of microbial nature have been used in clinical applications, such as polyketides, siderophores, antibacterial peptides, etc. These metabolites are produced to overcome competitors and vary greatly from one species to another (Baker et al. 2007). Attention is now being shifted to the unexplored niches of the earth's surface, which include the oceans and the polar regions (Dharmaraj 2010; Rocha-Martin et al. 2014) majorly. These are the untapped resources with unculturable, novel, unique microflora waiting to be discovered and exploited. The extreme conditions present in these regions offer hostile conditions for growth, such as temperature extremes, high pressures, oxygen unavailability, and oligotrophic conditions, resulting in well-adapted species of different flora found in other environments (Zhang et al. 2005). Hence the bioactive compounds produced by the marine microflora differ from compounds produced by terrestrial organisms, apropos to structure and function, being unique and diverse also.

Bioactive compounds in sponges are produced in response to competition for habitat, defense against predators, settlement inhibition, and symbiont protection (McClintock et al. 2005; Hellio et al. 2005; Lee et al. 2006). Marine sponges harbor a diverse range of microbes that make up 40% of their total biomass (Anbuchezhian and Li 2014), which are responsible for producing bioactive molecules. The sponges have been exhumed on a large scale to harness the bioactive compounds. Over-exploitation may lead to the extinction of sponge species and hampers industrial exploitation of this natural resource to its fullest. The answer is to start from the producer bacteria or its genes, encoding the biosynthetic pathways to ensure sustainable and sufficient levels of active molecules. The inability to mimic the environmental growth conditions in vitro has failed to culture the microbial symbionts of sponges (Kodzius and Gojobori 2015), and hence a lack of information on their functional characterization (Wu et al. 2009). The application of metagenomics, the cultivation-independent method, in biosynthetic compound gene identification and expression is gaining attention. Thus, using contemporary bioprospection techniques instead of conventional approaches, the biosynthetic machinery of non-cultivated bacteria can be more easily retrieved and utilized (Deane and Mitchell 2014).

2.2 Marine Sponges

Various reports indicate that sponges are, among metazoans, both benthic and sessile and are capable of filtering huge quantities of water through their many small holes (24,000 L seawater per kilogram sponge each day) (Vogel 1977). Sponges belong to the phylum Porifera meaning having many pores, and have been around for 700–800 million years. They are distributed throughout the world, including seas and inland waterbodies. Sponges in polar regions are abundant along the surfaces (Hentschel et al. 2012). About 1% of the species in freshwater habitats are sponges, whereas over half of the known species may be found in marine environments (Belarbi et al. 2003).

Marine sponges are divided into four classes: glass sponges, i.e., *Hexactinellida* found in deep marine habitats; demosponges, i.e., the *Demospongia* found in marine environments and freshwater; calcareous sponges, i.e., the *Calcarea* (calcium carbonate spicules), encrusting sponges *Homoscleromorpha* (Gazave et al. 2010). Marine sponges occur in various shapes, sizes, and colors, with environmental and biotic conditions influencing parameters (Pineda et al. 2015). They are multicellular invertebrates lacking nervous system and muscle cells, having a matrix called mesohyl, made up of freely moving amoeboid cells (Hochmuth and Piel 2009).

Various microorganisms, including marine animals, pass via the holes and enter the mesohyl layer of the sponge, which forms 40–60% of the sponge's total biomass (Kennedy et al. 2007). Symbiotic connection with the sponges, bacterial populations ranging between 10^8 and 10^{10} cells per gram of sponge wet weight, may be discovered in symbiosis with the sponges (Ouyang et al. 2010). These organisms fulfill all tasks that affect ecological balance by aiding the growth of other organisms while using controlled amounts of fertilizer and chemical defense against pathogens. Additionally, symbionts can enjoy a steady source of nutrients, a home for growth, and a trickle of nitrogen expelled as ammonia by marine sponges (Hentschel et al. 2012). Furthermore, sponges offer various species' habitats, impact their behavior by changing biotic and abiotic factors, and provide unique microhabitats (Maldonado et al. 2015). About half of the most often prescribed pharmaceuticals in the United States (Newman et al. 2000) are derived from natural compounds manufactured utilizing terrestrial microorganisms on an industrial scale. The number of marine-origin bioactive chemicals discovered in recent years from sponges, tunicates, mollusks, and bryozoans has been quite significant (Proksch et al. 2002). Bioactive natural products such as lipids and phytochemicals are found in marine sponges because they create these substances to protect against environmental rivals, predators, and diseases (Pimentel-Elardo et al. 2012).

The bioactive compounds have numerous applications to the producer organism in establishing and protecting itself in the niche (Uriz et al. 1996; Becerro et al.

Table 2.1 Synthesis of bioactive metabolites by sponge

Sponge species	Compound isolated	Applications	References
Callyspongia subarmigera		Inhibitory activity towards *Vibrio harveyi*	Dobretsov et al. (2005)
Callyspongia diffusa and *Sigmodocia carnosa*		Bioactivities	Lipton et al. (2006)
Discodermia calyx (*D. calyx*)	Calyculins	Cytotoxic compounds	Kato et al. (1986)
	Cyclodipeptides		He et al. (2013)
	Calyxamides		Kimura et al. (2012)
	Porphyrin pigments		He et al. (2012a)
	Alpha-hydroxyl fatty acids	Antibacterial	He et al. (2012b)
Halichondria and *Lissodendoryx* spp.	Eribulin, asynthetic analogue of halichondrin B	Metastatic breast cancer	Cortes et al. (2010)
Hippospongia communis	Untenospongin B	Antifungal extract	Rifai et al. (2004)
Plakortis simplex	Polyketide peroxides like plakortin	Antimalarial	Higgs and Faulkner (1978)
P. halichondrioides	Plakosides, simplexides, and plaxyloside	–	Costantino et al. (2001a, b)

1997). Apart from being a resource for secondary metabolites of pharmaceutical significance, marine benthic sponges also serve as a source for biomaterials such as biosilica, chitin, chitosan, ceramic, and collagen (Maldonado et al. 2015; van Soest et al. 2012) as well as natural products such as polyketides, nonribosomal peptides, and alkaloids (Proksch et al. 2002), anticancer compounds with cytotoxic activity (Zhang et al. 2003), and immunomodulators (Belarbi et al. 2003), antifouling agents (Armstrong et al. 1999). Some examples of bioactive compounds obtained from sponges are given in Table 2.1. The concentrations of the desired bioactive metabolites as high as 12% have been recovered in sponges though relatively low concentrations are present.

Few bioactive compounds have been commercialized even though many have been discovered (Armstrong et al. 1999; Proksch et al. 2002; Belarbi et al. 2003; Zhang et al. 2003). Monks et al. (2002) reported Brazilian sponge extracts' antibacterial, antitumor, and anti-chemotactic effects. Manoalide is recovered from the wild sponge *Luffariella variabilis* (Pomponi 1999). Numerous sponge-derived natural products exhibit promising activities against various diseases, most notably cancer (Hentschel et al. 2012).

Developing drugs for a specific target has improved due to the range of bioactive produced using sponges that can be applied at different stages of the disease. Additionally, several marine-derived bioactive chemicals in the preclinical assessment are in clinical trials, e.g., ET-743 and aplidine (Jimeno et al. 2004). Still, because of restricted supply, most natural marine sponges cannot be effectively exploited (Pomponi 1999). Many novel techniques have recently been used to show the numerous roles of the community of microbes and symbiotic groups found in sponges (Moitinho-Silva et al. 2017a).

2.3 Marine Microbes Associated with Marine Sponges

Marine sponges are multicellular animals harbouring different microbial forms (Gurgui and Piel 2010). Thomas et al. (2016) established the presence of 39 distinct microbial species in 81 sponge species in the ground-breaking research work. Research has shown that in sponge species *Plakortis halichondrioides*, microbial associated comprise 38–57% of total sponge biomass (Laroche et al. 2007). Shallow water habitat data from around the world's oceans provided insight into the symbiont sponge communities' diversity, variability, specificity, and correspondence to evolutionary patterns, as well as aided in explaining community complexity and composition.

Thaumarchaea, a class of archaea, is among the most dominant bacterial symbionts (Moitinho-Silva et al. 2017a). Crenarchaeata, another archaeal group identified from the sponge *Cenarchaeum symbiosum*, has also been described (Preston et al. 1996). Forty-two phyla of bacteria symbiotic to sponges are now known to exist (Webster and Thomas 2016). Due to the constant struggle for resources and worsening environmental conditions, many species create secondary metabolites to assist them in survival (Mehbub et al. 2014). *Plakortis* is known as a "High microbial abundance" (HMA) sponge due to its symbiotic microbial population richness (Gloeckner et al. 2014).

Sponge-bacteria symbiosis is common, and certain bacteria are linked with certain sponges for their whole lives, supporting the theory of sponge-bacteria symbiosis (Hentschel et al. 2002). Analyses of phylogenetic diversity indicated that the diversity of the bacterial communities in sponges, despite their distance in the taxonomic hierarchy and the diverse places of origin, is very homogeneous. Selvin et al. (2010) explain the functional connections and capabilities of linked microorganisms and sponge hosts well. The sea sponges are great models for microbial-macrofauna associations in that they are where scientists uncover many bioactive compounds (Faulkner 2002).

Bioactives produced by the sponge associates assist in nitrogen fixation, nitrification, photosynthesis, sulfate reduction, etc. (Mehbub et al. 2014). Many microorganisms found in sponges can help construct the intricate machinery needed for natural product biosynthesis, such as polyketides.

Research has proven that microbial associates of sponges play a significant role in the overall host fitness (Webster and Thomas 2016), for example, growth nutrient provision by synthesis of photosynthates and vitamins (Siegl et al. 2011), in-host geochemical cycling (Pita et al. 2013), nitrification (Fan et al. 2012), and polyphosphate production (Zhang et al. 2015), chemical defense by biosynthesis of bioactive such as polyketides (Mori et al. 2018) and removal of metabolic by-products (Webster and Thomas 2016) and sulphide (Hoffmann et al. 2005), avoid predation by producing feeding deterrent compounds (Rohde et al. 2015). Especially because of the distinctive phylogenetic makeup of the marine sponge microbiome, which has generated considerable research interest, and the vast natural product biosynthesis catalog, these symbiotic relationships are interesting to look into for possible applications in healthcare and pharmaceuticals.

However, due to methodological variations in sampling, sample processing, and molecular data analysis, it has not been easy to comprehensively compare the diverse sponge symbionts (Taylor et al. 2007). This consistency of microbial diversity and composition across large-scale initiatives, such as the Human Microbiome Project (2012) and the Earth Microbiome Project (Gilbert et al. 2014), can now be reliably and consistently defined. Large knowledge sets for host-associated microbiomes of vertebrates, notably humans, have been created due to these efforts, but comparable information for invertebrates is lacking. Microbial diversity, abundance, and location in sponge tissues have all been investigated using various methods. These techniques include restriction length fragment polymorphism (RFLP), fluorescence in situ hybridization (FISH), high-pressure liquid chromatography (HPLC), polymerase chain reaction–denaturing gradient gel electrophoresis (PCR-DGGE), epifluorescence microscopy, electron microscopy, direct sequencing, etc. (Taylor et al. 2007; Passarini et al. 2015).

2.4 Secondary Metabolites Produced by Sponge Microbiome

Marine microorganisms are the second major contributor to bioactive natural products. Evidence suggests that many bioactive metabolites, such as trace nutrients, generated by invertebrates like sponges, tunicates, and bryozoans are likely created by microbial symbionts in these animals. Archaea, bacteria, fungi, cyanobacteria, and microalgae are all significant symbionts in the sponges, and these organisms generate the bioactive compounds of relevance (Lee et al. 2001; Proksch et al. 2002; Selvin et al. 2010).

Many bioactive like antibiotics, antivirals, cytotoxics, cytotoxic agents, immunosuppressants, and enzymes, are found in spongobionts. A few of these metabolites, including antibiotics, antivirals, cytotoxics, cytotoxic agents, immunosuppressants, enzyme inhibitors, antimalarial, anti-inflammatory and siderophores (Taylor et al. 2007; Pimentel-Elardo et al. 2012) have been attributed to the microbiome (Laroche et al. 2007). Complex polyketides such as onnamide (Piel 2004), psymberin (Fisch et al. 2009), and Swingline A (Andrianasolo et al. 2005) make up the majority of the secondary metabolites viz. nonribosomal peptides, or polyketide-nonribosomal peptide hybrids which are generated by marine invertebrate symbiotic bacteria (Lane and Moore 2012).

Table 2.2 Biosynthesis of bioactive compounds by sponge-associated microbiome

Sponge	Microbial symbiont	Location or role	References
Baikalospongia bacillifera strain 28Bb-08	*Pseudomonas fluorescens*	PKS gene clusters	Lipko et al. (2012)
Dysidea herbacea	*Oscillatoria spongeliae*	Polybrominated biphenyl ether antibiotics	Osinga et al. (1998)
Demospongiae (e.g. *Theonella swinhoei, Discodermia dissoluta*)	*Candidatus Entotheonella* spp.	Filamentous bacteria living outside the sponge cells mostly in the mesohyl part	Gomes et al. (2013)
Dendrilla nigra	*Streptomyces* spp. BLT7	Antibacterial activity	Thomas et al. (2010)
Dictyoceratida	*S. officinalis*	Polyketides and terpenes	Manzo et al. (2011)
H. simulans	19 different genotypes of fungi belonging to *Agaricomycotina, Mucoromycotina, Saccharomycotina, and Pezizomycotina*	Antibacterial activity	Baker et al. (2009)
Haliclona spp.	*Streptomyces*	Largest number of natural products	Khan et al. (2011)
Ircinia fasciculata	*Penicillium chrysogenum*	Antileukemic Sorbicillactone A	Bringmann et al. (2007)
Lamellodysidea herbacea	Cyanobacterial symbionts	Fish feeding deterrent	Unson & Faulkner (1993)
Plakortis simplex	Symbiotic microbes	Bacteriohopanoids in cells of microbial symbionts	Laroche et al. (2007), Costantino et al. (2001b)
Pseudoceratina clavata	Novel *Actinomycete* genus *Salinispora*		Kim et al. (2005)
Rhabdastrella spp.	*Candidatus Entotheonella* spp.	Diverse pathways	Piel (2010)
Theonella swinhoei	*Candidatus Entotheonella*	Polyketides and modified peptides	Ueoka et al. (2015)
Tedania ignis	*Micrococcus* spp.	Biologically active secondary metabolites	Stierle et al. (1988)

The data in Table 2.2 depicts a small selection of the many bioactive chemicals in the sponge microbiome. For microorganisms to remain in the sponge, they must evade the host cells' phagocytosis. Sponge symbionts profusely produce eukaryotic-like protein domains (ELPs), such as ankyrin repeat proteins, tetratricopeptide repeat proteins, and leucine-rich repeat proteins (Díez-Vives et al. 2017). This vast resource has been unutilized by pharmaceutical businesses due to a lack of knowledge on culture methods and media formulations for cultivating marine microorganisms.

2.5 Culture-Dependent Methods

New pharmaceutical drug discovery by exploiting the enormous resources from the sponge microbiome requires isolating and cultivating microorganisms from sponges (Gomes et al. 2013). While the collection of marine sponges is difficult, their collection has to be done sustainably as it may lead to extinction. This will cause a hindrance in the development of bioactive from this source. Hence isolation of sponge microflora, it's screening, and maintenance is of priority.

However, culturing the sponge microbial symbionts has not succeeded, as more than 90% of microbes remain uncultured (Penesyan et al. 2010). The novel methods of culturing techniques include growth media modifications such as supplementing media with sponge extract as well as the addition of components such as acyl-homoserine lactones (AHLs) and growth regulators have also been recommended (Selvin et al. 2009). Growth parameters need careful attention because marine symbionts are fastidious organisms and slight variations in media composition or growth conditions can affect growth and, subsequently, metabolite synthesis qualitatively as well as quantitatively (Gomes et al. 2013; Scherlach and Hertweck 2009). Overall, the sponge sample processing methods, the growth medium composition, and the incubation conditions (Esteves et al. 2016) are important factors that determine the culturability of the microbiome.

For maximum productivity, active metabolites and cultures are grown in conditions mimicking the original environment (Knappe et al. 2008; Penesyan et al. 2010). According to Giovannoni and Stingl (2005), the microtiter dish dilution plate technique cultivates common marine bacteria. The coculturing method may prove to be very successful in helping scientists to discover a wide range of microbial secondary metabolites since this initiates the expression of dormant genes (Cueto et al. 2001), therefore increasing the complexity of natural products (Scherlach and Hertweck 2009). Several techniques, such as incubating sponges implanted with nutrient-containing diffusion growth chambers (DGCs) in the field (Kaeberlein et al. 2002), or the related use of solid or liquid media with and without antibiotics (Sipkema et al. 2011).

Several host species can reliably support known groups of Alphaproteobacteria, including the genera *Pseudovibrio* and *Ruegeria*. Culturable method-based studies have been conducted on the microbiome diversity of sponges that prove the association of these genera with sponge species (Sipkema et al. 2011). When the cultivable bacterial population is considered, it has been estimated that only 15% of the sponge microbiome is cultivable (Sipkema et al. 2011). For instance, the most prevalent sponge bacterial symbionts remain undiscovered (Taylor et al. 2007; Hardoim et al. 2014). The marine sponge *Discodermia dissolute* generates a potential anticancer chemical, discodermolide (Gunasekera et al. 1990). The culture of a microbe having discodermolide-producing ability might provide a continuous chemical supply. However, several bacteria linked with sponges have not been cultivated in vitro (Webster et al. 2001), indicating the need for alternative methods.

The advancement of metagenomics techniques gives an intriguing and culture-independent technique to explore marine sponges for their biosynthetic potential and overcome the limitation of culture-dependent approaches. The study of genome fragments from sponge-microbe associations will enable the detection of the biosynthetic gene clusters of bioactive compounds, enabling their utilization for industrial-scale production of the bioactive compounds.

2.6 Culture-Independent Methods

The marine symbiotic microbiome produces most of the secondary metabolites, which include compounds like complex polyketides, nonribosomal peptides, or polyketide-non-ribosomal peptide hybrids (Hochmuth and Piel 2009; Piel 2004). These metabolites range in activity from antimicrobial to antitumor, antimalarial, immunosuppressants, anti-inflammatory, antifouling, toxins, etc. (Pimentel-Elardo et al. 2012). Difficulties in culturing the sponge microbiome, which includes but are not limited to complex nutritional requirements, mimicking the environmental conditions, need to coculture, are the major constraint in studies related to the applicability of the sponge symbionts (Webster and Hill 2001). The inability to culture the symbiotic biosynthesizer hinders access to the biosynthetic pathways and subsequently harnesses their potential (Kurnia et al. 2017).

Metagenomics is an alternative tool to study the sponge symbionts by a non-culturable approach (Schloss and Handelsman 2003). These technologies have been applied to study the nature of the relationship between the sponge symbiotic microbiome and their abundance to identify the microbial communities within them (Hentschel et al. 2002; Webster et al. 2001), the discovery of new species (Preston et al. 1996) and exploit them for various uses (Thomas et al. 2010). Metagenomics can be applied to study and understand bioactive compound synthesis in sponge microbiomes. Various reports on the application of metagenome mining for identifying bioactive compounds are available. Sponge-derived polyketides and modified peptides (Matroodi et al. 2020), onnamide A (Piel et al. 2004), psymberin (Fisch et al. 2009), polytheonamides (Freeman et al. 2012), and misakinolide (Ueoka et al. 2015), have been identified prominently using these approaches. Bacteria genetically engineered to contain the necessary gene clusters for sponge-derived drug discovery might be grown on a massive scale to produce enormous quantities of therapeutic candidates (Piel 2004). There are discoveries to be made about the microbiome and its function in biotechnological applications. New genes, enzymes, pathways, and bioactive compounds are being discovered in this exploration. A new area of research is opened to exploration with metagenomics, which involves discovering novel biomaterials from uncultural microorganisms, constituting 99% of the microbial variety (Sleator et al. 2008).

Metagenomics has increased the whole genome availability, and studies using known sequence data for biosynthetic enzymes have revealed an overabundance of genes encoding bioactive secondary metabolites (Brakhage 2013). However, some studies indicate a difference in the species composition and the biosynthetic genes of symbiotic microbiome discovered by culture-based and culture-independent approaches. Thus, a combined strategy can be used to better and rapidly exploit the diversity in biosynthetic pathways associated with marine sponges.

Additionally, the holobiont of the *Cymbastella concentrica*, which is made up of two proteobacteria, a thaumarchaeon, and a diatom (Moitinho-Silva et al. 2017b), have recently been discovered using an analysis that has merged metagenomic binning, metatranscriptomics, and visualization methods. Several recent studies have focused on using fluorescence in situ hybridization (FISH) (Fieseler et al. 2004) or 16S rRNA sequencing (Hentschel et al. 2002) for the detection of microbial populations, which has shown that the majority of sponges exhibit uniform but diverse microbial populations. Due to this microbial diversity, sponges appear to synthesize numerous bioactive compounds (Davidson et al. 2001). Discoveries using technologies such as SCG and cultivation-independent genome binning using metagenomic data have been made, for example, *Chloroflexi*, *Actinobacteria*, and *Poribacteria* species (Bayer et al. 2013) have a high capacity for halogenation. *Synechococcus spongiarum* possesses various adaptive traits, some of which were discovered by analyzing its genome alone.

Functional gene diversity of the microbiome can be studied using previously unaccessible metagenomics. It gives an insight into various aspects of microbiomes by providing data on uncultivated microorganisms' evolutionary, functional, structural, and phylogenetic characteristics (Béjà et al. 2000a, b). Using metagenomics, it is possible to discover, isolate, and express the genes of interest in cultivable host cells, study microbial evolution, ecology, metabolism, and adaptations, and create and test metagenomic libraries. DNA of mixed populations of bacteria is extracted, cloned, and screened for gene sequences of novel bioactive compounds (Schloss and Handelsman 2003). Numerous biosynthetic pathways and enzymes have been thus identified using metagenomics libraries, clones and expressed using this approach (Piel et al. 2004). The application of metagenomics consists of taking total environmental DNA, referred to as eDNA, and fragmentation and inserting the fragments into a culture-ready host. Through the study of the expression of the genes, it is possible to explore the taxonomic diversity and biosynthetic potential of the uncultured species (Kennedy et al. 2010). Another approach involves direct sequencing of the metagenomic DNA libraries and deciphering the whole community's functions while ascertaining genes for natural product pathways (Katz et al. 2016).

2.6.1 The Methodology in Metagenomics Is as Follows

2.6.1.1 Extraction of eDNA

Metagenomic DNA is extracted from various sources, directly as well as by indirect methods. In the direct methods, highly pure DNA is obtained; however, some eukaryotic DNA also gets extracted (Ogram et al. 1987). A blending method can be used to avoid host DNA interference, which is high molecular weight DNA extraction after sample homogenization (Gabor et al. 2003). The DNA obtained is free of host DNA and can be used for metagenomics library construction. This technique was used to isolate the metagenomics DNA of the marine sponge *Mycale mytilorum* (Gabor et al. 2003).

2.6.1.2 eDNA Digestion

The restriction enzymes will partly digest the eDNA, making it much easier to handle. High molecular weight DNA is required for library creation as biosynthetic gene clusters encoding antimicrobial chemicals can vary from 10 to 100 kb in length (Singh and MacDonald 2010). Metagenomic libraries are created by cloning genetic material onto vectors, such as fosmid and cosmid. Functional screening of clones is done whereby cloning vectors play a critical role in determining the efficacy and fate of the screening methods (Courtois et al. 2003).

2.6.1.3 Library Construction and Transfection in Host

Based on the size of the insert, either fosmid, i.e., F1 origin-based cosmid vector or cosmid libraries, for DNA inserts of up to 40 kb, while BACs for inserts of up to 200 kb, can be employed (Ouyang et al. 2010). The highly stable fosmid vectors replicate using F (fertility)-factor replicon, requiring only *E. coli* as a host, because of the F factor, for use in functional screening. Aakvik et al. (2009) reported using a shuttle vector with an extended host range (Weiland et al. 2010) through the efficiency of replication and expression in the non-*E. coli* hosts are not good enough (Trindade et al. 2015). However, they can use multiple hosts for replication, facilitating the screening to be achieved (Trindade et al. 2015). *E. coli* as a host for expression has its advantages in being the bacteria used commonly in industrial fermentations. It has a well-characterized genome, is easily cultivable under standard laboratory conditions, and grows rapidly utilizing inexpensive substrates and high transfection efficiency (Uria et al. 2005). However, it lacks the machinery needed for bioactive metabolite biosynthesis of interest. Shuttle vectors can be used to construct libraries using multiple hosts, though the efficiency levels required are difficult to achieve in this non-*E.coli* hosts such as *Streptomyces* spp., *Pseudomonas* spp., and *Bacillus* spp. (Martinez et al. 2004). *Streptomyces* spp., for example, being a producer of a large number of secondary metabolites, can provide the precursors needed for new secondary metabolite synthesis and hence is a suitable host for metagenomics screening and expression involving bioactive compounds (Mocali and Benedetti 2010).

2.6.1.4 Replication of DNA Inserts in Host

Once the DNA insert with the vector of interest has been loaded, the vector multiplies in the host, creating several clones of the DNA fragments. Bioactive and enzymatic substances encoded by the metagenomic DNA fragments can be tested. To represent the entirety of the microbiome in an environmental sample, it is necessary to design metagenomic libraries as big as feasible with genetic data representing the species, strains, and populations that make up the microbiome, which may become as then include up to two million clones (Singh and MacDonald 2010).

2.6.2 The Detection of Novel Natural Products: Based on the Sequence-Based Mining

2.6.2.1 Sequence-Based Screening/PCR or Probe-Based Screening

This technique might be applied to investigate the composition of microbiome and bioproduct synthesis genes in environmental samples. In combination with hybridization probes, random sequencing can be used to test for certain genes. Using this approach eradicates the need for cloned gene expression in the host. Metagenomic sequencing, using a phylogenetic marker, can be done sequentially, whereby all clones are sequenced to provide an overall picture of the possible taxonomic source of DNA fragments; alternatively, using a random-DNA fragment-hunting approach, the genome around the gene of interest is sequenced, to find DNA markers for possible taxonomic sources. The replicated clones previously cultured on agar plates are now transferred to nylon membranes, which are hybridized with the tagged PCR products (Metsä-Ketelä et al. 1999). New genes, functions, and taxonomic groupings have been discovered through many applications of these strategies (Béjà et al. 2000a, b). Archaea-specific 16S rRNA gene sequencing by Stein et al. (1996) led to the identification of 16S rRNA genes from archaea along with helicase, glutamate semialdehyde aminotransferase, and novel proteorhodopsins. Typically, PKS and NRPS biosynthetic genes from the metagenomic library are identified using the same approach. These pathways are hard to identify due to the absence of expression and/or insufficient quantities in the laboratory, so it is necessary to conduct functional-based screening to find them (Wilson and Piel 2013).

Although the target sequence of the gene of interest, the biosynthetic genes, must be known to construct the primers, this technique has the advantage of allowing random access to any genes in the biosynthetic pathway (Singh and MacDonald 2010). Biosynthetic clusters accountable for the biosynthesis of an active metabolite may not be detected by the sequence screening approach (Yun and Ryu 2005). Even though sequence-based screening has certain limitations, it has proven highly successful when identifying and isolating polyketides and nonribosomal peptides from metagenomics DNA of marine invertebrate microbiomes, especially sponges. Gene clusters for theonamide and theopederin biosynthesis were identified from the metagenomic DNA of *Theonella swinhoei* by amplifying Ketoacyl synthase (KS) domains of Type I PKS (Piel et al. 2004).

2.6.2.2 Function-Based Screening

The presence of bioactive metabolite genes from metagenomic libraries of soil and marine invertebrate-associated microbiomes can be detected by functional metagenomics (Milshteyn et al. 2014). This strategy is an activity-based screening approach (Schmeisser et al. 2007). Instead of sequencing all of the genes necessary for an application's function ahead of time, it focuses on screening the millions of available metagenomic library clones for the existence of such genes. The reliability of the prediction is directly related to how the gene is expressed and if the outcome can be identified using biochemical testing. Initial studies on metagenomics used the

Sanger method for sequencing (Gillespie et al. 2002), which is now replaced by cost-effective and faster NGS technologies that simultaneously carry out the sequencing of multiple fragments (Klindworth et al. 2013).

Drug development requires the expression of the metagenomic gene sequences in hosts capable of transcription activation to synthesize bioactive chemicals (Katz et al. 2016). Novel shuttle vectors having a broad-host range, such as shuttle fosmid pCCERI, can be utilized in non-*E.coli* hosts such as *P. putida, P. bifermentans, S. lividans*, and *R. leguminosarum*. Thus the expression efficiency of functional screening can be improved significantly by providing the correct expression machinery from diverse host systems (Kennedy et al. 2008).

The clones with metagenomic DNA fragments for bioactive compounds are grown on agar media during the screening process. A mixture of soft molten agar and an indicator microbial strain is poured over the basal agar, overlaying it. Functional screening of metabolite producers is commonly performed using top agar overlay assays, commonly known as top agar overlay assays (Courtois et al. 2003). The clones indicate the antimicrobial activity around the colonies by a zone of clearance of the indicator strain or color generation. The fosmid DNA from each selected clone is extracted and sequenced. This sequencing identifies genes directly generating activity (Kennedy et al. 2010).

2.6.3 Advantages of Functional Screening Method

1. Prior knowledge of the gene of interest is not required.
2. DNA sequences of novel biosynthetic metabolites having useful functions in the pharmaceutical industry can be achieved can be detected, which is not possible with sequence-based methods alone (Kennedy et al. 2010).
3. The recovery of the biosynthetic product ensures its accurate synthesis by the host strain (Felczykowska et al. 2015).
4. The method enables the direct selection of clones with antimicrobial activity.

2.6.4 Limitation of Functional Screening Method

1. Essentially, to obtain a positive outcome, the expression of target genes in the host utilized for screening, phenotypic detection, and grouping all genes linked to function must be accomplished (Schloss and Handelsman 2003).
2. Using *E. coli* as an expression host may not be able to express genes from distantly related microorganisms, and fewer clones may be obtained during screening (Kennedy et al. 2010).
3. The biosynthetic pathway genes from uncultured microbial strains must be cloned in heterogeneous screening hosts for expression (Ginolhac et al. 2004).
4. Function-based screening can only be done through metagenomic libraries, while sequenced-based screening can be done using cloned metagenomic libraries.

2.7 Bioinformatics with Metagenomics

Metagenomic studies enable the detection of functional genes and the sequencing of whole genomes without the need to culture the microbes (Li et al. 2014; Singh and MacDonald 2010). Reduction in sequencing costs, as well as the development of bioinformatics tools, have accelerated the rate of gene cluster discovery in recent years (Ng et al. 2015), such as the identification of new NRPS and PKS gene clusters (Jenke-Kodama and Dittmann 2009). Based on microbial genome sequences, many algorithms, such as the Basic Local Alignment Search Tool (BLAST) algorithm on National Center for Biotechnology Information (NCBI), have been developed and applied in biometabolite biosynthesis studies (Nikolouli and Mossialos 2012). This algorithm detects protein-encoding genes having a short region in common with known NRPS and PKS enzymes. Thus, genes for various NRPS and PKS products can be detected from sequences of known and unknown organisms.

Another genome mining tool is the Antibiotics and Secondary Metabolite Analysis SHell (antiSMASH) (http://antismash.secondarymetabolites.org/) software that allows the identification of encoding secondary metabolites gene clusters from assembled contiguous sequences of microbial metagenomics DNA (Medema et al. 2011). From the DNA sequences uploaded in FASTA, EMBL, or GenBank format, information on genome function and predictable chemical structures of bioactive metabolites related to PKS and NRPS biosynthetic gene clusters, can be obtained (Medema et al. 2011). This is also termed genome annotation, involving gene sequence analyses, finding non-coding regions, identifying genes, i.e., gene prediction, and attaching biological or protein functions to the gene.

Several beneficial bioinformatics tools for investigating metagenomics data are available, including PRediction Informatics for Secondary Metabolomes (PRISM) (http://magarveylab.ca/prism/). PRISM helps identify bioactive compounds and prediction of its chemical structure, especially of NRPs and polyketides (Skinnider et al. 2015). A sequencing dataset may be uploaded as FASTA or GenBank format and studied for bioactive metabolite production characteristics connected to bioactive metabolite synthesis (Lane and Moore 2012). The identified biosynthetic gene sequence can be cloned and expressed in easily culturable hosts and used for functional screening to confirm its presence. The move to make widespread use of high-throughput sequencing, competent bioinformatics tools, and heterologous gene expression marks the start of a whole new era in research using bacterial communities that had not been thoroughly studied until now.

2.8 Conclusions

The metagenomics approach provides quick and reliable means for biosynthetic bioactive product discovery and also offers a better understanding of the biochemistry in unculturable microbiomes. Recent developments in co-cultivation show the possibility of culturing uncultured bacteria. Different cultivation methods, such as coculture using semi-porous culture chambers, where the cultures grow in different compartments but an exchange of nutrients occurs through a semipermeable membrane; natural seawater for culturing marine bacteria; and use of microfluidics, have been employed successfully for novel bacterial species thought to be uncultivable.

Incorporation of metagenomics with the modern tools used in single-cell analysis, meta-transcriptomics, and meta-proteomics will certainly benefit the understanding potential of metabolite biosynthesis in uncultivated microbes and aid in their expression. An individual laboratory-generated volume of data can be massive with the development of advanced and more affordable NGS technology. Large data sets need automated genome mining techniques, like anti-SMASH or ClusterMine360. While these methods use homology-based searches, new approaches for bioinformatic genome mining will be mandated to identify novel chemistry. The next step in novel drug discovery and its development is genomic annotation, which effectively performs functional screening of the genome and its expression in different host systems. The pattern recognition algorithms are used to search for previously overlooked ORFs linked with biosynthetic genes that are conserved throughout several species. However, several obstacles must be addressed to realize the full potential of metagenomics for discovering bioactive. So, comprehensive whole-genome sequencing suggests that only a small fraction of the biosynthetic potential has been revealed from the DNA of well-studied producers, showing that a much greater amount of previously unknown metabolic potential remains to be discovered. We may reasonably infer that the number of biocatalysts and biosynthetic gene clusters to be discovered among the non-cultural majority is vast.

Conflict of Interest The authors declare no conflict of interest, financial or otherwise.

References

Aakvik T, Degnes KF, Dahlsrud R, Schmidt F, Dam R, Yu L, Völker U, Ellingsen TE, Valla S (2009) A plasmid RK2-based broad-host-range cloning vector useful for transfer of metagenomic libraries to a variety of bacterial species. FEMS Microbiol Lett 296:149–158. https://doi.org/10.1111/j.1574-6968.2009.01639.x. PMID 19459950

Abdelmohsen UR, Pimentel-Elardo SM, Hanora A, Radwan M, Abou-El-Ela SH, Ahmed S, Hentschel U (2010) Isolation, phylogenetic analysis and anti-infective activity screening of marine sponge-associated actinomycetes. Mar Drugs 8(3):399–412. https://doi.org/10.3390/md8030399. PMID 20411105

Anbuchezhian R, Li Z (2014) Marine sponge *Craniella australiensis*-associated bacterial diversity. In: Encyclopedia of metagenomics. Springer, Berlin, pp 1–9

Andrianasolo EH, Gross H, Goeger D, Musafija-Girt M, McPhail K, Leal RM, Mooberry SL, Gerwick WH (2005) Isolation of swinholide A and related glycosylated derivatives from two field collections of marine cyanobacteria. Org Lett 7(7):1375–1378. https://doi.org/10.1021/ol050188x. PMID 15787510

Armstrong E, McKenzie JD, Goldsworthy GT (1999) Aquaculture of sponges on scallops for natural products research and antifouling. J Biotechnol 70:163–174

Baker DD, Chu M, Oza U, Rajgarhia V (2007) The value of natural products to future pharmaceutical discovery. Nat Prod Rep 24(6):1225–1244. https://doi.org/10.1039/b602241n. PMID 1 8033577

Baker PW, Kennedy J, Dobson AD, Marchesi JR (2009) Phylogenetic diversity and antimicrobial activities of fungi associated with *Haliclona simulans* isolated from Irish coastal waters. Mar Biotechnol (NY) 11(4):540–547. https://doi.org/10.1007/s10126-008-9169-7. PMID 19083060

Bayer K, Scheuermayer M, Fieseler L, Hentschel U (2013) Genomic mining for novel FADH2-dependent halogenases in marine sponge-associated microbial consortia. Mar Biotechnol 15: 63–72

Becerro MA, Turon X, Uriz MJ (1997) Multiple functions for secondary metabolites in encrusting marine invertebrates. J Chem Ecol 23(6):1527–1547. https://doi.org/10.1023/B:JOEC.0000006420.04002.2e

Béjà O, Aravind L, Koonin EV, Suzuki MT, Hadd A, Nguyen LP, Jovanovich SB, Gates CM, Feldman RA, Spudich JL, Spudich EN, DeLong EF (2000a) Bacterial rhodopsin: evidence for a new type of phototrophy in the sea. Science 289(5486):1902–1906. https://doi.org/10.1126/science.289.5486.1902. PMID 10988064

Béjà O, Suzuki MT, Koonin EV, Aravind L, Hadd A, Nguyen LP, Villacorta R, Amjadi M, Garrigues C, Jovanovich SB, Feldman RA, DeLong EF (2000b) Construction and analysis of bacterial artificial chromosome libraries from a marine microbial assemblage. Environ Microbiol 2(5):516–529. https://doi.org/10.1046/j.1462-2920.2000.00133.x. PMID 11233160

Belarbi EH, Contreras Gómez A, Chisti Y, García Camacho F, Molina Grima E (2003) Producing drugs from marine sponges. Biotechnol Adv 21(7):585–598. https://doi.org/10.1016/s0734-9750(03)00100-9. PMID 14516872

Brakhage AA (2013) Regulation of fungal secondary metabolism. Nat Rev Microbiol 11(1):21–32. https://doi.org/10.1038/nrmicro2916. PMID 23178386

Bringmann G, Gulder TA, Lang G, Schmitt S, Stöhr R, Wiese J, Nagel K, Imhoff JF (2007) Large-scale biotechnological production of the antileukemic marine natural product sorbicillactone A. Mar Drugs 5(2):23–30. https://doi.org/10.3390/md502023. PMID 18463724

Cortes J, Vahdat L, Blum JL, Twelves C, Campone M, Roché H, Bachelot T, Awada A, Paridaens R, Goncalves A, Shuster DE, Wanders J, Fang F, Gurnani R, Richmond E, Cole PE, Ashworth S, Allison MA (2010) Phase II study of the halichondrin B analog eribulin mesylate in patients with locally advanced or metastatic breast cancer previously treated with an anthracycline, a taxane, and capecitabine. J Clin Oncol 28(25):3922–3928. https://doi.org/10.1200/JCO.2009.25.8467. PMID 20679609

Costantino V, Fattorusso E, Imperatore C, Mangoni A (2001a) A biosynthetically significant new bacteriohopanoid present in large amounts in the Caribbean sponge *Plakortis simplex*. Tetrahedron 57(18):4045–4048. https://doi.org/10.1016/S0040-4020(01)00286-1

Costantino V, Fattorusso E, Imperatore C, Mangoni A (2001b) Plaxyloside from the marine sponge *Plakortis simplex*: an improved strategy for NMR structural studies of carbohydrate chains. Eur J Org Chem 2001:4457–4462

Courtois S, Cappellano CM, Ball M, Francou FX, Normand P, Helynck G, Martinez A, Kolvek SJ, Hopke J, Osburne MS, August PR, Nalin R, Guérineau M, Jeannin P, Simonet P, Pernodet JL (2003) Recombinant environmental libraries provide access to microbial diversity for drug discovery from natural products. Appl Environ Microbiol 69(1):49–55. https://doi.org/10.1128/AEM.69.1.49-55.2003. PMID 12513976

Cueto M, Jensen PR, Kauffman C, Fenical W, Lobkovsky E, Clardy J (2001) Pestalone, a new antibiotic produced by a marine fungus in response to bacterial challenge. J Nat Prod 64(11): 1444–1446. https://doi.org/10.1021/np0102713. PMID 11720529

Davidson SK, Allen SW, Lim GE, Anderson CM, Haygood MG (2001) Evidence for the biosynthesis of bryostatins by the bacterial symbiont '*Candidatus Endobugula sertula*' of the bryozoan *Bugula neritina*. Appl Environ Microbiol 67(10):4531–4537. https://doi.org/10.1128/AEM.67.10.4531-4537.2001. PMID 11571152

Deane CD, Mitchell DA (2014) Lessons learned from the transformation of natural product discovery to a genomedriven endeavor. J Ind Microbiol Biotechnol 41(2):315–331. https://doi.org/10.1007/s10295-013-1361-8. PMID 24142337

Dharmaraj S (2010) Marine streptomyces as a novel source of bioactive substances. World J Microbiol Biotechnol 26(12):2123–2139. https://doi.org/10.1007/s11274-010-0415-6

Díez-Vives C, Moitinho-Silva L, Nielsen S, Reynolds D, Thomas T (2017) Expression of eukaryotic-like protein in the microbiome of sponges. Mol Ecol 26(5):1432–1451. https://doi.org/10.1111/mec.14003. PMID 28036141

Dobretsov S, Dahms HU, Qian PY (2005) Antibacterial and anti-diatom activity of Hong Kong sponges. Aquat Microb Ecol 38(2):191–201. https://doi.org/10.3354/ame038191

Esteves AIS, Amer N, Nguyen M, Thomas T (2016) Sample processing impacts the viability and cultivability of the sponge microbiome. Front Microbiol 7:499. https://doi.org/10.3389/fmicb.2016.00499. PMID 27242673

Fan L, Reynolds D, Liu M, Stark M, Kjelleberg S, Webster NS, Thomas T (2012) Functional equivalence and evolutionary convergence in complex communities of microbial sponge symbionts. Proc Natl Acad Sci U S A 109(27):E1878–E1887. https://doi.org/10.1073/pnas.1203287109. PMID 22699508

Faulkner DJ (2002) Marine natural products. Nat Prod Rep 19(1):1–48. https://doi.org/10.1039/b009029h. PMID 11902436

Felczykowska A, Krajewska A, Zielińska S, Łoś JM, Bloch SK, Nejman-Faleńczyk B (2015) The most widespread problems in the function-based microbial metagenomics. Acta Biochim Pol 62(1):161–166. https://doi.org/10.18388/abp.2014_917. PMID 25680374

Fieseler L, Horn M, Wagner M, Hentschel U (2004) Discovery of the novel candidate phylum Poribacteria in marine sponges. Appl Environ Microbiol 70(6):3724–3732

Fisch KM, Gurgui C, Heycke N, van der Sar SA, Anderson SA, Webb VL, Taudien S, Platzer M, Rubio BK, Robinson SJ, Crews P, Piel J (2009) Polyketide assembly lines of uncultivated sponge symbionts from structure-based gene targeting. Nat Chem Biol 5(7):494–501. https://doi.org/10.1038/nchembio.176. PMID 19448639

Freeman MF, Gurgui C, Helf MJ, Morinaka BI, Uria AR, Oldham NJ, Sahl HG, Matsunaga S, Piel J (2012) Metagenome mining reveals polytheonamides as post-translationally modified ribosomal peptides. Science 338(6105):387–390. https://doi.org/10.1126/science.1226121. PMID 22983711

Gabor EM, de Vries EJ, Janssen DB (2003) Efficient recovery of environmental DNA for expression cloning by indirect extraction methods. FEMS Microbiol Ecol 44(2):153–163. https://doi.org/10.1016/S0168-6496(02)00462-2. PMID 19719633

Gazave E, Lapébie P, Renard E, Vacelet J, Rocher C, Ereskovsky AV, Lavrov DV, Borchiellini C (2010) Molecular phylogeny restores the supra-generic subdivision of homoscleromorph sponges (Porifera, Homoscleromorpha). PLoS One 5(12):e14290. https://doi.org/10.1371/journal.pone.0014290. PMID 21179486

Gilbert JA, Jansson JK, Knight R (2014) The earth microbiome project: successes and aspirations. BMC Biol 12:69. https://doi.org/10.1186/s12915-014-0069-1. PMID 25184604

Gillespie DE, Brady SF, Bettermann AD, Cianciotto NP, Liles MR, Rondon MR, Clardy J, Goodman RM, Handelsman J (2002) Isolation of antibiotics turbomycin A and B from a metagenomic library of soil microbial DNA. Appl Environ Microbiol 68(9):4301–4306. https://doi.org/10.1128/AEM.68.9.4301-4306.2002. PMID 12200279

Ginolhac A, Jarrin C, Gillet B, Robe P, Pujic P, Tuphile K, Bertrand H, Vogel TM, Perrière G, Simonet P, Nalin R (2004) Phylogenetic analysis of polyketide synthase I domains from soil metagenomic libraries allows selection of promising clones. Appl Environ Microbiol 70(9):5522–5527. https://doi.org/10.1128/AEM.70.9.5522-5527.2004. PMID 15345440

Giovannoni SJ, Stingl U (2005) Molecular diversity and ecology of microbial plankton. Nature 437(7057):343–348. https://doi.org/10.1038/nature04158. PMID 16163344

Gloeckner V, Wehrl M, Moitinho-Silva L, Gernert C, Schupp P, Pawlik JR, Lindquist NL, Erpenbeck D, Wörheide G, Hentschel U (2014) The HMA-LMA dichotomy revisited: an electron microscopical survey of 56 sponge species. Biol Bull 227(1):78–88. https://doi.org/10.1086/BBLv227n1p78. PMID 25216505

Gomes ES, Schuch V, de Macedo Lemos EG (2013) Biotechnology of polyketides: new breath of life for the novel antibiotic genetic pathways discovery through metagenomics. Braz J Microbiol 44(4):1007–1034. https://doi.org/10.1590/s1517-83822013000400002. PMID 24688489

Gunasekera SP, Gunasekera M, Longley RE, Schulte GK (1990) Discodermolide: a new bioactive polyhydroxylated lactone from the marine sponge *Discodermia dissoluta*. J Org Chem 55(16): 4912–4915. https://doi.org/10.1021/jo00303a029

Gurgui C, Piel J (2010) Metagenomic approaches to identify and isolate bioactive natural products from the microbiota of marine sponges. Methods Mol Biol 668:247–264. https://doi.org/10.1007/978-1-60761-823-2_17. PMID 20830569

Hardoim CCP, Cardinale M, Cúcio AC, Esteves AI, Berg G, Xavier JR, Cox CJ, Costa R (2014) Effects of sample handling and cultivation bias on the specificity of bacterial communities in keratose marine sponges. Front Microbiol 5:611. https://doi.org/10.3389/fmicb.2014.00611. PMID 25477868

He R, Wakimoto T, Egami Y, Kenmoku H, Ito T, Asakawa Y, Abe I (2012a) Heterologously expressed β-hydroxyl fatty acids from a metagenomic library of a marine sponge. Bioorg Med Chem Lett 22(24):7322–7325. https://doi.org/10.1016/j.bmcl.2012.10.082. PMID 23127887

He R, Wakimoto T, Takeshige Y, Egami Y, Kenmoku H, Ito T, Wang BC, Asakawa Y, Abe I (2012b) Porphyrins from a metagenomic library of the marine sponge *Discodermia calyx*. Mol Biosyst 8(9):2334–2338. https://doi.org/10.1039/c2mb25169h. PMID 22735778

He R, Wang B, Wakimoto T, Wang M, Zhu L, Abe I (2013) Cyclodipeptides from metagenomics library of a Japanese marine sponge. J Braz Chem Soc 24:1926–1932. https://doi.org/10.5935/0103-5053.20130240

Hellio C, Tsoukatou M, Maréchal JP, Aldred N, Beaupoil C, Clare AS, Vagias C, Roussis V (2005) Inhibitory effects of Mediterranean sponge extracts and metabolites on larval settlement of the barnacle *Balanus amphitrite*. Mar Biotechnol (NY) 7(4):297–305. https://doi.org/10.1007/s10126-004-3150-x. PMID 15971089

Hentschel U, Hopke J, Horn M, Friedrich AB, Wagner M, Hacker J, Moore BS (2002) Molecular evidence for a uniform microbial community in sponges from different oceans. Appl Environ Microbiol 68(9):4431–4440. https://doi.org/10.1128/AEM.68.9.4431-4440.2002PMID 122002 97

Hentschel U, Piel J, Degnan SM, Taylor MW (2012) Genomic insights into the marine sponge microbiome. Nat Rev Microbiol 10(9):641–654. https://doi.org/10.1038/nrmicro2839. PMID 22842661

Higgs MD, Faulkner DJ (1978) Plakortin, an antibiotic from *Plakortis halichondrioides*. J Org Chem 43(18):3454–3457. https://doi.org/10.1021/jo00412a006

Hochmuth T, Piel J (2009) Polyketide synthases of bacterial symbionts in sponges—evolution-based applications in natural products research. Phytochemistry 70(15–16):1841–1849. https://doi.org/10.1016/j.phytochem.2009.04.010. PMID 19443000

Hoffmann F, Larsen O, Thiel V, Rapp HT, Pape T, Michaelis W, Reitner J (2005) An anaerobic world in sponges. Geomicrobiol J 22(1–2):1–10. https://doi.org/10.1080/01490450590922505

Human Microbiome Project (2012) Structure, function and diversity of the healthy human microbiome. Nature 486(7402):207–214. https://doi.org/10.1038/nature11234. PMID 22 699609

Jenke-Kodama H, Dittmann E (2009) Evolution of metabolic diversity: insights from microbial polyketide synthases. Phytochemistry 70(15–16):1858–1866. https://doi.org/10.1016/j.phytochem.2009.05.021. PMID 19619887

Jimeno J, López-Martín JA, Ruiz-Casado A, Izquierdo MA, Scheuer PJ, Rinehart K (2004) Progress in the clinical development of new marine-derived anticancer compounds. Anti Cancer Drugs 15(4):321–329. https://doi.org/10.1097/00001813-200404000-00003. PMID 15057135

Kaeberlein T, Lewis K, Epstein SS (2002) Isolating "uncultivable" microorganisms in pure culture in a simulated natural environment. Science 296(5570):1127–1129. https://doi.org/10.1126/science.1070633. PMID 12004133

Kato Y, Fusetani N, Matsunaga S, Hashimoto K, Fujita S, Furuya T (1986) Bioactive marine metabolites. Part 16. Calyculin A. A novel antitumor metabolite from the marine sponge *Discodermia calyx*. J Am Chem Soc 108(10):2780–2781. https://doi.org/10.1021/ja00270a061

Katz M, Hover BM, Brady SF (2016) Culture-independent discovery of natural products from soil metagenomes. J Ind Microbiol Biotechnol 43(2–3):129–141. https://doi.org/10.1007/s10295-015-1706-6. PMID 26586404

Kennedy J, Marchesi JR, Dobson ADW (2007) Metagenomic approaches to exploit the biotechnological potential of the microbial consortia of marine sponges. Appl Microbiol Biotechnol 75(1): 11–20. https://doi.org/10.1007/s00253-007-0875-2. PMID 17318533

Kennedy J, Marchesi JR, Dobson ADW (2008) Marine metagenomics: strategies for discovering novel enzymes with biotechnological applications from marine environments. Microb Cell Fact 7(27):27. https://doi.org/10.1186/1475-2859-7-27. PMID 18717988

Kennedy J, Flemer B, Jackson SA, Lejon DPH, Morrissey JP, O'Gara F, Dobson ADW (2010) Marine metagenomics: new tools for studying and exploiting marine microbial metabolism. Mar Drugs 8(3):608–628. https://doi.org/10.3390/md8030608. PMID 20411118

Khan ST, Komaki H, Motohashi K, Kozone I, Mukai A, Takagi M, Shin-Ya K (2011) Streptomyces associated with a marine sponge *Haliclona* sp.; biosynthetic genes for secondary metabolites and products. Environ Microbiol 13(2):391–403. https://doi.org/10.1111/j.1462-2920.2010.02337.x. PMID 20849448

Kim TK, Garson MJ, Fuerst JA (2005) Marine actinomycetes related to the 'Salinospora' group from the Great Barrier Reef sponge *Pseudoceratina clavata*. Environ Microbiol 7(4):509–518. https://doi.org/10.1111/j.1462-2920.2005.00716.x. PMID 15816928

Kimura M, Wakimoto T, Egami Y, Tan KC, Ise Y, Abe I (2012) Calyxamides A and B, cytotoxic cyclic peptides from the marine sponge *Discodermia calyx*. J Nat Prod 75(2):290–294. https://doi.org/10.1021/np2009187. PMID 22276742

Klindworth A, Pruesse E, Schweer T, Peplies J, Quast C, Horn M, Glöckner FO (2013) Evaluation of general 16S ribosomal RNA gene PCR primers for classical and next-generation sequencing-based diversity studies. Nucleic Acids Res 41(1):e1. https://doi.org/10.1093/nar/gks808. PMID 22933715

Knappe TA, Linne U, Zirah S, Rebuffat S, Xie X, Marahiel MA (2008) Isolation and structural characterization of capistruin, a lasso peptide predicted from the genome sequence of *Burkholderia thailandensis* E264. J Am Chem Soc 130(34):11446–11454. https://doi.org/10.1021/ja802966g. PMID 18671394

Kodzius R, Gojobori T (2015) Marine metagenomics as a source for bioprospecting. Mar Genomics 24(1):21–30. https://doi.org/10.1016/j.margen.2015.07.001. PMID 26204808

Kurnia NM, Uria AR, Kusnadi Y, Dinawati L, Zilda DS, Hadi TA, Setyahadi S, Felix F (2017) Metagenomic survey of potential symbiotic bacteria and polyketide synthase genes in an Indonesian marine sponge. HAYATI J Biosci 4(1):6–15. https://doi.org/10.1016/j.hjb.2017.04.004

Lane AL, Moore BS (2012) A sea of biosynthesis: marine natural products meet the molecular age. Nat Prod Rep 28(2):411–428

Laroche M, Imperatore C, Grozdanov L, Costantino V, Mangoni A, Hentschel U, Fattorusso E (2007) Cellular localisation of secondary metabolites isolated from the Caribbean sponge *Plakortis simplex*. Mar Biol 151(4):1365–1373. https://doi.org/10.1007/s00227-006-0572-1

Lee YK, Lee JH, Lee HK (2001) Microbial symbiosis in marine sponges. J Microbiol 39:254–264

Lee OO, Lau SCK, Qian PY (2006) Defense against epibiosis in the sponge *Mycale adhaerens*: modulating the bacterial community associated with its surface. Aquat Microb Ecol 43:55–65. https://doi.org/10.3354/ame043055

Li ZY, Wang YZ, He LM, Zheng HJ (2014) Metabolic profiles of prokaryotic and eukaryotic communities in deep-sea sponge *Neamphius huxleyi* indicated by metagenomics. Sci Rep 4: 3895. https://doi.org/10.1038/srep03895. PMID 24463735

Lipko IA, Kalyuzhnaya OV, Kravchenko OS, Parfenova VV (2012) Identification of polyketide synthase genes in genome of *Pseudomonas fluorescens* strain 28Bb-06 from freshwater sponge

Baikalospongia bacillifera. Mol Biol 46(4):609–611. https://doi.org/10.1134/S0026893312040073

Lipton AP, Pramitha VS, Selvin J, Huxley AJ (2006) Bioactivity profile of sponge-associated bacteria isolated from marine sponges, *Callyspongia* sp. and *Sigmadocia* sp. In: 7th international sponge symposium, vol 2006, Buzios-Rio de Janeiro, Brazil, p 250

Maldonado M, Aguilar R, Bannister R, Bell J, Conway K, Dayton P et al (2015) Sponge grounds as key marine habitats: a synthetic review of types, structure, functional roles and conservation concerns. In: Rossi S, Bramanti L, Gori A, Orejas C (eds) Marine animal forests. Springer, Cham, pp 1–39

Manzo E, Ciavatta ML, Villani G, Varcamonti M, Sayem SM, van Soest R, Gavagnin M (2011) Bioactive terpenes from *Spongia officinalis*. J Nat Prod 74(5):1241–1247. https://doi.org/10.1021/np200226u. PMID 21548580

Martinez A, Kolvek SJ, Yip CLT, Hopke J, Brown KA, MacNeil IA, Osburne MS (2004) Genetically modified bacterial strains and novel bacterial artificial chromosome shuttle vectors for constructing environmental libraries and detecting heterologous natural products in multiple expression hosts. Appl Environ Microbiol 70(4):2452–2463. https://doi.org/10.1128/AEM.70.4.2452-2463.2004. PMID 15066844

Matroodi S, Siitonen V, Baral B, Yamada K, Akhgari A, Metsä-Ketelä M (2020) Genotyping-guided discovery of Persiamycin a from sponge-associated halophilic *Streptomonospora* sp. PA3. Front Microbiol 11:1237. https://doi.org/10.3389/fmicb.2020.01237. PMID 32582127

McClintock JB, Amsler CD, Baker BJ, Van Soest RWM (2005) Ecology of Antarctic marine sponges: an overview. Integr Comp Biol 45(2):359–368. https://doi.org/10.1093/icb/45.2.359. PMID 21676781

Medema MH, Blin K, Cimermancic P, de Jager V, Zakrzewski P, Fischbach MA, Weber T, Takano E, Breitling R (2011) AntiSMASH: rapid identification, annotation and analysis of secondary metabolite biosynthesis gene clusters in bacterial and fungal genome sequences. Nucleic Acids Res 39(Web Server issue):W339–W346. https://doi.org/10.1093/nar/gkr466. PMID 21672958

Mehbub MF, Lei J, Franco C, Zhang W (2014) Marine sponge derived natural products between 2001 and 2010: trends and opportunities for discovery of bioactives. Mar Drugs 12(8):4539–4577. https://doi.org/10.3390/md12084539. PMID 25196730

Metsä-Ketelä M, Salo V, Halo L, Hautala A, Hakala J, Mäntsälä P, Ylihonko K (1999) An efficient approach for screening minimal PKS genes from Streptomyces. FEMS Microbiol Lett 180(1):1–6. https://doi.org/10.1111/j.1574-6968.1999.tb08770.x. PMID 10547437

Milshteyn A, Schneider JS, Brady SF (2014) Mining the metabiome: identifying novel natural products from microbial communities. Chem Biol 21(9):1211–1223. https://doi.org/10.1016/j.chembiol.2014.08.006. PMID 25237864

Mocali S, Benedetti A (2010) Exploring research frontiers in microbiology: the challenge of metagenomics in soil microbiology. Res Microbiol 161(6):497–505. https://doi.org/10.1016/j.resmic.2010.04.010. PMID 20452420

Moitinho-Silva L, Díez-Vives C, Batani G, Esteves AI, Jahn MT, Thomas T (2017a) Integrated metabolism in sponge-microbe symbiosis revealed by genome-centered metatranscriptomics. ISME J 11(7):1651–1666. https://doi.org/10.1038/ismej.2017.25. PMID 28338677

Moitinho-Silva L, Nielsen S, Amir A, Gonzalez A, Ackermann GL, Cerrano C, Astudillo-Garcia C, Easson C, Sipkema D, Liu F, Steinert G, Kotoulas G, McCormack GP, Feng G, Bell JJ, Vicente J, Björk JR, Montoya JM, Olson JB, Reveillaud J, Steindler L, Pineda MC, Marra MV, Ilan M, Taylor MW, Polymenakou P, Erwin PM, Schupp PJ, Simister RL, Knight R, Thacker RW, Costa R, Hill RT, Lopez-Legentil S, Dailianis T, Ravasi T, Hentschel U, Li Z, Webster NS, Thomas T (2017b) The sponge microbiome project. Gigascience 6(10):1–7. https://doi.org/10.1093/gigascience/gix077. PMID 29020741

Monks NR, Lerner C, Henriques AT, Farias FM, Schapoval EES, Suyenaga ES, da Rocha AB, Schwartsmann G, Mothes B (2002) Anticancer, antichemotactic and antimicrobial activities of marine sponges collected off the coast of Santa Catarina, southern Brazil. J Exp Mar Biol Ecol 281(1–2):1–12. https://doi.org/10.1016/S0022-0981(02)00380-5

Mori T, Cahn JKB, Wilson MC, Meoded RA, Wiebach V, Martinez AFC, Helfrich EJN, Albersmeier A, Wibberg D, Dätwyler S, Keren R, Lavy A, Rückert C, Ilan M, Kalinowski J, Matsunaga S, Takeyama H, Piel J (2018) Single-bacterial genomics validates rich and varied specialized metabolism of uncultivated *Entotheonella* sponge symbionts. Proc Natl Acad Sci U S A 115(8):1718–1723. https://doi.org/10.1073/pnas.1715496115. PMID 29439203

Newman DJ, Cragg GM, Snader KM (2000) The influence of natural products upon drug discovery. Nat Prod Rep 17(3):215–234. https://doi.org/10.1039/a902202c. PMID 10888010

Ng TB, Cheung RCF, Wong JH, Bekhit AA, Ael-D B (2015) Antibacterial products of marine organisms. Appl Microbiol Biotechnol 99(10):4145–4173. https://doi.org/10.1007/s00253-015-6553-x. PMID 25874533

Nikolouli K, Mossialos D (2012) Bioactive compounds synthesized by nonribosomal peptide synthetases and type-I polyketide synthases discovered through genome-mining and metagenomics. Biotechnol Lett 34(8):1393–1403. https://doi.org/10.1007/s10529-012-0919-2. PMID 22481301

Ogram A, Sayler GS, Barkay T (1987) The extraction and purification of microbial DNA from sediments. J Microbiol Methods 7(2–3):57–66. https://doi.org/10.1016/0167-7012(87)90025-X

Osinga R, Tramper J, Wijffels RH (1998) Cultivation of marine sponges for metabolite production: applications for biotechnology? Trends Biotechnol 16(3):130–134. https://doi.org/10.1016/S0167-7799(97)01164-5

Ouyang Y, Dai S, Xie L, Ravi Kumar MSR, Sun W, Sun H, Tang D, Li X (2010) Isolation of high molecular weight DNA from marine sponge bacteria for BAC library construction. Mar Biotechnol (NY) 12(3):318–325. https://doi.org/10.1007/s10126-009-9223-0. PMID 19685098

Passarini MRZ, Miqueletto PB, de Oliveira VM, Sette LD (2015) Molecular diversity of fungal and bacterial communities in the marine sponge *Dragmacidon reticulatum*. J Basic Microbiol 55(2): 207–220. https://doi.org/10.1002/jobm.201400466. PMID 25213208

Penesyan A, Kjelleberg S, Egan S (2010) Development of novel drugs from marine surface associated microorganisms. Mar Drugs 8(3):438–459. https://doi.org/10.3390/md8030438. PMID 20411108

Piel J (2004) Metabolites from symbiotic bacteria. Nat Prod Rep 21(4):519–538. https://doi.org/10.1039/b310175b. PMID 15282634

Piel J (2010) Biosynthesis of polyketides by trans-AT polyketide synthases. Nat Prod Rep 27(7): 996–1047. https://doi.org/10.1039/b816430b. PMID 20464003

Piel J, Hui D, Wen G, Butzke D, Platzer M, Fusetani N, Matsunaga S (2004) Antitumor polyketide biosynthesis by an uncultivated bacterial symbiont of the marine sponge *Theonella swinhoei*. Proc Natl Acad Sci U S A 101(46):16222–16227. https://doi.org/10.1073/pnas.0405976101. PMID 15520376

Pimentel-Elardo SM, Grozdanov L, Proksch S, Hentschel U (2012) Diversity of nonribosomal peptide synthetase genes in the microbial metagenomes of marine sponges. Mar Drugs 10(6): 1192–1202. https://doi.org/10.3390/md10061192. PMID 22822366

Pineda MC, Duckworth A, Webster N (2015) Appearance matters: sedimentation effects on different sponge morphologies. J Mar Biol Assoc 96:481–492

Pita L, Turon X, López-Legentil S, Erwin PM (2013) Hosts. Host rules: spatial stability of bacterial communities associated with marine sponges (*Ircinia* spp.) in the Western Mediterranean Sea. FEMS Microbiol Ecol 86(2):268–276. https://doi.org/10.1111/1574-6941.12159. PMID 2383 7533

Pomponi SA (1999) The bioprocess—technological potential of the sea. J Biotechnol 70(1–3): 5–13. https://doi.org/10.1016/S0168-1656(99)00053-X

Preston CM, Wu KY, Molinski TF, DeLong EF (1996) A psychrophilic crenarchaeon inhabits a marine sponge: *Cenarchaeum symbiosum* gen. nov., sp. nov. Proc Natl Acad Sci U S A 93(13): 6241–6246. https://doi.org/10.1073/pnas.93.13.6241. PMID 8692799

Proksch P, Edrada RA, Ebel R (2002) Drugs from the sea—current status and microbiological implications. Appl Microbiol Biotechnol 59(2–3):125–134. https://doi.org/10.1007/s00253-002-1006-8. PMID 12111137

Rifai S, Fassouane AF, Kijjoa A, Van Soest R (2004) Antimicrobial activity of Untenospongin B, a metabolite from the marine sponge *Hippospongia communis* collected from the Atlantic Coast of Morocco. Mar Drugs 2(3):147–153. https://doi.org/10.3390/md203147

Rocha-Martin J, Harrington C, Dobson ADW, O'Gara F (2014) Emerging strategies and integrated systems microbiology technologies for biodiscovery of marine bioactive compounds. Mar Drugs 12(6):3516–3559. https://doi.org/10.3390/md12063516. PMID 24918453

Rohde S, Nietzer S, Schupp PJ (2015) Prevalence and mechanisms of dynamic chemical defenses in tropical sponges. PLoS One 10(7):e0132236. https://doi.org/10.1371/journal.pone.0132236. PMID 26154741

Scherlach K, Hertweck C (2009) Triggering cryptic natural product biosynthesis in microorganisms. Org Biomol Chem 7(9):1753–1760. https://doi.org/10.1039/b821578b. PMID 19590766

Schloss PD, Handelsman J (2003) Biotechnological prospects from metagenomics. Curr Opin Biotechnol 14(3):303–310. https://doi.org/10.1016/s0958-1669(03)00067-3. PMID 12849784

Schmeisser C, Steele H, Streit WR (2007) Metagenomics, biotechnology with non-culturable microbes. Appl Microbiol Biotechnol 75(5):955–962. https://doi.org/10.1007/s00253-007-0945-5. PMID 17396253

Selvin J, Gandhimathi R, Kiran GS, Priya SS, Ravji TR, Hema TA (2009) Culturable heterotrophic bacteria from the marine sponge *Dendrilla nigra*: isolation and phylogenetic diversity of actinobacteria. Helgol Mar Res 63(3):239–247. https://doi.org/10.1007/s10152-009-0153-z

Selvin J, Ninawe AS, Kiran GS, Lipton AP (2010) Sponge microbial interactions: ecological implications and bioprospecting avenues. Crit Rev Microbiol 36(1):82–90. https://doi.org/10.3109/10408410903397340. PMID 20088685

Siegl A, Kamke J, Hochmuth T, Piel J, Richter M, Liang C, Dandekar T, Hentschel U (2011) Single-cell genomics reveals the lifestyle of Poribacteria, a candidate phylum symbiotically associated with marine sponges. ISME J 5(1):61–70. https://doi.org/10.1038/ismej.2010.95. PMID 20613790

Singh BK, MacDonald CA (2010) Drug discovery from uncultivable microorganisms. Drug Discov Today 15(17–18):792–799. https://doi.org/10.1016/j.drudis.2010.07.002. PMID 20656054

Sipkema D, Schippers K, Maalcke WJ, Yang Y, Salim S, Blanch HW (2011) Multiple approaches to enhance the cultivability of bacteria associated with the marine sponge *Haliclona* (*Gellius*) sp. Appl Environ Microbiol 77(6):2130–2140. https://doi.org/10.1128/AEM.01203-10. PMID 21296954

Skinnider MA, Dejong CA, Rees PN, Johnston CW, Li H, Webster ALH, Wyatt MA, Magarvey NA (2015) Genomes to natural products PRediction informatics for secondary metabolomes (PRISM). Nucleic Acids Res 43(20):9645–9662. https://doi.org/10.1093/nar/gkv1012. PMID 2 6442528

Sleator RD, Shortall C, Hill C (2008) Metagenomics. Lett Appl Microbiol 47(5):361–366. https://doi.org/10.1111/j.1472-765X.2008.02444.x. PMID 19146522

Stein JL, Marsh TL, Wu KY, Shizuya H, DeLong EF (1996) Characterization of uncultivated prokaryotes: isolation and analysis of a 40-kilobase-pair genome fragment from a planktonic marine archaeon. J Bacteriol 178(3):591–599. https://doi.org/10.1128/jb.178.3.591-599.1996. PMID 8550487

Stierle AC, Cardellina JH, Singleton FL (1988) A marine *micrococcus* produces metabolites ascribed to the sponge *Tedania ignis*. Experientia 44(11–12):1021. https://doi.org/10.1007/BF01939910. PMID 3197807

Taylor MW, Radax R, Steger D, Wagner M (2007) Sponge-associated microorganisms: evolution, ecology, and biotechnological potential. Microbiol Mol Biol Rev 71(2):295–347. https://doi.org/10.1128/MMBR.00040-06. PMID 17554047

Thomas T, Rusch D, DeMaere MZ, Yung PY, Lewis M, Halpern A, Heidelberg KB, Egan S, Steinberg PD, Kjelleberg S (2010) Functional genomic signatures of sponge bacteria reveal unique and shared features of symbiosis. ISME J 4(12):1557–1567. https://doi.org/10.1038/ismej.2010.74. PMID 20520651

Thomas T, Moitinho-Silva L, Lurgi M, Björk JR, Easson C, Astudillo-García C, Olson JB, Erwin PM, López-Legentil S, Luter H, Chaves-Fonnegra A, Costa R, Schupp PJ, Steindler L, Erpenbeck D, Gilbert J, Knight R, Ackermann G, Victor Lopez J, Taylor MW, Thacker RW, Montoya JM, Hentschel U, Webster NS (2016) Diversity, structure and convergent evolution of the global sponge microbiome. Nat Commun 7:11870. https://doi.org/10.1038/ncomms11870. PMID 27306690

Trindade M, van Zyl LJ, Navarro-Fernández J, Abd Elrazak A (2015) Targeted metagenomics as a tool to tap into marine natural product diversity for the discovery and production of drug candidates. Front Microbiol 6:890. https://doi.org/10.3389/fmicb.2015.00890. PMID 26379658

Ueoka R, Uria AR, Reiter S, Mori T, Karbaum P, Peters EE, Helfrich EJN, Morinaka BI, Gugger M, Takeyama H, Matsunaga S, Piel J (2015) Metabolic and evolutionary origin of actin-biding polyketides from diverse organisms. Nat Chem Biol 11:705.e12

Unson MD, Faulkner DJ (1993) Cyanobacterial symbiont biosynthesis of chlorinated metabolites from *Dysidea herbacea* (Porifera). Experientia 49(4):349–353. https://doi.org/10.1007/BF01923420

Uria AR, Fawzya YN, Chasanah E (2005) Novel molecular methods for discovery and engineering of biocatalysts from uncultured marine microorganisms. J Coast Dev 8(2):53–71

Uriz MJ, Turon X, Becerro MA, Galera J (1996) Feeding deterrence in sponges. The role of toxicity, physical defenses, energetic contents, and life-history stage. J Exp Mar Biol Ecol 205(1–2):187–204. https://doi.org/10.1016/S0022-0981(96)02609-3

van Soest RWM, Boury-Esnault N, Vacelet J, Dohrmann M, Erpenbeck D, De Voogd NJ et al (2012) Global diversity of sponges (Porifera). PLoS One 7(4):e35105. https://doi.org/10.1371/journal.pone.0035105. PMID 22558119

Vogel S (1977) Current-induced flow through living sponges in nature. Proc Natl Acad Sci U S A 74(5):2069–2071. https://doi.org/10.1073/pnas.74.5.2069. PMID 266728

Webster NS, Hill RT (2001) The culturable community of the Great Barrier Reef sponge *Rhopaloeidesodorabile* is dominated by an proteobacterium. Mar Biol 138(4):843–851. https://doi.org/10.1007/s002270000503

Webster NS, Thomas T (2016) The sponge hologenome. mBio 7(2):e00135-16. https://doi.org/10.1128/mBio.00135-16. PMID 27103626

Webster NS, Wilson KJ, Blackall LL, Hill RT (2001) Phylogenetic diversity of bacteria associated with the marine sponge *Rhopaloeidesodorabile*. Appl Environ Microbiol 67(1):434–444. https://doi.org/10.1128/AEM.67.1.434-444.2001. PMID 11133476

Weiland N, Löscher C, Metzger R, Schmitz R (2010) Construction and screening of marine metagenomic libraries. Methods Mol Biol 668:51–65

Wilson MC, Piel J (2013) Metagenomic approaches for exploiting uncultivated bacteria as a resource for novel biosynthetic enzymology. Chem Biol 20(5):636–647. https://doi.org/10.1016/j.chembiol.2013.04.011. PMID 23706630

Wu D, Hugenholtz P, Mavromatis K, Pukall R, Dalin E, Ivanova NN, Kunin V, Goodwin L, Wu M, Tindall BJ, Hooper SD, Pati A, Lykidis A, Spring S, Anderson IJ, D'haeseleer P, Zemla A, Singer M, Lapidus A, Nolan M, Copeland A, Han C, Chen F, Cheng JF, Lucas S, Kerfeld C, Lang E, Gronow S, Chain P, Bruce D, Rubin EM, Kyrpides NC, Klenk HP, Eisen JA (2009) A phylogeny-driven genomic encyclopaedia of bacteria and archaea. Nature 462(7276): 1056–1060. https://doi.org/10.1038/nature08656. PMID 20033048

Yun J, Ryu S (2005) Screening for novel enzymes from metagenome and SIGEX, as a way to improve it. Microb Cell Factories 4(1):8. https://doi.org/10.1186/1475-2859-4-8. PMID 1 5790425

Zhang W, Zhang X, Cao X, Xu J, Zhao Q, Yu X, Jin M, Deng M (2003) Optimizing the formation of in vitro sponge primmorphs from the Chinese sponge *Stylotella agminata* (Ridley). J Biotechnol 100(2):161–168. https://doi.org/10.1016/s0168-1656(02)00255-9. PMID 12423910

Zhang L, An R, Wang J, Sun N, Zhang S, Hu J, Kuai J (2005) Exploring novel bioactive compounds from marine microbes. Curr Opin Microbiol 8(3):276–281. https://doi.org/10.1016/j.mib.2005.04.008. PMID 15939350

Zhang F, Blasiak LC, Karolin JO, Powell RJ, Geddes CD, Hill RT (2015) Phosphorus sequestration in the form of polyphosphate by microbial symbionts in marine sponges. Proc Natl Acad Sci U S A 112(14):4381–4386. https://doi.org/10.1073/pnas.1423768112. PMID 25713351

3 Biomedical Applications of Marine Biopolymers in Tissue Engineering and Regenerative Medicine

Prakash Narayana Reddy, Harish Babu Kolla, Chanda Vikrant Berde, Rohini Krishna Kota, and Pallaval Veera Bramhachari

Abstract

Marine ecosystems harbor incredibly rich biodiversity, which houses an extensive variety of biological substances with abundant economic value. Tissue engineering and regenerative medicine (TERM) employ engineering and natural biotic laws for fabricating novel tissues and organs, encouraging the repair and renewal of tissues and organs impaired by trauma or disease. Biopolymers were studied extensively in regenerative medicine due to their nontoxicity, low antigenicity, feasibility to complicated shapes and scaffolds, appropriate porosity, and mechanical strength with the ability to support cell growth and proliferation. Biopolymers derived from marine sources form a part of the growing number of materials reported as suitable options in regenerative medicine due to their natural availability, biodegradability, biocompatibility, and ease of isolation. The biomedical industry is investigating marine biomaterials with specific properties for application in bone regeneration, wound dressing, wound healing, regenerative

P. N. Reddy (✉)
Microbiology Department, Dr. V.S. Krishna Government Degree College (Autonomous), Visakhapatnam, Andhra Pradesh, India

H. B. Kolla
Department of Biotechnology, Vignan's Foundation for Science, Technology and Research (Deemed to be University), Guntur, Andhra Pradesh, India

C. V. Berde
Ocean and Atmospheric Sciences, Goa University, Panaji, Goa, India

R. K. Kota
Development Quality Control (DQC), Bharat Biotech, Genome Valley, Shameerpet, Hyderabad, India

P. Veera Bramhachari
Department of Biotechnology, Krishna University, Machilipatnam, Andhra Pradesh, India

P. Veera Bramhachari, C. V. Berde (eds.), *Marine Bioactive Molecules for Biomedical and Pharmacotherapeutic Applications*,
https://doi.org/10.1007/978-981-99-6770-4_3

medicine, and drug delivery. In the present review, we have discussed some recent developments in marine biopolymers focusing on their promise for biomedical applications, specifically in TERM.

Keywords

Marine biopolymers · Biomedicine · Tissue engineering · Drug delivery · Carrageenan · Fucoidan

3.1 Introduction

Damage to tissues and organs is a serious problem caused due to accidents, infections, or illnesses. It remained an unresolved challenge to anneal the damaged organs and restore their functionality due to the shortage of organs for transplantation. In the United States, an average of 18 deaths are recorded each day due to the shortage of organs (Liu et al. 2013). The increase in organ transplantation surgeries' failure due to graft rejections has imposed another challenge to surpass this serious issue. With the failures in organ transplantation surgeries and the shortage of organs, tissue engineering strategies have become eminent in developing new alternatives to repair and replace defective organs (Mandrycky et al. 2017). With advances in material sciences, biomedicine and bioengineering technologies, researchers are fostering various approaches to provide solutions to address this serious issue. In 1993, regenerative medicine through tissue engineering (TE) was proposed to overcome the body's shortage of organs and graft rejections (Vacanti and Vacanti 2000).

Tissue engineering and regenerative medicine employ the principles of engineering and biology to fabricate fresh tissues and organs and induce the renewal of new tissues and organs destroyed by trauma or any disease. It focuses on developing supporting matrices or scaffolds where cells are seeded and cultured with certain growth regulators and bioactive compounds. These supplements promote cell growth, and the supporting scaffolds point to new organs as implants to replace the damaged tissues or organs (Kaihara and Vacanti 1999). The engineered implants adequately integrate at the implantation location, followed by cells undergoing multiplication and differentiation to produce a new extracellular matrix. Then the implanted scaffolds undergo degradation and excrete out of the body (Velasco et al. 2015). With this perception, various biodegradable polymers such as collagen, silk fibroin and chondroitin sulfate were used to develop the scaffold materials. But it is believed that these materials would interfere with the body's growth, physiology and biological roles by interacting with other cells (Grabska-Zielińska et al. 2020). The marine ecosystem harbors abundant natural substances with great economic and medicinal importance. Marine biopolymers were considered ideal polymers for developing scaffolds in supporting tissue and organ matrices. Their inertness towards the cellular processes, porosity, ability to support cell growth, nontoxicity, low antigenicity, biodegradability, and flexibility to complex shapes are suitable

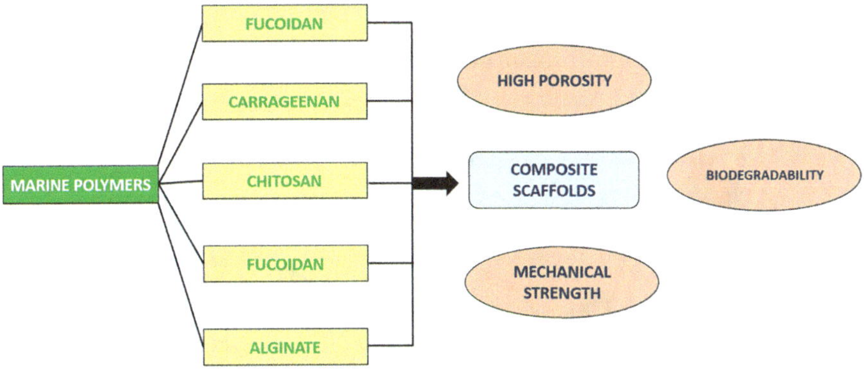

Fig. 3.1 Popular marine polymers and their desirable properties for use in TERM

materials in developing scaffolds for TERM. Chitin/chitosan, alginic acid, fucoidan, and carrageenan are marine biopolymers currently used in modern medicine and applied in TERM. Recent advancements in biomedicine focusing on potential marine biopolymers in TERM were reviewed in the following sections of this article. Some of the marine biopolymers used in TERM and their desirable properties are depicted in Fig. 3.1.

3.2 Marine Biopolymers in TERM

Regenerative medicine is an interdisciplinary field of biomedical research that integrates the laws of medicine, materials science and biology that enables the generation of organs having better function and biological structures. It combines tissue engineering and drug delivery approaches (Murphy and Atala 2014). Regenerative medicine helps implant the framework material that is the basis of recruiting the native cells into the scaffold, thereby resulting in extracellular matrix (ECM) deposition. Cell scaffold provides artificial ECM in a temporary environment that will help support the cells in adherence, infiltration, proliferation, and differentiation (Hussey et al. 2018). Cell scaffolds perform a major part in TE as they can provide early mechanical support and maintain cells present in the damaged areas aiding in cell division, metabolism and production of ECM (Mantha et al. 2019).

The properties of an ideal scaffolding material include desired mechanical strength, suitable microstructure, and biocompatibility, along with the capability to host cells and retain their biochemical processes (Salvatore et al. 2021). Cell-supporting matrices are made of both natural and synthetic biomaterials. Polymers from the natural origin are preferred as they share similarities with ECM, are chemically versatile, and have improved biological performance (Boni et al. 2018). Marine biopolymers were used in regenerative medicine for varied tissue applications, which is depicted in Fig. 3.2.

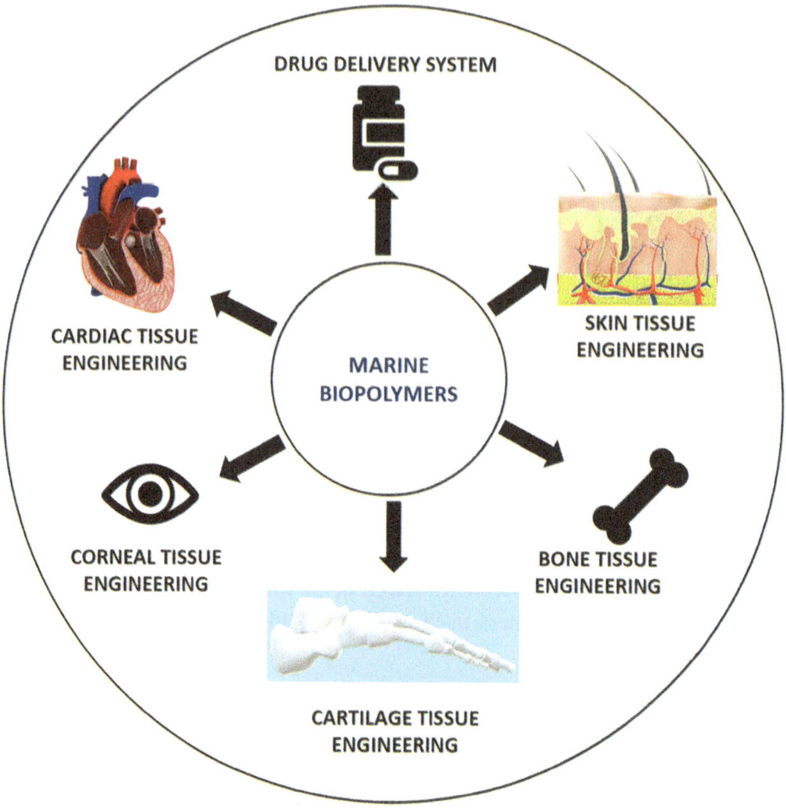

Fig. 3.2 Medical applications of marine biopolymers in tissue engineering. (Figure adapted from Kalirajan et al. 2021 and redrawn for this article)

3.3 Tissue Engineering: Perspectives of Marine Biopolymers

3.3.1 Fucoidan

Fucoidan is a polysaccharide composed of fucose and sulfate groups. It is present in various pheophycaea (brown algae) members. Fucoidan is also a component in the cell wall of red algae to protect them from external stress conditions. The commercial form of fucoidan was obtained from the seaweeds like *Cladosiphon okamuranus*, *Fucus vesiculosus*, *Laminaria japonica,* and *Undaria pinnatifida* (Fitton 2011). Other fucoidan forms were also obtained from invertebrate species, such as sea hedgehogs and sea cucumbers (Atashrazm et al. 2015). Fucoidan is primarily comprised of galactose, sulfated fucose, xylose, and uronic acids etc. (Rioux et al. 2007). Fucoidan was studied for its therapeutic activities like anti-cancer, anti-coagulant, anti-inflammatory, anti-oxidant, anti-viral, and immunomodulatory functions (Jin and Kim 2011). Fucoidan has been used in engineering various damaged tissues and organs in the past decade. The anionic property of fucoidan, owing to the fucose and sulfate group esters, tends to induce osteoblast formations in the mesenchymal stem cells (Li et al. 2008). A composite from the chitosan-alginate-fucoidan combined improvised the bioactivity, mineralization and adsorption of proteins for bone regeneration in MG63 cells (Venkatesan et al. 2014). In another study, a fucoidan-incorporated scaffold Tcp-Fu-Ch (Tricalcium phosphate-fucoidan and chitosan) exhibited osteogenic differentiation of bone marrow cells that aided treatment for bone ailments (Puvaneswary et al. 2016). Besides the regeneration of bones, fucoidans were extensively investigated for their potential role in wound healing and dressing. Fucoidan alleviates and stimulates the heparin-binding cytokines that help to promote new blood vessel formation during wound repair. Fucoidans interact with cell signaling factors such as fibroblast growth factor (FGF) and transforming growth factor-β (TGF-β) to induce fibroblast proliferation resulting in the reconstruction of skin epidermis (O'Leary et al. 2004; Song et al. 2014). Therefore, it facilitates wound healing with nontoxic action on the tissues.

Several reports have proved that fucoidans are key regulators for wound healing. For instance, a low-molecular-weight fucoidan with 5 kDa derived from the seaweed *Undaria pinnatifida* has thickened dermal excision in a wound-healing rat model (Obluchinsksya et al. 2015). This was a dose-dependent effect where wounded animals treated with this solution showed accelerated angiogenesis and remodeled collagen. A combination of fucoidan and polyvinyl alcohol (PVA) in a 70:30 ratio showed enhanced wound healing effects within 8 days post-treatment. Higher collagen content was observed in the treatment group compared to the control with complete re-epithelialization and renewal of the epidermis at the wounded site (Feki et al. 2019). A micro chitosan-fucoidan biodegradable hydrogel created by Nakamura et al. enhanced the fibroblast growth factor 2 (FGF-2) function and secured it from deactivation. Mice injected with this hydrogel subcutaneously developed significant neovascularization at the injection site (Nakamura et al. 2008). Micro and macro 3D scaffolds based on fucoidan filled with vascular endothelial growth factor (VEGF) facilitate neovascularization by ensuring angiogenic activity in ischemic tissues (Purnama et al. 2013).

3.3.2 Carrageenan

Carrageenan (CRG) is a seaweed polysaccharide isolated from marine red algae (Rhodophyceae) bearing distinguishing structural and functional characteristics. CRG is an anionic, sulfated polygalacton comprising of interchanging linear polymers of α-1,3 D-galactose and β-(1-4)-3,6-anhydro-D-galactose bearing ester sulfates (15–40%). CRGs closely resemble glycosaminoglycans of ECM (Yegappan et al. 2018). CRGs are categorized into six types based on the percent of sulfation and their origin, such as Kappa (κ)-, Iota (ι)-, Lambda (λ)-, Mu (μ)-, Nu (ν)-, and Theta (θ)-CRGs. Kappa, Iota, and Lambda are commercially important (Cunha and

Grenha 2016). CRG is popularly utilized as gelling, emulsifying, and stabilizing agent in medical and industrial applications (Liu et al. 2015). CRG was studied for its anti-inflammatory and anti-cancer properties. CRG was further identified as a herpes and human papillomavirus inhibitor (Zhou et al. 2004; Panlasigui et al. 2003; Carlucci et al. 1999; Buck et al. 2006). Recently hydrogels based on marine biopolymers (chitosan, alginate, agarose, carrageenan) have been utilized in tissue engineering due to their abundance (Varoni et al. 2012; Vignesh et al. 2018). Some reports where CRG was utilized in regenerative medicine are presented in Table 3.1.

Hydrogels are complex and intricate framework of hydrophilic long biopolymers comprising 90–99% of water. These materials gained a reputation for their applications in tissue engineering due to excellent biological behavior like biodegradability, biocompatibility, minimum cytotoxicity, immunogenicity and ease of tuning their physical and chemical properties (Saul and Williams 2011). The 3D structure closely mimics the tissue extracellular matrix environment, providing a biocompatible environment for cell growth and proliferation (Geckil et al. 2010). Out of the CRGs available in markets, Kappa CRG, upon cooling and under the availability of cations such as K^+ and Ca^{2+} undergoes structural and mechanical transitions and aggregations led to form thermotropic and ionotropic hydrogels. CRG hydrogels are brittle under physiological conditions and have a high swelling ratio with poor mechanical stability (Thakur et al. 2016). CRG polymeric backbone is amenable to chemical alterations because hydroxyl or sulfate functional moieties are available. Some chemical modifications on CRG, such as oxidation, over sulfation, carboxymethylation, methylacrylation, and phosphorylation, have endowed the polymer with new functionalities and properties (Yegappan et al. 2018).

When mixing with functional bioactive molecules, CRGs can form an apatite layer (Kim et al. 2011). CRG hydrogels incorporated with varied proportions of rod-shaped *n*-hydroxyapatite (*n*HAP) demonstrated enhanced cell-cell communication and adherence (Morais et al. 2013; González et al. 2017). Poly (hydroxybutyrate) (PHB) and poly(hydroxybutyrate valerate) (PHBV) nanofilaments engineered by superficial application of κ-CRG hydrogel led to improved osteoblast cell specialization and biomineralization. Compared to non-CRG fibers, CRG-coated fibers exhibit nano–microscale apatite crystals. Furthermore, CRG-coated fibers improved biomineralization upon culturing in osteosarcoma cells (Goonoo et al. 2017). Selenium nanoparticles (SeNPs) in the presence of hydroxyl, carboxyl and sulfate groups of CRG achieved excellent stability. Moreover, λ-CRG/SeNPs hydrogel showed no toxicity and displayed calcium deposits indicating biomineralization when cultured in the presence of osteoblast D1 cells (Kim et al. 2016). A blended scaffolding designed using ι-CRG, chitosan, and gelatin resulted in osteogenic delineation and rescue of adipose-derived mesenchymal stem cells (AdMSCs) because the availability of several chemical moieties such as hydroxyl, carboxyl, amine, and sulfate in the scaffolds favored cell survival (Li et al. 2015). Microvascular-like endothelial cells encapsulated inside κ-CRG hydrogel coated with the chitosan retained their functionality and phenotype, and survived a prolonged period (Mihaila et al. 2014). Collagen, hydroxyapatite and

Table 3.1 Recent biomedical applications of CRG in combination with other biomaterials in tissue engineering

S. no.	Carrageenan + other polymers	Properties observed	References
1.	Nanocomposite films of aldehyde-modified carrageenan/gelatin/halloysite nanotubes (AD-Carr/Gel/HNTs)	AD-Carr/Gel/HNTs nanocomposite films are found to be hemocompatible and nontoxic by in vitro assays	Akrami-Hasan-Kohal et al. (2020)
2.	Nano hydroxyapatite-silver—3% wt carrageenan	3% CRG addition led to higher stiffness of the hydrogel scaffold, induced better cell adhesion and proliferation and biodegradable rate	Swain et al. (2021)
3.	Poly(vinyl alcohol) (PVA)/methacrylate kappa-carrageenan (κ-CaMA) composite hydrogel encapsulated with chitooligosaccharides (COS) [PVA/κ-CaMA/COS]	PVA/κ-CaMA/COS hydrogel showed uniformly thick, highly porous 3D architecture with uniformly distributed pores, a high fluid absorption and retention capacity, mechanical stability and in vitro biocompatibility. Wound dressing resulted in tissue re-epithelization and skin tissue remodeling	Chandika et al. (2021)
4.	Carboxymethyl kappa-carrageenan + poly(vinyl alcohol)	Enhanced fibrinogen adsorption, platelet activation and blood clotting. Displayed antibacterial activity against *S. aureus* and *P. aeruginosa*; showed promise as wound dressing	Madruga et al. (2021b)
5.	Kappa carrageenan (κCA)-coated starch/cellulose nanofiber (CNF)	Surface coating of starch/CNF hydrogel with κCA improved mechanical properties, reduced swelling ability and degradation rate of starch/CNF, and showed blood compatibility and blood clotting ability	Tavakoli et al. (2021)
6.	Poly(vinyl alcohol) (PVA) and carboxymethyl-kappa-carrageenan (CMKC) blend poly(vinyl alcohol) (PVA) and carboxymethyl-kappa-carrageenan (CMKC) blends poly (vinyl alcohol) (PVA) and carboxymethyl-kappa-carrageenan (CMKC) blends poly(vinyl alcohol) and carboxymethyl-kappa-carrageenan (CMKC) blends	CMKC improved cytocompatibility, biodegradability, cell growth and adhesion compared to PVA fibers alone. PVA/CMKC nanofibers enhance human adipose-derived stem cells' response to osteogenic differentiation signals	Madruga et al. (2021a)
7.	Porous bioactive alginate/carrageenan/calcium silicate scaffolds	In BTE, the formation of hydroxyapatite crystals was observed on the surface of the scaffold after soaking in SBF solution; the presence of calcium	Sathain et al. (2021)

(continued)

Table 3.1 (continued)

S. no.	Carrageenan + other polymers	Properties observed	References
		silicate enhanced the bioactivity and mechanical properties of the scaffolds; in vitro drug release behavior of diclofenac-loaded scaffold was suitable for acute inflammation post-surgery	

κ-CRG complexed scaffold was fabricated bearing excellent bone-like innate structure. CRG components in the composite protect the nerve growth factor (NGF) by preventing its degradation and allowing sustained release to overcome frequent administration of NGF, which is critical for the repair, preservation and growth of nerves (Wang et al. 2009, 2010).

CRG contains several sulfate moieties in its structure, which imitates the biologically existing sulfated glycosaminoglycans in the ECM portion of cartilage, thus making CRG an excellent choice for the engineering of cartilage tissue (Popa et al. 2016). κ-CRG hydrocolloid in combination with iron oxide magnetic nanoparticles showed feasible properties such as increased stability, chondrocyte differentiation, cellular responses, and increase in expression of chondrocyte gene markers (Popa et al. 2012, 2016). Human adipose stem cells (HASCs) encapsulated in κ-CRG scaffold were noncytotoxic, and supported chondrocytes differentiation and proliferation in the presence of chondrogenic growth factors (Popa et al. 2015). CRG hydrogels have excellent mechanical properties and freeze-thaw stability during cryopreservation—thawing process making them excellent materials for encapsulation systems (Popa et al. 2013). κ-CRG hydrogel in the presence of human adipose-derived stem cells (hADSC) and TGF-β1 promoted chondrogenic differentiation combined with chondrocyte-related gene expression (Rocha et al. 2011). CRG, fibrin, and hyaluronic acid (HA) composite scaffold encapsulated human articular chondrocytes (HACs) have displayed better elasticity in comparison with fibrin/HA only besides promoting cartilage ECM accumulation (Fuss et al. 2000). Gelatin methacryloyl (GelMA) and methacrylated (Mκ-CRG) incorporated with nano silicates possessed multiple applications in tissue engineering. Seeding cells in GelMA and Mκ-CRG regions displayed round and diffused arrangement, a characteristic feature of chondrocytes and osteoblasts (Pereira et al. 2009). Despite several advantages of CRG in tissue engineering, few reports of adverse effects of CRG is an important drawback that requires attention when utilizing CRG for medical purposes. More in vivo research reports are needed to support the regenerative efficacy and medical application of CRG as a valid biopolymer for tissue engineering applications.

3.3.3 Chitosan

Chitosan is derived through the deacetylation of chitin which is part of the crustacean exoskeleton of Arthropod members such as crustaceans and insects. Chitosan is a natural polymer with a linear structure consisting of D-glucosamine units bonded through β(1-4)glycosidic linkages and varied *N*-acetyl-D-glucosamine (NAG) groups. The molecular mass of chitosan falls in the range of 300 to above 1000 kDa, subject to the degree of deacetylation, preparation method, and source of the polymer. Several reports have evidenced that chitosan displays increased cell attachment, division, osteoblast specialization, and mineralization (Seol et al. 2004; Di Martino et al. 2005; Şenel et al. 2004). In bone tissue engineering (BTE), chitosan could be fabricated to create 3D scaffoldings with numerous ceramics and polymers of varied porosity and possessing properties such as being bioactive, biodegradable, and biocompatible with a hydrophilic surface.

These composites can mimic the natural bone function and promote cell attachment, multiplication and differentiation. Chitosan could be chemically modified by attaching functional groups, peptides or amino acids on primary amines and secondary hydroxyl groups that aid in optimization and functionalization for use in BTE (Amidi et al. 2010). Chitosan was also used extensively in promoting wound healing, and it also possesses antimicrobial activity owing to its cationic property. Chitosan activates and modulates the inflammatory cells, thereby facilitating the development of tissues. As a wound dressing component, chitosan promotes blood clotting (Alexander et al. 1996). The role of chitosan in cell division, osteoblast specialization and bone mineralization has been well recognized (Van Tienen et al. 2002). Some of the research reports where chitosan was used for TERM applications are presented in Table 3.2.

Due to its physical properties, it is possible to fabricate chitosan into several structures, such as membranes, scaffolds, 3D structures, and fibers. Chitosan-based scaffolds are produced through lyophilization or freeze drying, 3D printing, salt leaching, and electrospinning methods. Most of these fabrication techniques have no control over the scaffold geometry and architecture. The application of CS scaffolds, irrespective of fabrication methodology on load-bearing sites, is highly doubtful. Therefore, the amalgamation of CS with other natural, synthetic, and ceramic polymers has displayed increased biological activity and structural properties compared to natural CS. Scaffold porosity is crucial for cellular penetration, attachment, release of extracellular matrix elements, and bone growth (Van Tienen et al. 2002; Yang et al. 2001). Therefore, the essential criterion for optimal cell infiltration and adhesion is controlling the required pore size. The porosity of the CS scaffold is influenced by CS percentage, type of crosslinkers, temperature, and addition of nanomaterials etc., including crosslinkers like tripolyphosphate and ethyl-3 [3-dimethylaminopropyl] carbodiimide hydrochloride (EDAC) improved the pore structure and pore connections inside the scaffolds (Liu et al. 1989; Lie et al. 2003; Park et al. 2002). Adding nanoparticles to chitosan scaffolds or CS amalgamated matrices may or may not affect the pore size. Adding the Hap material has

Table 3.2 Applications of chitosan in tissue engineering and biomedical applications when combined with different polymers for improved biological properties

S. no.	Chitosan + other polymer(s)	Properties observed	References
1	Chitosan: gelatin (Ch:Gel) cryogels	• Good biocompatibility with L929 mouse fibroblasts • Neither necrosis nor foreign body rejection was witnessed • Nonirritant and nontoxic for bone tissue	Öfkeli et al. (2021)
2	Hydroxyapatite (HAP), chitosan (CS), and antibiotic gentamicin (Gent) composite reinforced with graphene (Gr)	• Coatings had porous, uniform, fracture-free surfaces • Large specific areas of graphene enabled strong bonding with chitosan • Gentamicin addition strongly improved the antibacterial activity • Exhibited low cytotoxicity and good biocompatibility	Stevanović et al. (2020)
3	Chitosan (CS) + polyethylene oxide (PEO) nanofibers + Coral	• Plasma treatment of composite enhanced osteoblast performance and cell adhesion • Thick layers of flake-like carbonate apatite nanocrystals were covering plasma-treated coral/CS/PEO fibers	Tabaei et al. (2021)
4	Chitosan + silver nanoparticles (AgNPs)	• CS-Ag polymeric scaffolds are highly porous and biodegradable • Found to be cyto-compatible and supported growth of osteoblasts • Showed enhanced osteogenic differentiation and matrix mineralization	Vaidhyanathan et al. (2021)
5	Bioglass + chitosan (BG/CH) beads loaded with streptomycin sulfate (STRS)	• STRS loaded BG/CH beads displayed inhibition against Gram-positive and Gram-negative organisms • Beads displayed the biological activity of deposition of a layer of hydroxyapatite on their surface	Al-esnawy et al. (2021)
6	Injected hydrogels containing poly(*N*-isopropylacrylamide) (PNIPAAm)-based copolymer/graphene oxide (GO) composite with different ratios to chitosan (CS)	• Displayed enhanced deposition of minerals and activity of alkaline phosphatase (ALP) • Upregulate the expression of the Runt-related transcription factor 2 and osteocalcin in the human dental pulp stem cells (hDPSCs)	Amiryaghoubi et al. (2020)

(continued)

Table 3.2 (continued)

S. no.	Chitosan + other polymer(s)	Properties observed	References
7	Chitosan (CS) + polycaprolactone (PCL) + zinc divalent ions	• CS/PCL/Zn scaffolds were compatible with mesenchymal cells from human ovarian follicular fluid (HOFF) and promoted osteoblast differentiation	Chandramohan et al. (2021)
		• The construct improved calcium deposition and ALP activity	
		• Osteoblast marker expression increased, such as Runx2, type-1 collagen, osteonectin (ON) and osteocalcin (OC)	
8	Chitosan (CS) + mesoporous zinc sulfate (mZS)	• Scaffolds showed improved porosity, degradation rate and biomineralization	Jindal et al. (2020)
		• Displayed antibacterial activity against *Staphylococcus aureus*	
		• Scaffold showed biocompatibility and allowed cellular attachment	
9	Chitosan (CS) + regenerated cellulose (rCL) nanofibers	• The composite showed unique pore morphology, higher compressive strength, improved biomineralization and increased pre-osteoblast cell (MC3T3-E1) viability, attachment and proliferation	Maharjan et al. (2021)

significantly reduced the porosity. Adding Zirconium to the CS scaffolds yielded better pore connectivity (Lima et al. 2013; Teimouri et al. 2015).

The connectivity among the pores of the CS scaffold occupies a critical role in transporting oxygen and nutrients, removing wastes and neovascularization. Water retention ability denotes tissue engineering scaffolds' capacity to absorb and retain water. This property is also referred to as swelling ability. The swelling ability is influenced by ionizable chemical moieties inside the scaffold and the neighboring surroundings (Chapekar 2000). CS swells through protonation of amine/imine groups which determine the swelling ability. Adding $n\mathrm{SiO_2}$ and $n\mathrm{ZrO_2}$ significantly inhibited the swelling ability of the CS scaffold (Pattnaik et al. 2011). Attachment of proteins or peptides to the TE scaffolds for a basis for ensuing cell communications, cell proliferation and spreading upon introduction into a biological system (Misra et al. 2012; Rogina et al. 2015). NH_2 and COOH functional groups of chitosan are crucial for protein adsorption. Since these functional groups are hydrophilic, they associate with proteins using their hydrophilic side chains through noncovalent interactions such as Vander Waals and electrostatic forces. Nanoparticles are frequently added to CS scaffold to increase the protein adsorption capacity, including

$nSiO_2$, and $nZrO_2$ chitosan/alginate of chitosan scaffold improved the protein adsorption capacity of biocomposite scaffolds (Sowjanya et al. 2013; Pattnaik et al. 2011). Biomineralization is a process of overlay of minerals and ions accumulated on the exterior of TE scaffolds after coming in contact with a biological medium. The deposition of bone apatite material on the scaffold exterior is critical for properly integrating the polymer scaffold with the bone. Therefore, the biomineralization ability of a composite scaffold is established by exposing the materials to simulated body fluids (SBF). Chitosan is an excellent matrix for biomineralization owing to the availability of amine and carboxyl side chains.

Chitosan alone showed poor biomineralization capacity compared to CS biocomposite scaffolds which showed excellent biomineralization potential—including $nZrO_2$ to CS/$nSiO_2$ scaffolds and including nBGC to CS/Gn particles increased nucleation sites, leading to enhanced apatite formation (Pattnaik et al. 2011; Peter et al. 2010). CS only exhibited marginal biomineralization, while CS combined with diopside particles enhanced the deposition of apatite due to the involvement of silanol groups (Kumar et al. 2014). CS/Alg scaffolds demonstrated enhanced biomineralization potential upon adding $nSiO_2$ particles (Sowjanya et al. 2013; Jongwattanapisan et al. 2011). Biodegradation is another hallmark of tissue bioengineering scaffolds which involves the slow and gradual breakdown of biomaterial in a living system through the action of enzymes and additional bioactive elements. Biodegradation is an important property for BTE scaffolds, and the amount of degradation must be equal to that of creating new bone.

3.3.4 Role of Alginate in Wound Healing

Alginate is a hydrophilic polysaccharide found naturally and accounted for one of the copiously biosynthesized materials derived from bacteria and red seaweed. Alginate has widespread applications in the field of biomedicine. It has excellent biocompatibility and biodegradability (Hinchliffe et al. 2021). Tissue engineering utilizes alginate-based biomaterials as successful drug delivery systems and cell carriers. The alginate derivatives can be obtained by modifying various morphological and chemical properties of the native alginate (Abasalizadeh et al. 2020). Advancements in the delivery systems resulted in the advancements in scaffold materials that are often utilized to deliver drugs, growth factors, and cells for the therapeutic purpose. Such composite scaffolds will help to keep the biologically active substances and the surrounding cells intact (Rodrigo-Navarro et al. 2021). Based on the site of the implant, biomaterials will be subjected to various pH that might affect the mechanical and degradation properties along with its swelling behavior. Hence, alginate stands predominantly for its long-term stability and performance as a biomaterial in vitro. Degradation rate is directly related to the molecular weight of the biomaterial (Nicu et al. 2021). Therefore, the molecular weight of alginate-based biomaterials strongly influences the structural rigidity and degradation rate. Increased molecular weight results in decreased availability of reactive groups for hydrolysis, resulting in a much slower rate of degradation. The degradation rate, in turn, will affect the mechanical properties inherently, which will impact the structural changes at molecular levels (Kreller et al. 2021).

U.S. FDA-approved alginate has gained popularity for several applications in nutrition supplements, regenerative medicine, and semi-permeable separation. There is an immense requirement for alginate-based biopolymers in drug delivery and TE (Thomas 2021). Understanding stem cells' relationship with alginate-based biomaterials is crucial, as stem cells are important components of regenerative medicine. Data from in vitro implantation and in vitro cytotoxicity assays suggest that alginate-based scaffolds and microcapsules have minimal or very less cytotoxicity and are histocompatible (Prasathkumar and Sadhasivam 2021). These in vitro studies have revealed information on the interactions that are adjustable between the biocomposites and platelet release-associated growth factors that will enhance the repair of the hematoma-like fracture. In addition, rat tail vertebrate was used to demonstrate the minimally invasive delivery approach through inoculation using microcomputed tomography for in situ rejuvenation of implant materials (Xue et al. 2021). These studies have proved that alginate-based biomaterials undergo degradation, thereby allowing vascularization and having negligible levels of inflammatory responses post-transplantation. Hence, alginate-based scaffolds have proved as potential carriers for cells and drugs that aid tissue regeneration (Sahoo and Biswal 2021).

The application of dressing to open wounds has been followed for ages as it can prevent wounds from getting injured further or from bacterial invasion. Here, gauze is the simplest material that is used for dressing. Advantages of gauze include low cost, easy handling and great absorbent capacity (Tang 2021). But the only disadvantage of the gauze is the secondary injury it might cause while peeling off. Recent findings include the development of high-quality wound dressings that will help create an occlusive environment that is moist and will promote a rapid healing process. Such dressings include gels, sponges, and semi-occlusive or occlusive dressings (Kim et al. 2021). Alginate has found use in several wound dressings, including hydrogels, sponges and electrospun mats. These will act as prospective substrates for the repair process offering wide benefits such as hemostatic ability and the ability to form a gel by absorbing the wound exudates (Bonferoni et al. 2021).

Alginate possesses all the properties that will promote wound healing, such as optimal water vapor transmission rate, good water absorptivity, gentle anti-septic properties, conformability, biodegradability, and nontoxicity. Various studies on alginate-based wound dressings suggest that alginate has enhanced properties of wound healing by monocyte stimulation which in turn produces increased concentrations of cytokines such as TNF-α and IL-6 (Miguel et al. 2021). Cytokine generation at the site of a wound will result in the generation of pro-inflammatory factors that will result in rapid wound healing. Enhanced bioactivity of the alginate could be due to the occurrence of endotoxins inherently. It was proposed that a hydrogel formed in situ from oxidized gelatin and alginate, along with a trace of borax, displayed excellent wound-healing properties when used as a wound dressing. In this wound dressing, alginate stands for wound healing, and borax stands for its anti-septic property (Balakrishnan et al. 2021).

In addition, the alginate backbone lacks a signal sequence responsible for adherence; thereby, alginate-based wound dressings will not result in secondary injury while peeling off. Wounds having large volume exudations can be dressed with an alginate-based sponge. A moist environment will promote wound healing, and alginate-based gels will help prevent the wound from drying out and keep it moist throughout the healing process (Tunçer 2021). As the wounds provide the most favorable environment for bacterial growth by encouraging the colonization of microbes, thereby leading to delayed healing and infection, the antimicrobial properties of the wound dressings prevent bacterial growth. A combination of alginate with silver nanoparticles and chitosan will be done for making antibacterial wound dressings. A composite polysaccharide sponge comprising alginate and water-soluble chitosan is an antimicrobial and anti-adhesive wound dressing (Kharaziha et al. 2021).

3.4 Conclusions

TARE technology is a suitable substitute for conventional tissue or organ grafting techniques. Among the marine biopolymers, chitosan, carrageenan, alginate, and fucoidan were promising and are researched extensively by in situ, in vitro, and in vivo methods. Although numerous scaffolds were designed using the above polymers and several biocompatible materials, they have drawbacks such as consistent reproducibility, mechanical strength, and neovascularization issues. Utilization of novel technologies for the fabrication of scaffolds containing the above bioactive materials and inclusion of bioactive molecules followed by testing their applications in promoting the tissue engineering and repair processes in vivo systems are needed.

Acknowledgments Prakash Narayana Reddy and Chintha Lalitha thank the Principal, Dr. V.-S. Krishna Government Degree College (Autonomous) for his support and encouragement. Rohini Krishna Kota thanks the administration of DRILS for funding her fellowship during the study period. Veera Brahmachari Pallaval thanks the administration of Krishna University for their support and encouragement. Chanda V. Berde thanks the administration of Goa University for their support and encouragement.

Conflict of Interest The authors declare no conflicts of interest.

References

Abasalizadeh F, Moghaddam SV, Alizadeh E, Kashani E, Fazljou SMB, Torbati M, Akbarzadeh A (2020) Alginate-based hydrogels as drug delivery vehicles in cancer treatment and their applications in wound dressing and 3D bioprinting. J Biol Eng 14(1):1–22. https://doi.org/10.1186/s13036-020-0227-7

Akrami-Hasan-Kohal M, Ghorbani M, Mahmoodzadeh F, Nikzad B (2020) Development of reinforced aldehyde-modified kappa-carrageenan/gelatin film by incorporation of halloysite nanotubes for biomedical applications. Int J Biol Macromol 160:669–676. https://doi.org/10.1016/j.ijbiomac.2020.05.222

Al-esnawy AA, Ereiba KT, Bakr AM, Abdraboh AS (2021) Characterization and antibacterial activity of streptomycin sulfate loaded bioglass/chitosan beads for bone tissue engineering. J Mol Struct 1227:129715. https://doi.org/10.1016/j.molstruc.2020.129715

Alexander H, Brunski JB, Cooper SL, Hench LL, Hergenrother RW, Hoffman AS, Yannas IV (1996) Classes of materials used in medicine. In: Biomaterials science. Elsevier, Amsterdam, pp 37–130. https://doi.org/10.1016/B978-0-08-050014-0.50007-9

Amidi M, Mastrobattista E, Jiskoot W, Hennink WE (2010) Chitosan-based delivery systems for protein therapeutics and antigens. Adv Drug Deliv Rev 62(1):59–82. https://doi.org/10.1016/j.addr.2009.11.009

Amiryaghoubi N, Pesyan NN, FathiM OY (2020) Injectable thermosensitive hybrid hydrogel containing graphene oxide and chitosan as dental pulp stem cells scaffold for bone tissue engineering. Int J Biol Macromol 162:1338–1357. https://doi.org/10.1016/j.ijbiomac.2020.06.138

Atashrazm F, Lowenthal RM, Woods GM, Holloway AF, Dickinson JL (2015) Fucoidan and cancer: a multifunctional molecule with anti-tumor potential. Mar Drugs 13(4):2327–2346. https://doi.org/10.3390/md13042327

Balakrishnan SB, Kuppu S, Thambusamy S (2021) Biologically important alumina nanoparticles modified polyvinylpyrrolidone scaffolds in vitro characterizations and it is in vivo wound healing efficacy. J Mol Struct 1246:131195. https://doi.org/10.1016/j.molstruc.2021.131195

Bonferoni MC, Caramella C, Catenacci L, Conti B, Dorati R, Ferrari F, Genta I, Modena T, Perteghella S, Rossi S, Sandri G, Sorrenti M, Torre ML, Tripodo G (2021) Biomaterials for soft tissue repair and regeneration: a focus on Italian research in the field. Pharmaceutics 13(9): 1341. https://doi.org/10.3390/pharmaceutics13091341

Boni R, Ali A, Shavandi A, Clarkson AN (2018) Current and novel polymeric biomaterials for neural tissue engineering. J Biomed Sci 25(1):1–21. https://doi.org/10.1186/s12929-018-0491-8

Buck CB, Thompson CD, Roberts JN, Müller M, Lowy DR, Schiller JT (2006) Carrageenan is a potent inhibitor of papillomavirus infection. PLoS Pathog 2(7):671–680. https://doi.org/10.1371/journal.ppat.0020069

Carlucci MJ, Scolaro LA, Damonte EB (1999) Inhibitory action of natural carrageenans on herpes simplex virus infection of mouse astrocytes. Chemotherapy 45(6):429–436. https://doi.org/10.1159/000007236

Chandika P, Kim MS, Khan F, Kim YM, Heo SY, Oh GW, Kim NG, Jung WK (2021) Wound healing properties of triple cross-linked poly(vinyl alcohol)/methacrylate kappa-carrageenan/chitooligosaccharide hydrogel. Carbohydr Polym 269:118272. https://doi.org/10.1016/j.carbpol.2021.118272

Chandramohan Y, Jeganathan K, Sivanesan S, Koka P, Amritha TMS, Vimalraj S, Dhanasekaran A (2021) Assessment of human ovarian follicular fluid derived mesenchymal stem cells in chitosan/PCL/Zn scaffold for bone tissue regeneration. Life Sci 264:118502. https://doi.org/10.1016/j.lfs.2020.118502

Chapekar MS (2000) Tissue engineering: challenges and opportunities. J Biomed Mater Res 53(6): 617–620. https://doi.org/10.1002/1097-4636(2000)53:6<617::AID-JBM1>3.0.CO;2-C

Cunha L, Grenha A (2016) Sulfated seaweed polysaccharides as multifunctional materials in drug delivery applications. Mar Drugs 14(3):42. https://doi.org/10.3390/md14030042

Di Martino A, Sittinger M, Risbud MV (2005) Chitosan: a versatile biopolymer for orthopaedic tissue-engineering. Biomaterials 26(30):5983–5990. https://doi.org/10.1016/j.biomaterials.2005.03.016

Feki A, Bardaa S, Hajji S, Ktari N, Hamdi M, Chabchoub N, Kallel R, Boudawara T, Nasri M, Ben AI (2019) *Falkenbergia rufolanosa* polysaccharide - poly(vinyl alcohol) composite films: a promising wound healing agent against dermal laser burns in rats. Int J Boil Macromol 144:954–966. https://doi.org/10.1016/j.ijbiomac.2019.09.173

Fitton JH (2011) Therapies from fucoidan; multifunctional marine polymers. Mar Drugs 9(10): 1731–1760. https://doi.org/10.3390/md9101731

Fuss M, Ehlers EM, Russlies M, Rohwedel J, Behrens P (2000) Characteristics of human chondrocytes, osteoblasts and fibroblasts seeded onto a type I/III collagen sponge under different culture conditions. A light, scanning and transmission electron microscopy study. Ann Anat 182(4):303–310. https://doi.org/10.1016/S0940-9602(00)80002-3

Geckil H, Xu F, Zhang X, Moon S, Demirci U (2010) Engineering hydrogelsas extracellular matrix mimics. Nanomedicine 5(3):469–484. https://doi.org/10.2217/nnm.10.12

González JI, Ossa CPO (2017) Injectability evaluation of bone-graft substitutes based on carrageenan and hydroxyapatite nanorods. In: Meyers M et al (eds) Proceedings of the 3rd Pan American materials congress. The minerals, metals & materials series. Springer, Cham, pp 33–46. https://doi.org/10.1007/978-3-319-52132-9_4

Goonoo N, Khanbabaee B, Steuber M, Bhaw-Luximon A, Jonas U, Pietsch U, Jhurry D, Schönherr H (2017) κ-carrageenan enhances the biomineralization and osteogenic differentiation of electrospun polyhydroxybutyrate and polyhydroxybutyrate valerate fibers. Biomacromolecules 18(5):1563–1573. https://doi.org/10.1021/acs.biomac.7b00150

Grabska-Zielińska S, Sionkowska A, Reczyńska K, PamułaE. (2020) Physico-chemical characterization and biological tests of collagen/silk fibroin/chitosan scaffolds cross-linked by dialdehyde starch. Polymers 12(2):372. https://doi.org/10.3390/polym12020372

Hinchliffe JD, Parassini Madappura A, Syed Mohamed SMD, Roy I (2021) Biomedical applications of bacteria-derived polymers. Polymers 13(7):1081. https://doi.org/10.3390/polym13071081

Hussey GS, Dziki JL, Badylak SF (2018) Extracellular matrix-based materials for regenerative medicine. Nat Rev Mater 3(7):159–173. https://doi.org/10.1038/s41578-018-0023-x

Jin G, Kim GH (2011) Rapid-prototyped PCL/fucoidan composite scaffolds for bone tissue regeneration: design, fabrication, and physical/biological properties. J Mater Chem 21(44): 17710–17718. https://doi.org/10.1039/C1JM12915E

Jindal A, Mondal T, Bhattacharya J (2020) An *in vitro* evaluation of zinc silicate fortified chitosan scaffolds for bone tissue engineering. Int J Biol Macromol 164:4252–4262. https://doi.org/10.1016/j.ijbiomac.2020.09.018

Jongwattanapisan P, Charoenphandhu N, Krishnamra N, Thongbunchoo J, Tang IM, Hoonsawat R, Smith SM, Pon-On W (2011) *In vitro* study of the SBF and osteoblast-like cells on hydroxyapatite/chitosan–silica nanocomposite. Mater Sci Eng C 31(2):290–299. https://doi.org/10.1016/j.msec.2010.09.009

Kaihara S, Vacanti JP (1999) Tissue engineering: toward new solutions for transplantation and reconstructive surgery. Arch Surg 134(11):1184–1188. https://doi.org/10.1001/archsurg.134.11.1184

Kalirajan C, Dukle A, Nathanael AJ, Oh TH, Manivasagam G (2021) A critical review on polymeric biomaterials for biomedical applications. Polymers 13(17):3015. https://doi.org/10.3390/polym13173015

Kharaziha M, Baidya A, Annabi N (2021) Rational design of immunomodulatory hydrogels for chronic wound healing. Adv Mater 33:e2100176. https://doi.org/10.1002/adma.202100176

Kim IY, Iwatsuki R, Kikuta K, Morita Y, Miyazaki T, Ohtsuki C (2011) Thermoreversible behavior of κ-carrageenan and its apatite-forming ability in simulated body fluid. Mater Sci Eng C 31(7): 1472–1476. https://doi.org/10.1016/j.msec.2011.05.015

Kim J, Lee KY, Lee CM (2016) Selenium nanoparticles formed by modulation of carrageenan enhance osteogenic differentiation of mesenchymal stem cells. J Nanosci Nanotechnol 16(3): 2482–2487. https://doi.org/10.1166/jnn.2016.10764

Kim K, Mahajan A, Patel K, Syed S, Acevedo-Jake AM, Kumar VA (2021) Materials and cytokines in the healing of diabetic foot ulcers. Adv Ther 4:2100075. https://doi.org/10.1002/adtp.202100075

Kreller T, Distler T, Heid S, Gerth S, Detsch R, Boccaccini AR (2021) Physico-chemical modification of gelatine for the improvement of 3D printability of oxidized alginate-gelatine hydrogels towards cartilage tissue engineering. Mater Des 208:109877. https://doi.org/10.1016/j.matdes.2021.109877

Kumar JP, Lakshmi L, Jyothsna V, Balaji DR, Saravanan S, Moorthi A, Selvamurugan N (2014) Synthesis and characterization of diopside particles and their suitability along with chitosan matrix for bone tissue engineering *in vitro* and *in vivo*. J Biomed Nanotechnol 10(6):970–981. https://doi.org/10.1166/jbn.2014.1808

Li B, Lu F, Wei X, Zhao R (2008) Fucoidan: structure and bioactivity. Molecules (Basel, Switzerland) 13(8):1671–1695. https://doi.org/10.3390/molecules13081671

Li J, Yang B, Qian Y, Wang Q, Han R, Hao T, Shu Y, Zhang Y, Yao F, Wang C (2015) Iota-carrageenan/chitosan/gelatin scaffold for the osteogenic differentiation of adipose-derived MSCs *in vitro*. J Biomed Mater Res Part B Appl Biomater 103(7):1498–1510. https://doi.org/10.1002/jbm.b.33339

Lie GC, Mao Z, Shen J, Hu X, Han C (2003) Thermal dehydration treatment and glutaraldehyde cross-linking to increase the biostability of collagen–chitosan porous scaffolds used as dermal equivalent. J Biomater Sci Polym Ed 14(8):861–874. https://doi.org/10.1163/156856203768366576

Lima PAL, Resende CX, de Almeida Soares GD, Anselme K, Almeida LE (2013) Preparation, characterization and biological test of 3D-scaffolds based on chitosan, fibroin and hydroxyapatite for bone tissue engineering. Mater Sci Eng C 33(6):3389–3395. https://doi.org/10.1016/j.msec.2013.04.026

Liu B, Harrell R, Davis RH, Dresden MH, Spira M (1989) The effect of gamma irradiation on injectable human amnion collagen. J Biomed Mater Res 23(8):833–844. https://doi.org/10.1002/jbm.820230803

Liu Y, Lim J, Teoh SH (2013) Development of clinically relevant scaffolds for vascularised bone tissue engineering. Biotechnol Adv 31(5):688–705. https://doi.org/10.1016/j.biotechadv.2012.10.003

Liu J, Zhan X, Wan J, Wang Y, Wang C (2015) Review for carrageenan-based pharmaceutical biomaterials: favourable physical features versus adverse biological effects. Carbohydr Polym 121:27–36. https://doi.org/10.1016/j.carbpol.2014.11.063

Madruga LY, Balaban RC, Popat KC, Kipper MJ (2021a) Biocompatible crosslinked nanofibers of poly(vinyl alcohol)/carboxymethyl-kappa-carrageenan produced by a green process. Macromol Biosci 21(1):2000292. https://doi.org/10.1002/mabi.202000292

Madruga LY, Popat KC, Balaban RC, Kipper MJ (2021b) Enhanced blood coagulation and antibacterial activities of carboxymethyl-kappa-carrageenan-containing nanofibers. Carbohydr Polym 273:118541. https://doi.org/10.1016/j.carbpol.2021.118541

Maharjan B, Park J, Kaliannagounder VK, Awasthi GP, Joshi MK, Park CH, Kim CS (2021) Regenerated cellulose nanofiber reinforced chitosan hydrogel scaffolds for bone tissue engineering. Carbohydr Polym 251:117023. https://doi.org/10.1016/j.carbpol.2020.117023

Mandrycky C, Phong K, Zheng Y (2017) Tissue engineering toward organ-specific regeneration and disease modeling. MRS Commun 7(3):332–347. https://doi.org/10.1557/mrc.2017.58

Mantha S, Pillai S, Khayambashi P, Upadhyay A, Zhang Y, Tao O, Pham HM, Tran SD (2019) Smart hydrogels in tissue engineering and regenerative medicine. Materials 12(20):3323. https://doi.org/10.3390/ma12203323

Miguel SP, Ribeiro MP, Coutinho P (2021) Biomedical applications of biodegradable polymers in wound care. In: Kumar P, Kothari V (eds) Wound healing research. Springer, Singapore, pp 509–597. https://doi.org/10.1007/978-981-16-2677-7_17

Mihaila SM, Popa EG, Reis RL, Marques AP, Gomes ME (2014) Fabrication of endothelial cell-laden carrageenan microfibers for microvascularized bone tissue engineering applications. Biomacromolecules 15(8):2849–2860. https://doi.org/10.1021/bm500036a

Misra RDK, Girase B, Nune VKC, Xu W (2012) Cellular interactions and modulated osteoblasts functions mediated by protein adsorption. Adv Eng Mater 14(5):B247–B257. https://doi.org/10.1002/adem.201180071

Morais DS, Rodrigues MA, Silva TI, Lopes MA, Santos M, Santos JD, Botelho CM (2013) Development and characterization of novel alginate-based hydrogels as vehicles for bone substitutes. Carbohydr Polym 95(1):134–142. https://doi.org/10.1016/j.carbpol.2013.02.067

Murphy SV, Atala A (2014) 3D bioprinting of tissues and organs. Nat Biotechnol 32(8):773–785. https://doi.org/10.1038/nbt.2958
Nakamura S, Nambu M, Ishizuka T, Hattori H, Kanatani Y, Takase B, Kishimoto S, Amano Y, Aoki H, Kiyosawa T et al (2008) Effect of controlled release of fibroblast growth factor-2 from chitosan/fucoidanmicro complex-hydrogel on in vitro and in vivo vascularization. J Biomed Mater Res A 85(3):619–627. https://doi.org/10.1002/jbm.a.31563
Nicu R, Ciolacu F, Ciolacu DE (2021) Advanced functional materials based on nanocellulose for pharmaceutical/medical applications. Pharmaceutics 13(8):1125. https://doi.org/10.3390/pharmaceutics13081125
O'Leary R, Rerek M, Wood EJ (2004) Fucoidan modulates the effect of transforming growth factor (TGF)-beta1on fibroblast proliferation and wound repopulation in in vitro models of dermal wound repair. Biol Pharm Bull 27(2):266–270. https://doi.org/10.1248/bpb.27.266
Obluchinsksya ED, Makarova MN, Pozharitskaya ON, Shikov AN (2015) Effects of ultrasound treatment on the chemical composition and anticoagulant properties of dry fucus extract. Pharm Chem J 49(3):183–186. https://doi.org/10.1007/s11094-015-1250-8
Öfkeli F, Demir D, Bölgen N (2021) Biomimetic mineralization of chitosan/gelatin cryogels and in vivo biocompatibility assessments for bone tissue engineering. J Appl Polym Sci 138(14): 50337. https://doi.org/10.1002/app.50337
Panlasigui LN, Baello OQ, Dimatangal JM, Dumelod BD (2003) Blood cholesterol and lipid-lowering effects of carrageenan on human volunteers. Asia Pac J Clin Nutr 12(2): 209–214. PMID: 12810413
Park SN, Park JC, Kim HO, Song MJ, Suh H (2002) Characterization of porous collagen/hyaluronic acid scaffold modified by 1-ethyl-3-(3-dimethylaminopropyl) carbodiimide cross-linking. Biomaterials 23(4):1205–1212. https://doi.org/10.1016/S0142-9612(01)00235-6
Pattnaik S, Nethala S, Tripathi A, Saravanan S, Moorthi A, Selvamurugan N (2011) Chitosan scaffolds containing silicon dioxide and zirconia nano particles for bone tissue engineering. Int J Biol Macromol 49(5):1167–1172. https://doi.org/10.1016/j.ijbiomac.2011.09.016
Pereira RC, Scaranari M, Castagnola P, Grandizio M, Azevedo HS, Reis RL, Cancedda R, Gentili C (2009) Novel injectable gel (system) as a vehicle for human articular chondrocytes in cartilage tissue regeneration. J Tissue Eng Regen Med 3(2):97–106. https://doi.org/10.1002/term.145
Peter M, Binulal NS, Nair SV, Selvamurugan N, Tamura H, Jayakumar R (2010) Novel biodegradable chitosan–gelatin/nano-bioactive glass ceramic composite scaffolds for alveolar bone tissue engineering. Chem Eng J 158(2):353–361. https://doi.org/10.1016/j.cej.2010.02.003
Popa E, Reis R, Gomes M (2012) Chondrogenic phenotype of different cells encapsulated in κ-carrageenan hydrogels for cartilage regeneration strategies. Biotechnol Appl Biochem 59(2):132–141. https://doi.org/10.1002/bab.1007
Popa EG, Rodrigues MT, Coutinho DF, Oliveira MB, Mano JF, Reis RL, Gomes ME (2013) Cryopreservation of cell laden natural origin hydrogels for cartilage regeneration strategies. Soft Matter 9(3):875–885. https://doi.org/10.1039/C2SM26846A
Popa EG, Caridade SG, Mano JF, Reis RL, Gomes ME (2015) Chondrogenic potential of injectable κ-carrageenan hydrogel with encapsulated adipose stem cells for cartilage tissue-engineering applications. J Tissue Eng Regen Med 9(5):550–563. https://doi.org/10.1002/term.1683
Popa EG, Santo VE, Rodrigues MT, Gomes ME (2016) Magnetically responsive hydrogels for modulation of chondrogenic commitment of human adipose-derived stem cells. Polymers 8(2): 28. https://doi.org/10.3390/polym8020028
Prasathkumar M, Sadhasivam S (2021) Chitosan/hyaluronic acid/alginate and an assorted polymers loaded with honey, plant, and marine compounds for progressive wound healing—know-how. Int J Biol Macromol 186:656–685. https://doi.org/10.1016/j.ijbiomac.2021.07.067
Purnama A, Aid-Launais R, Haddad O, Maire M, Mantovani D, Letourneur D, Hlawaty H, Le Visage C, Maire M (2013) Fucoidan in a 3D scaffold interacts with vascular endothelial growth factor and promotes neovascularization in mice. Drug Deliv Transl Res 5(2):187–197. https://doi.org/10.1007/s13346-013-0177-4

Puvaneswary S, Raghavendran HB, Talebian S, Murali MR, Mahmod SA, Singh S, Kamarul T (2016) Incorporation of fucoidan in β-tricalcium phosphate-chitosan scaffold prompts the differentiation of human bone marrow stromal cells into osteogenic lineage. Sci Rep 6(1): 1–11. https://doi.org/10.1038/srep24202

Rioux LE, Turgeon SL, Beaulieu M (2007) Rheological characterization of polysaccharides extracted from brown seaweeds. J Sci Food Agric 87(9):1630–1638. https://doi.org/10.1002/jsfa.2829

Rocha PM, Santo VE, Gomes ME, Reis RL, Mano JF (2011) Encapsulation of adipose-derived stem cells and transforming growth factor-β1 in carrageenan-based hydrogels for cartilage tissue engineering. J Bioact Compat Polym 26(5):493–507. https://doi.org/10.1177/0883911511420700

Rodrigo-Navarro A, Sankaran S, Dalby MJ, del Campo A, Salmeron-Sanchez M (2021) Engineered living biomaterials. Nat Rev Mater 6(12):1175–1190. https://doi.org/10.1038/s41578-021-00350-8

Rogina A, Rico P, Ferrer GG, Ivanković M, Ivanković H (2015) Effect of in situ formed hydroxyapatite on microstructure of freeze-gelled chitosan-based biocomposite scaffolds. Eur Polym J 68:278–287. https://doi.org/10.1016/j.eurpolymj.2015.05.004

Sahoo DR, Biswal T (2021) Alginate and its application to tissue engineering. SN Appl Sci 3(1): 1–19. https://doi.org/10.1007/s42452-020-04096-w

Salvatore L, Gallo N, Natali ML, Terzi A, Sannino A, Madaghiele M (2021) Mimicking the hierarchical organization of natural collagen: toward the development of ideal scaffolding material for tissue regeneration. Front Bioeng Biotechnol 9:258. https://doi.org/10.3389/fbioe.2021.644595

Sathain A, Monvisade P, Siriphannon P (2021) Bioactive alginate/carrageenan/calcium silicate porous scaffolds for bone tissue engineering. Mater Today Commun 26:102165. https://doi.org/10.1016/j.mtcomm.2021.102165

Saul JM, Williams DF (2011) Hydrogels in regenerative medicine. Handbook of polymer applications in medicine and medical devices. William Andrew Publishing, Norwich, pp 279–302. https://doi.org/10.1016/B978-0-323-22805-3.00012-8

Şenel S, McClure SJ (2004) Potential applications of chitosan in veterinary medicine. Adv Drug Deliv Rev 56(10):1467–1480. https://doi.org/10.1016/j.addr.2004.02.007

Seol YJ, Lee JY, Park YJ, Lee YM, Rhyu IC, Lee SJ, Han SB, Chung CP (2004) Chitosan sponges as tissue engineering scaffolds for bone formation. Biotechnol Lett 26(13):1037–1041. https://doi.org/10.1023/B:BILE.0000032962.79531.fd

Song YS, Li H, Balcos MC, Yun HY, Baek KJ, Kwon NS, Choi HR, Park KC, Kim DS (2014) Fucoidan promotes the reconstruction of skin equivalents. Korean J Physiol Pharmacol 18(4): 327–331. https://doi.org/10.4196/kjpp.2014.18.4.327

Sowjanya JA, Singh J, Mohita T, Sarvanan S, Moorthi A, Srinivasan N, Selvamurugan N (2013) Biocomposite scaffolds containing chitosan/alginate/nano-silica for bone tissue engineering. Colloids Surf B Biointerfaces 109:294–300. https://doi.org/10.1016/j.colsurfb.2013.04.006

Stevanović M, Djošić M, Janković A, Kojić V, Vukašinović-Sekulić M, Stojanović J, Odović J, Sakač MC, Yop RK, Mišković-Stanković V (2020) Antibacterial graphene-based hydroxyapatite/chitosan coating with gentamicin for potential applications in bone tissue engineering. J Biomed Mater Res A 108(11):2175–2189. https://doi.org/10.1002/jbm.a.36974

Swain S, Kwon TY, Rautray TR (2021) Fabrication of silver doped nano hydroxyapatite-carrageenan hydrogels for articular cartilage applications. bioRxiv. https://doi.org/10.1101/2020.12.31.424664

Tabaei PSE, Asadian M, Ghobeira R, Cools P, Thukkaram M, Derakhshandeh PG, Abednatanzi S, Voort PVD, Verbeken K, Vercruysse C, Declerch H, Morent R, De Geyter N (2021) Combinatorial effects of coral addition and plasma treatment on the properties of chitosan/polyethylene oxide nanofibers intended for bone tissue engineering. Carbohydr Polym 253:117211. https://doi.org/10.1016/j.carbpol.2020.117211

Tang SLP (2021) Dressings for advanced wound care. CRC Press, Boca Raton

Tavakoli S, Kharaziha M, Nemati S, Kalateh A (2021) Nanocomposite hydrogel based on carrageenan-coated starch/cellulose nanofibers as a hemorrhage control material. Carbohydr Polym 251:117013. https://doi.org/10.1016/j.carbpol.2020.117013

Teimouri A, Ebrahimi R, Emadi R, Beni BH, Chermahini AN (2015) Nanocomposite of silk fibroin–chitosan/nano ZrO2 for tissue engineering applications: fabrication and morphology. Int J Biol Macromol 76:292–302. https://doi.org/10.1016/j.ijbiomac.2015.02.023

Thakur A, Jaiswal MK, Peak CW, Carrow JK, Gentry J, Dolatshahi-Pirouz A, Gaharwar AK (2016) Injectable shear-thinning nanoengineered hydrogels for stem cell delivery. Nanoscale 8(24): 12362–12372. https://doi.org/10.1039/C6NR02299E

Thomas LV (2021) Polymeric biomaterials in tissue engineering: retrospect and prospects. In: Bhaskar B, Sreenivasa Rao P, Kasoju N, Nagarjuna V, Baadhe RR (eds) Biomaterials in tissue engineering and regenerative medicine. Springer, Singapore, pp 89–118. https://doi.org/10.1007/978-981-16-0002-9_4

Tunçer S (2021) Biopolysaccharides: properties and applications. In: Polysaccharides: properties and applications. Wiley, New York, pp 95–134. https://doi.org/10.1002/9781119711414.ch6

Vacanti CA, Vacanti JP (2000) The science of tissue engineering. Orthop Clin North Am 31(3): 351–356. https://doi.org/10.1016/s0030-5898(05)70155-3

Vaidhyanathan B, Vincent P, Vadivel S, Karuppiah P, Al-Dhabi NA, Sadhasivam DR, Vimalraj S, Saravanan S (2021) Fabrication and investigation of the suitability of chitosan-silver composite scaffolds for bone tissue engineering applications. Process Biochem 100:178–187. https://doi.org/10.1016/j.procbio.2020.10.008

Van Tienen TG, Heijkants RG, Buma P, de Groot JH, Pennings AJ, Veth RP (2002) Tissue ingrowth and degradation of two biodegradable porous polymers with different porosities and pore sizes. Biomaterials 23(8):1731–1738. https://doi.org/10.1016/S0142-9612(01)00280-0

Varoni E, Tschon M, Palazzo B, Nitti P, Martini L, Rimondini L (2012) Agarose gel as biomaterial or scaffold for implantation surgery: characterization, histological and histomorphometric study on soft tissue response. Connect Tissue Res 53(6):548–554. https://doi.org/10.3109/03008207.2012.712583

Velasco MA, Narváez-Tovar CA, Garzón-Alvarado DA (2015) Design, materials, and mechanobiology of biodegradable scaffolds for bone tissue engineering. Biomed Res Int 2015:729076. https://doi.org/10.1155/2015/729076

Venkatesan J, Bhatnagar I, Kim SK (2014) Chitosan-alginate biocomposite containing fucoidan for bone tissue engineering. Mar Drugs 12(1):300–316. https://doi.org/10.3390/md12010300

Vignesh S, Gopalakrishnan A, Poorna MR, Nair SV, Jayakumar R, Mony U (2018) Fabrication of micropatterned alginate-gelatin and k-carrageenan hydrogels of defined shapes using simple wax mould method as a platform for stem cell/induced Pluripotent Stem Cells (iPSC) culture. Int J Biol Macromol 112:737–744. https://doi.org/10.1016/j.ijbiomac.2018.02.031

Wang L, Cao J, Lei D, Cheng X, Yang Y, Hou R, Cui F (2009) Effects of nerve growth factor delivery via a gel to inferior alveolar nerve in mandibular distraction osteogenesis. J Craniofac Surg 20(6):2188–2192. https://doi.org/10.1097/SCS.0b013e3181bf04c7

Wang L, Cao J, Lei DL, Cheng XB, Zhou HZ, Hou R, Cui FZ (2010) Application of nerve growth factor by gel increases formation of bone in mandibular distraction osteogenesis in rabbits. Br J Oral Maxillofac Surg 48(7):515–519. https://doi.org/10.1016/j.bjoms.2009.08.042

Xue X, Hu Y, Deng Y, Su J (2021) Recent advances in design of functional biocompatible hydrogels for bone tissue engineering. Adv Funct Mater 31(19):2009432. https://doi.org/10.1002/adfm.202009432

Yang S, Leong KF, Du Z, Chua CK (2001) The design of scaffolds for use in tissue engineering. Part I. Traditional factors. Tissue Eng 7(6):679–689. https://doi.org/10.1089/107632701753337645

Yegappan R, Selvaprithiviraj V, Amirthalingam S, Jayakumar R (2018) Carrageenan based hydrogels for drug delivery, tissue engineering and wound healing. Carbohydr Polym 198: 385–400. https://doi.org/10.1016/j.carbpol.2018.06.086

Zhou G, Sun YP, Xin H, Zhang Y, Li Z, Xu Z (2004) *In vivo* antitumor and immunomodulation activities of different molecular weight lambda-carrageenans from *Chondrus ocellatus*. Pharmacol Res 50(1):47–53. https://doi.org/10.1016/j.phrs.2003.12.002

4 Metagenome Mining Approaches for the Discovery of Marine Microbial Natural Products

Devika N. Nagar, Kabilan Mani, and Judith M. Bragança

Abstract

Microbial natural products (NPs) are a group of specialized metabolites produced by microorganisms such as Archaea, Bacteria, Fungi, and Microalgae conferring these organisms with adaptive advantages in certain environments. Owing to the bio-activity potential of these small molecules, their possible use as antibiotics, chemotherapeutics, or pesticides is currently being explored. Accessing the least or unexplored ecosystems, such as marine environments, will unearth a rich, unique microbiome. These microorganisms can potentially yield novel natural products (NPs) with favorable functionalities, which can be further engineered for use in the pharmaceutical and healthcare industry. Although oceans cover more than 70% of our planet's surface and possess the possibility of having the highest species diversity, the majority of marine microbes are yet to be cultured. Our inability to decipher the culture conditions and culture the bulk of marine microbes has hindered us in discovering and exploiting a possibly large collection of marine microbes-derived NPs. The advent of the genomic era and our ability to directly sequence the environmental DNA (eDNA), has expanded research in Microbial ecology significantly. Metagenomics, direct sequencing, and DNA analysis from microbial assemblages have revolutionized our understanding of the diversity and ecological processes involving marine archaea, bacteria, and eukaryotic microorganisms. Further, the strides made in omics-based technology

D. N. Nagar · J. M. Bragança (✉)
Department of Biological Sciences, Birla Institute of Technology and Science, Pilani, K K Birla Goa Campus, Pilani, Goa, India
e-mail: judith@goa.bits-pilani.ac.in

K. Mani
Center for Molecular Medicine & Therapeutics, PSG Institute of Medical Sciences and Research, Coimbatore, India

P. Veera Bramhachari, C. V. Berde (eds.), *Marine Bioactive Molecules for Biomedical and Pharmacotherapeutic Applications*,
https://doi.org/10.1007/978-981-99-6770-4_4

have rendered the uncultured marine microorganisms accessible for function-based metagenomics, allowing researchers to mine the hidden NPs and their pathways from the complex biological big data. This chapter presents various methods of eDNA extraction, and various next-generation sequencing technologies available for generating metagenomic data and strategies for mining the generated data for identifying new NPs.

Keywords

Natural products (NPs) · Marine microorganisms · Metagenomics · Genome mining

4.1 Introduction

Natural products are crucial metabolites essential for many biological activities like signaling, regulation, enzyme activity, scavenging, self-defense etc. Microorganisms with a potential for producing these natural products with bioactive properties have been of recent focus. Marine microorganisms, in particular, are known for producing these compounds due to their high diversity and need to thrive in various environmental conditions. The capability of these bioactive compounds has caught an eye for the welfare of humanity for various applications such as antibiotics, antioxidants, antimicrobials, and chemotherapeutics (Kenshole et al. 2021). The marine environment is the largest type of ecosystem, occupying around 70% of the earth's surface and inhabiting many microorganisms with larger marine biomass and metabolism than any other environment. Thus, it possesses high biological and chemical diversity to explore (Trindade et al. 2015).

The marine ecosystem consists largely of biotic components such as plants, mammals, and other marine organisms. Also, it has a rich microbial community of bacteria, archaea, fungi, algae, protozoans, plankton etc. (Solanki et al. 2008). Marine microorganisms contribute a large portion of marine biomass and metabolism have grabbed attention in recent times for the synthesis of bioactive natural compounds (Sogin et al. 2006) compared to that of larger marine organisms like sponges, corals etc., from which varieties of natural products have been reported previously (Hu et al. 2011; Leal et al. 2012; Trindade et al. 2015). It is reported that the oceanic environment contains approximately 3.6×10^{30} microorganisms (Mondol et al. 2013). However, very little, i.e., around 0.001–1% of marine microorganisms, have been successfully cultured and scale-up to the required amount of biomass is still time-consuming and difficult to achieve (Vollmers et al. 2017). Even after investing time and effort, the challenging environmental conditions like varying temperatures, pressure, salinity, pH, and light sources dictate the type of bioactive natural product synthesis (Das and Mangwani 2015).

In the last decade, the initial culture-dependent techniques for studying metagenomics have been refocused towards culture-independent high throughput metagenomics that offers a great level of information on the genetic constituency of the microorganisms. Further, this modern approach has provided us with information on the plethora of genes and pathways involved in the metabolism and synthesis of metabolites, allowing us to reconstruct the entire intracellular process. Recently, marine natural products with potential biological activities are being evaluated through direct cloning of the target gene from metagenomic libraries while exploring possible avenues for scaling up for mass production (Mahapatra et al. 2020; König et al. 2006). This chapter discusses metagenomic approaches for identification, isolation and the applications of natural products derived from marine microorganisms.

4.2 Marine Sources of Natural Products

Marine natural products have been gaining attention steadily, and the studies concerning their extraction and characterization are varied. Considering the huge biodiversity of marine microbes like marine bacterial, archaeal, fungal and algal communities and their potential for synthesis of a variety of bioactive natural compounds, they are valuable reservoirs (Imhoff et al. 2011). These natural products from marine environments are useful assets to society once accepted as commercial products. The most noticeable application of natural products is their therapeutic activities, thus often identified with pharmacokinetic properties for clinical usages and in a few other socio-economic applications (Trindade et al. 2015). The beneficial aspects include antimicrobial activity (Donia and Hamann 2003, Paul et al. 2021; Bérdy 2005), antitumor effect (Feling et al. 2003; Taori et al. 2008; Rath et al. 2011), anti-inflammatory property (Jenab et al. 2020), antiviral activity (Sayed 2000), anti-coagulant properties, anti-parasitic activity (Manivasagan et al. 2014), antidiabetic property, as well as effect on cardiovascular, nervous, and immune systems (Mayer et al. 2010; Waters et al. 2010; Imhoff et al. 2011). Antioxidant activity possessed by pigments and phenolic compounds like flavonoids, carotenoids, gallic acid, and other marine secondary metabolites like mycosporines etc. is documented (Corinaldesi et al. 2017). The low molecular weight marine secondary metabolites like mycosporines and mycosporine-like amino acids are accumulated and produced by marine microorganisms functioning in chemical defenses, UV protection, radical scavenging, reproductive signaling etc., in response to various stress conditions like salt and temperature besides serving as accessory light-harvesting pigments in photosynthesis. These compounds can be exploited as potential UV protectants in sunscreens, as well as antioxidants and anti-inflammatory products (Oren and Gunde-Cimerman 2007; Claveric et al. 2020). Other than therapeutics or pharmacological applicative products, marine microbial bio-compounds also include enzymes like amylase, protease, lipase etc., therefore proving to be potential bioactive compounds in biotechnology, agriculture, cosmetic, food, textile industries, and other commercial fields (Paul et al. 2021). Concerning such a broad and potential range of applications, marine natural products from microorganisms are still largely unexplored. There are challenges and limitations to isolating and screening bioactive natural products from microorganisms (Poli et al. 2017). Various examples of natural products from marine microorganisms and their potential applications are summarized in Table. 4.1.

4.2.1 Marine Archaea and Bacteria

Marine archaea and bacteria like thermophiles, halophiles, psychrophiles, acidophiles, alkalophiles, piezophiles, and other extremophiles are capable of adapting to extreme conditions due to certain cellular mechanisms which help them not only tolerate but also thrive in such conditions. Therefore, this unique and novel survival mechanism involves novel genes encoding for microbial metabolites and biologically active compounds of desired interest.

Marine archaea are mostly extremophilic, therefore, are reservoirs of novel and sustainable metabolites. For example, compared to regular enzymes, the enzymes derived from extremophiles have supremacy for industrial purposes because they are stable even at different extreme and harsh conditions (Egorova and Antranikian 2005). The thermostable archaeal enzymes derived from hyperthermophiles are stable at high temperatures at high pressure in the presence of solvents, detergents and proteolytic degradation (Rotter et al. 2021). Thermophilic archaea have been a source of hydrolytic enzymes like chitinases (Chen et al. 2019). Similarly, halophilic archaea prevailing at high salinities also show photocatalytic properties (Singh and Gabani 2011). These properties are due to the inherent synthesis of carotenoids; both C40 and C50 are best known for their higher antioxidant properties, specifically Bacterioruberin type of C50 carotenoids (Giani et al. 2019; Rodrigo-Baños et al. 2015). Other than antioxidant properties, archaeal carotenoids are reported to have anticancer and antihemolytic activities (Hou and Cui 2018).

Marine bacteria are interesting and have been highlighted for their potential to produce various natural products. A few bacterial natural products have been characterized based on their activities. From 1997 to 2008, around 660 marine bacterial compounds were detected. After that, approximately 115 marine bacterial compounds were found from 2010 to 2012, 161 from 2013 to 2015, and then almost 179 in year 2016 were reported. Among these compounds, Class Actinobacteria contributes up to 40%, Cyanobacteria up to 33%, whereas Proteobacteria up to 12% and members of Firmicutes and Bacteroidetes contribute up to 5% (Williams 2009). Natural product depsipeptide largazole, produced by the marine cyanobacterium *Symploca* sp., has shown selectivity against transformed human mammary epithelial and osteosarcoma cells (Taori et al. 2008; Bowers et al. 2008). Metabolites of marine actinomycetes are the sources of novel antimicrobials (Subramani and Aalbersberg 2013) and anticancer compounds (Olano et al. 2009). Proximicins, the bioactive metabolite isolated from *Verrucosispora* sp. MG-37 and *Verrucosispora maris* AB-18-032, are weak antibacterial agents with strong cytostatic effects on human tumors (Schneider et al. 2008). Recent reports of bioactive compounds include isolates have been reported from *Bacillus aerius*, *B. safensis, B. altitudinis, B. boroniphilus, B. oryzicola*, and *Virgibacillus senegalensis*, having profound antimicrobial activity against food poisonings causative agents like *Vibrio*

Table 4.1 Natural products produced by marine microorganisms and their potential applications

Marine microorganisms	Natural products	Activities	Reference
Haloplanus vescus	Carotenoids (C50)	Anticancer	Hou and Cui (2018)
Halobacterium halobium	Carotenoidsextract	Antitumor	Abbes et al. (2013)
Halogeometricum limi	Carotenoids (C50)	Anticancer	Hou and Cui (2018)
Nocardia dassonvillei	Phenazine derivative	Anticancer	Gao et al. (2012)
Dermacoccus abyssi	Dermacozines	Antitumor	Abdel-Mageed et al. (2010)
Streptomyces sp.	Elaiomycins B and C	Antitumor	Helaly et al. (2011)
Streptomyces sp. WBF16	Aureolic acids	Antitumor	Lu et al. (2012)
Streptomyces sp. and *Bacillus* sp. (co-culture)	Dentigerumycin E	Antitumor	Shin et al. (2018)
Nocardiopsis sp.	cyclo-(l-Pro-l-Met)	Anti-angiogenesis	Shin et al. (2010)
Aspergillusniger AKV-MKBU	L-asparaginase	Anticancer	Vala et al. (2018)
Hansfordia sinuosae	Polysaccharide	Antitumoral	Li et al. (2018a)
Penicillium minioluteum ZZ1657	Purpuride G	Antitumoral	Ma et al. (2020)
Aspergillus sp.	Asperphenin A	Antitumoral	Bae et al. (2020)
Prorocentrum hoffmannianum, Prorocentrum arenarium, Prorocentrum reticulatum, Alexandrium tamarensis and *Gambierdiscus australes*	Microalgal extracts	Apoptotic	de Vera et al. (2018)
Streptomyces sp. NPS853	Usabamycins	Cytotoxicity	Sato et al. (2011)
Amycolatopsis alba.	Pyridinium	Cytotoxicity	Dasari et al. (2012)
Streptomyces sp.S2A	Bacterial metabolites	Cytotoxicity	Siddharth and Vittal (2018)
Chaetomium sp. NA-S01-R1	Chaetovirides, Chaephilone	Cytotoxicity	Wang et al. (2018)
Fusarium equiseti	(11S)-1,3,6-trihydroxi-7-(1-hydroxyethyl)-anthracene-9,10-dione	Cytotoxic	Zhao et al. (2018)

(continued)

Table 4.1 (continued)

Marine microorganisms	Natural products	Activities	Reference
Curvularia sp. IFB-Z10	Spirocurvulaide	Cytotoxic	An et al. (2019)
Amphidinium carterae	Amphidinol 22	Cytotoxicity	Martínez et al. (2019)
Halobacterium halobium	Carotenoid extract	Antioxidant	Abbes et al. (2013)
Haloferax volcanii, Halogranum rubrum, Halopelagius inordinatus and *Halogeometricum rufum*	Carotenoids (C50)	Antioxidant	Hou and Cui (2018)
Streptomyces sp. strain MS1/7	2-allyloxyphenol	Antioxidant	Arumugam et al. (2010)
Dermacoccus abyssi	Dermacozines	Antioxidant	Abdel-Mageed et al. (2010)
Janibacter melonis and *Pseudomonas stutzeri*	Bacterial extracts	Antioxidant	Shivale et al. (2018)
Streptomyces sp.S2A	Bacterial metabolites	Antioxidant	Siddharth and Vittal (2018)
Novosphingobium sp. PP1Y	Bacterial extracts	Antioxidant	Petruk et al. (2019)
Microbacterium aurantiacum FSW-25	Exopolysaccharide	Antioxidant	Sran et al. (2019)
Nostoc sphaericum	Glycosylated mycosporine-like amino acid	UV protection and antioxidant	Ishihara et al. (2017)
Penicillium citrinum HDN-152–088	Dicitrinones	Antioxidant	Wang et al. (2019)
Penicillium sclerotiorin	Azaphilone derivative	Anti-inflammatory	Liu et al. (2019)
Navicula sp.	Sulfated polysaccharides	Antioxidant	Fimbres-Olivarria et al. (2018)
B. amyloliquefaciens	Macrolactins V	Antibacterial	Gao et al. (2010)
Streptomyces sp. *strain MS1/7*	2-allyloxyphenol	Antibacterial	Arumugam et al. (2010)
Nocardiopsis sp.	Thiopeptide	Antibacterial	Kerstin et al. (2010)
Streptomyces sp. *GU045544.1*	Anthraquinone	Antibacterial	Ravikumar et al. (2012)
Amycolatopsis alba.	Pyridinium	Antibacterial	Dasari et al. (2012)
Bacillus aerius, Bacillus oryzicola, Bacillus safensis, Bacillus boroniphilus,	Bacterial extracts	Antibacterial	Galaviz-Silva et al. (2018)

(continued)

Table 4.1 (continued)

Marine microorganisms	Natural products	Activities	Reference
Bacillus altitudinis and Virgibacillus senegalensis			
Streptomyces sp.*S2A*	Bacterial metabolites	Antibacterial	Siddharth and Vittal (2018)
Streptomyces sp. *ZZ745*	Bagremycins F and G	Antibacterial	Zhang et al. (2018)
Streptomyces sp. *Al Dhabi-90*	Bacterial extract	Antibacterial	Al-Dhabi et al. (2019)
Simplicillium lamellicola, Leptosphaerulina sp., *Penicillium citrinum, Penicillium chrysogenum, Aspergillus sydowii, Hansfordia sinuosae*	Fungal extracts	Antibacterial	Agrawal et al. (2018)
Penicillium sp. *ZZ901*	(+)-Scleroderolide	Antibacterial	Li et al. (2018b)
Aspergillus alabamensis EN-547	Diketomorpholine derivatives	Antibacterial	Yang et al. (2018)
Penicillium sp. *IMB17-046*	Trypilepyrazinol and (+)-neocitreoviridin	Antibacterial	Li et al. (2019)
Aspergillus sp. *LS116*	Aspergillsteroid A	Antibacterial	Xu et al. (2020)
Penicillium minioluteum ZZ1657	N-acetyl-L-valine conjugated drimarane sesquiterpenoids	Antibacterial	Ma et al. (2020)
Chaetomium sp. NA-S01-R1	Chaetovirides, Chaephilone	Anti-methicillin resistant Staphylococcus aureus (anti-MRSA) and antibacterial	Wang et al. (2018)
Fusarium sp. 152	Equisetin	Anti-methicillin resistant Staphylococcus aureus (anti-MRSA)	Luo et al. (2018)
Skeletonema costatum, Chaetoceros pseudocurvisetus	Microalgal extract	Antibacterial	Lauritano et al. (2018)
Verrucosispora strain MS100137	Abyssomicin	Antiviral	Zhang et al. (2020)
Diaporthe sp. SCSIO 41011	Polyketides	Antiviral	Luo et al. (2018)
Trichoderma sp. SCSIO41004	5-acetyl-2-methoxy-1,4,6-trihydroxy-anthraquinone	Antiviral	Pang et al. (2018)

(continued)

Table 4.1 (continued)

Marine microorganisms	Natural products	Activities	Reference
Truncatella angustata	Isoprenylated cyclohexanols	Antiviral	Zhao et al. (2018)
Penicillium sp. IMB17-046	Trypilepyrazinol and 3b-Hydroxyergosta-8,14,24(28)-trien-7-one	Antiviral	Li et al. (2019)
Nocardia dassonvillei.	Phenazine derivative	Antifungal	Gao et al. (2012)
Streptomyces sp.S2A	Bacterial metabolites	Antifungal	Siddharth and Vittal (2018)
Streptomyces chumphonensis	Bacterial extracts	Antifungal	Hu et al. (2019)
Penicillium minioluteum ZZ1657	N-acetyl-L-valine conjugated drimarane sesquiterpenoids	Antifungal	Ma et al. (2020)
Amphidinium carterae	Amphidinol 22	Antifungal	Martínez et al. (2019)
	Other industrial applications		
Haloferax volcanii	Caspase	Detergent	Seth-Pasricha et al. (2013)
Natronococcus sp.	Lipase	Detergent	Corral et al. (2013)
Martelella mediterranea	Beta-glucosidase	Cellulose conversion; detergent	Mao et al. (2010)
Halomonas sp. AAD21	Alpha-amylase	Starch conversion; detergent	Uzyol et al. (2012)
Deinococcus geothermalis	Peptidase	Detergent	Pietrow et al. (2013)
Zunongwangia profunda	Alpha-amylase	Starch conversion; detergent	Qin et al. (2014)
Halothermothrix orenii	Amylase Endoamylase	Starch conversion; detergent	Bhattacharya and Pletschke (2014)
Amphibacillus sp. *NM-Ra2*	Amylase	Starch conversion; detergent	Mesbah and Wiegel (2014)

parahaemolyticus and *Staphylococcus aureus* (Galaviz-Silva et al. 2018). Co-cultivation of marine strains of *Bacillus* sp. and *Streptomyces* sp. resulted in the detection of novel metabolite dentigerumycin, having antiproliferative and antimetastatic properties against human carcinoma (Shin et al. 2018).

Exopolysaccharide, isolated and purified from marine bacterium *Aerococcus uriaeequi*, exhibited biocompatible antioxidant activity in mice, thus safe for topical and oral applications (Wang et al. 2018). The novel bargremycin analogs with antibacterial activity against *E. coli* were identified from marine *Streptomyces* sp. ZZ745 (Zhang et al. 2018). Marine-derived actinomycete *Streptomyces* sp. Al-Dabhi 90 exhibited significant antimicrobial activity against drug-resistant microorganisms like *S. aureus, Klebsiella pneumoniae, E. coli, Pseudomonas aeruginosa, Proteus mirabilis*, and *Enterococcus faecium* (Al-Dhabi et al. 2019). Antagonistic activity against prochloraz-resistant *Penicillium digitatum* strains (causes green mould in post-harvested citrus) was identified in the isolates from a marine actinomycete, *Streptomyces chumphonensis* (Hu et al. 2019). Bioactive extracts from the marine Gram-negative bacterium *Novosphingobium* sp. PP1Y have been reported to show remarkable antioxidant property (Petruk et al. 2019). More recently, a novel abyssomicin polyketide and other six known abyssomicin and proximicin analogs were detected in the extracts isolated from the marine bacterium *Verrucosispora* sp. MS100137 exhibits antiviral properties against Influenza A viruses (IAV) (Zhang et al. 2020; Ameen et al. 2021).

4.2.2 Marine Fungi

The fundamental information about fungi's ecological function and administration is yet to be explored. They colonize in different ecological niches in the marine environment or symbiotic or parasitic associations with other marine organisms (Poli et al. 2018; Rotter et al. 2021). Previously in 1992, only 15 marine fungal metabolites were identified (Fenical and Jensen 1993) and further increased to 270 in 2002 (Bugni and Ireland 2004), and until mid-2010, around 690 fungal bio-compounds from marine habitats were isolated and reported (Rateb and Ebel 2011). However, compared to other marine microorganisms, marine fungi are under explored, though the natural products from fungi are reported to be applicable for various purposes in medicine, cosmetics, food, biofertilizers, etc. (Tisthammer et al. 2016). Prominent examples of bioactive compounds isolated from fungi are cyclodepsipeptides such as scopularide from *Scopulariropsis breviccaulis* (Yu et al. 2008), as an antiproliferative agent and has been patented as a potential antitumor drug (Imhoff et al. 2011). The most recent examples include the isolation of purpuride from a marine fungus, *Penicillium minioluteum* ZZ1657. One of the purpuride was identified as a new drimane sesquiterpenoid, exhibiting antiproliferative activity against human glioma cells and the other two purpurides were identified as N acetyl-L-valine conjugated drimarane sesquiterpenoids, having antimicrobial activities against methicillin-resistant *Staphylococcus aureus* (MRSA), *C. albicans* and *E. coli* (Ma et al. 2020). Biologically active compound exhibiting antibacterial activity against aquatic pathogen *Vibro harveyi* was isolated from marine sponge-associated fungus *Aspergillus* sp. LS116. This antibacterial compound was identified as a C23 steroid with biocyclo [4.4.1] A/B ring and was named Aspergillsteroid A (Xu et al. 2020).

4.2.3 Marine Microalgae

Microalgae with shorter generation times and high growth rates can be considered cell factories for synthesizing natural bioactive compounds for biotechnological and industrial purposes (Rotter et al. 2021). Microalgae, due to their biomass and secondary metabolites, phenols, lipids, bioactive peptides, enzymes etc., are potential sources of renewable and sustainable energy; nutritional components, colorants in the food industry and stabilizing antioxidant agents as well as antiviral, antimicrobial, anticancer agents (Suganya et al. 2016; Moreno-Garcia et al. 2017). They produce value-added components like enzymes, polymers, vitamins, toxins, sterols etc. (Rotter et al. 2021). Their biomass can be used as bioplastics, biosurfactants, bio-emulsifier, and bio-fertilizers (Skjånes et al. 2013). Recent reports on microalgal natural products from marine microalgal dioflagellates *Prorocentrum hoffmannianum, P. arenarium, P. reticulatum, Gambierdiscus australes* and *Alexandrium tamarensis* expressed apoptotic properties in cell lines (de Vera et al. 2018). The organic extract from marine diatoms *Chaetoceros pseudocurvisetus* and *Skeletonema costatum,* when grown under controlled and phosphate starvation conditions, shows anti-tuberculosis activity (Lauritano et al. 2018). Polysaccharides from the marine diatom, *Navicula* sp., exhibited potential antioxidant activity (Fimbres-Olivarria et al. 2018). A new marine natural compound amphidinol 22, synthesized by marine microalgae *Amphidinium carterae*, had cytotoxic properties as well as antimicrobial activity against *C. albicans* (Martínez et al. 2019; Ameen et al. 2021).

These bioactive natural products are relevant to ecological and biomedical functions (König et al. 2006). Though there are such wide and potential applications of bioactive natural products derived from marine microorganisms, these products are yet to be explored to their full potential (Corinaldesi et al. 2017). Therefore metagenomics is acknowledged as a powerful and futuristic tool for understanding the probable prospects of marine microbially derived natural compounds.

4.3 Metagenomics for Marine Natural Product Discovery

Prokaryotes are most genetically diverse, with exceptional physiological, phenotypical and metabolic diversity. They prevail in varying and unique environments, thus, are rich sources of biologically applicable novel metabolites, enzymes, and natural products etc. (Nováková and Farkašovský 2013; Imhoff et al. 2011). To explore the ability of the microorganisms for the possible natural products, traditionally, the target microorganisms were cultured and screened for applications (Tringe et al. 2005; Fortman and Sherman 2005). A marine ecosystem has unique and vast biodiversity in itself. Still, only a fraction of it (0.001-1%) has been explored and cultured successfully under in-vitro environments to the extent possible (Kennedy et al. 2008). Since it is difficult to design media and mimic the natural microenvironment of microorganisms, cultivating most marine microbes with available standard laboratory facilities is challenging (Simon and Daniel 2009). Therefore, with the available nominal cultured marine microorganisms, the discovery of natural products from marine microbes is severely impacted (Barone et al. 2014). Concerning these limitations, metagenomics has emerged as a propitious tool for exploring novel gene clusters involved in biosynthetic pathways for bioproducts like enzymes, natural products etc., from the mixed marine microbial community, without culturing them (Fortman and Sherman 2005). Metagenomic approaches (illustrated in Fig. 4.1) are now being used for exploring various microbial communities; however, these approaches need to be further refined for the discovery of rare marine microorganisms with novel genes not yet represented in metagenomic libraries (Kennedy et al. 2010).

4.4 Metagenomics as a Strategy

Metagenomic strategies and other culture-independent approaches can be employed as potential tools for identifying and retrieving the natural products from marine microbial communities. Metagenomics approaches previously used to study non-cultured microbial communities for bioproducts have provided a better understanding of the global microbial ecology of different environmental niches and marine ecosystems. These approaches allow us to unleash various natural products with biotechnological potential (Madhavan et al. 2017). Screening the natural products of microorganisms from marine environments through functional-driven metagenomic strategies by next-generation sequencing generates large sequencing data from diverse environments, including marine water bodies (Kennedy et al. 2008). Sequence-driven screen strategies like Shotgun analysis and Next generation sequencing have been employed to explore, probe and retrieve novel genes or gene clusters for potential natural products (Madhavan et al. 2017; Mahapatra et al. 2020).

As mentioned above, metagenomic approaches bypass the need for the cultivation of microorganisms retrieved from the environment. The genomic DNA is extracted or isolated from the desired econiche, fragmented and cloned into the vector. The metagenomic libraries are constructed depending upon the desired gene size into vectors such as plasmids, cosmids, fosmids, bacterial artificial chromosomes (BACs) etc., thus allowing replication, identification, expression and analysis of genes for natural products. Likewise, potential genes such as polyketide synthases, non-ribosomal peptide synthetase, and post-translationally modified peptides were detected by screening, identification, and analysis of environmentally isolated DNA and many other genomic groups (Nazir 2016).

4.4.1 Environmental DNA (eDNA) Extraction/Isolation

With the vast array of mixed environmental genomic clusters, extraction or isolation of DNA for the metagenomic studies is the key primary step, followed by cloning transformation and library screening for the genes of biosynthesized natural products

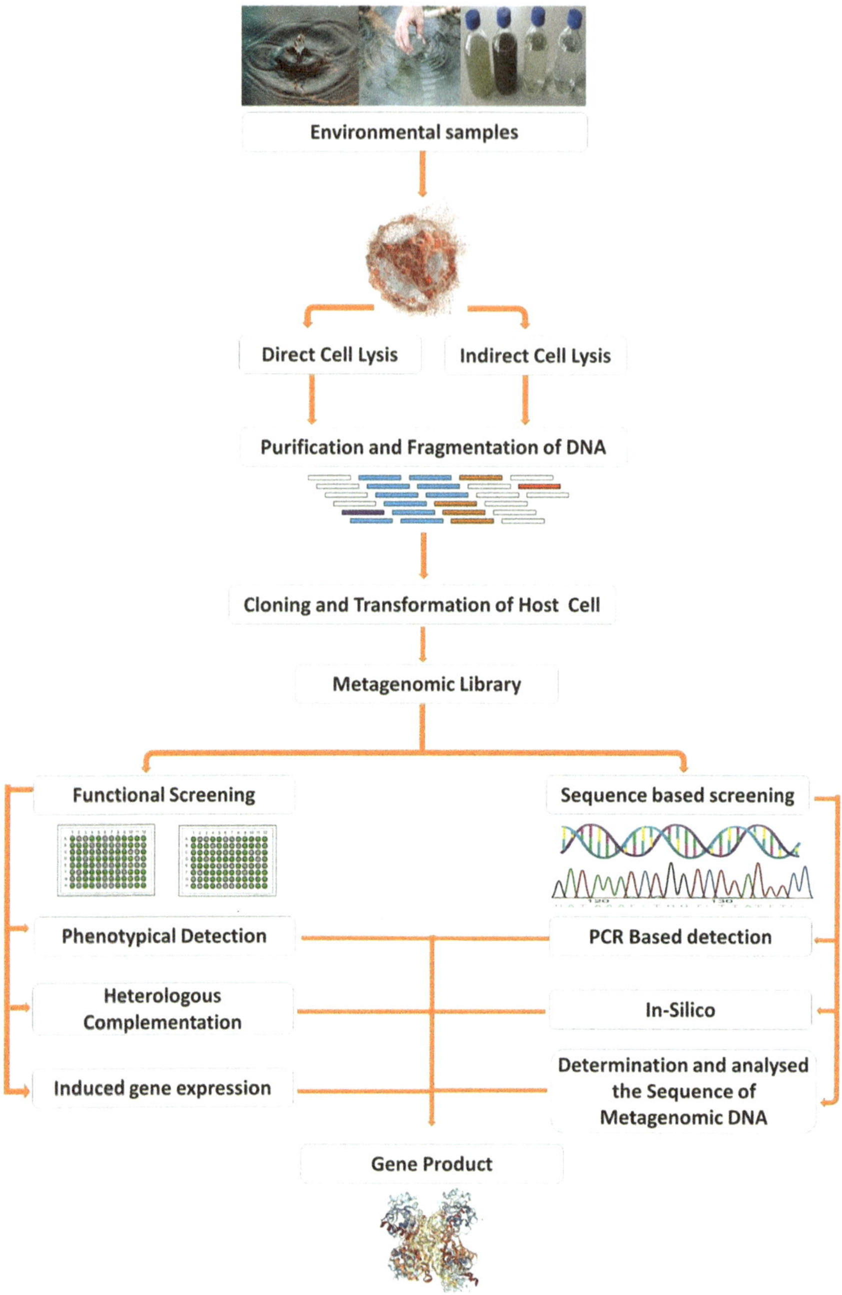

Fig. 4.1 Metagenomic approches for identification of natural products from marine ecosystem

(Fortman and Sherman 2005). The initial step of DNA extraction involves two methods, direct DNA extraction and indirect DNA extraction (Nazir 2016).

Direct genomic DNA isolation involves extraction of DNA by cell lysis directly from environmental samples availed and separation of DNA for further direct cloning into a vector (Brady 2007). This technique of DNA isolation is more robust for exploring genetic diversity in environmental samples than constructing libraries from isolated genomic material (Biddle et al. 2008). These libraries have paved the way for the biochemical profiling of environmental microbes and identification of novel natural products such as bioactive compounds, enzymes, and metabolites Steele et al. 2008).

The DNA from the environment can also be extracted via an indirect method involving segregating cells from any natural resources by utilizing various sedimentation techniques before DNA extraction. The cells are treated and lysed with enzymes and detergents, thus via gel electrophoresis retrieving the pure DNA (Liles et al. 2008). Though the DNA can be extracted through both approaches, high-quality DNA is key to constructing complex metagenomic libraries from marine environmental samples. This is because successful cloning and the subsequent library construction hinges on pure DNA, free from contamination.

4.4.2 Metagenomic Techniques for Creating and Screening Libraries

Metagenomic libraries are created from the extracted DNA from environmental samples. These libraries are constructed considering the DNA fragment insert size; thus, suitable vectors are selected. Small insertions less than 10 kb in plasmid vectors are required for single genes or small operons. While for example, a group of genes involved in a biosynthetic pathway or environmental DNA fragment requires cloning of an entire cluster of pathway genes of around 40 kb; hence, it is cloned into vectors like fosmids and BACs for gene fragments larger than 100 kb (Nováková and Farkašovský 2013). Thus, these extracted environmental DNAs are cloned into respective vectors creating libraries. The constructed metagenomic libraries are investigated or screened at the interrelated levels, sample processing, gene sequencing, and activity analysis to understand marine microbial functioning and recover natural products from these environmental samples (Nazir 2016).

Metagenomic techniques for screening these libraries include:

1. **Function or activity-based screening** involves the detection or looking for the desired phenotype or biological activities (Daniel 2005; Madhavan et al. 2017; Barone et al. 2014).
2. **Sequence or Homology based screening** is PCR amplification and hybridization sequencing techniques to identify genes coding for target proteins.

4.4.2.1 Function-Based Metagenomic Screening

Function-based metagenomics is a targeted approach involving screening functional gene products for the detection of bioactive compounds, therefore also called activity-based screening. Genomic libraries are created from the gene extracted from environmentally retrieved samples. Libraries are then screened directly for the phenotypic expressions of natural products with desired biological activity. Novel marine natural bioactive products with antioxidant, antimicrobial, antiviral and anticancer activities can be recognized with ease by functional or activity-based screening of the genomic libraries (Nováková and Farkašovský 2013). Metagenomic library screening based on functional analysis implied different strategies.

1. Enzyme activity screening of clones or phenotypes like a cellulose screening test for cellulase. This involves direct detection of phenotypes based on enzyme-substrate reactions expressing colored compounds to reveal visual or spectroscopic detection of the desired phenotype examples are that of starch iodine test for amylase or lipolytic enzymes and phosphatase from marine bioproducts (Staley and Sadowsky 2016). For activity-based screening, prior knowledge of gene sequence is unnecessary; thus, identifying desired genes for bioactive compounds is feasible and convenient (Ferrer et al. 2008; Riesenfeld et al. 2004).
2. Heterologous expression of desired genes requires heterologous compatibility or complementation between the host strain and cloned targeted gene sequence for growth and expression under selective conditions. The heterologous expression-based approach is mainly helpful for determining attributes like phylogenetic origin, promoters in gene sequences, gene toxicity, resistance mechanism, host physiology, GC content, and energy required in mRNA folding. Genes encoding for lipid hydrolysis enzymes, antimicrobial resistance, lysine racemases etc., are a few gene-encoding bioproducts identified by this strategy (Mahapatra et al. 2020).
3. Substrate-induced gene expression (SIGEX) wherein the novel catabolic genes, which are difficult to procure via conventional methods, can be identified and isolated by SIGEX. Environmental samples obtained are restriction digested into DNA fragments, ligated into the operon-trap expression vector, then transformed into a suitable cloning host, thus creating a metagenomic library (Uchiyama and Watanabe 2008). Substrate-specific positive clones are identified through the co-expression of green fluorescent protein (GFP) as a reporter gene in the vector. These positive clones are segregated by fluorescent-activated cell sorting (FACS), which helps rapidly screen the desired library (Uchiyama et al. 2005). Several challenges are met in functional-based metagenomic approaches via SIGEX for marine-derived samples, such as activation of multiple transcriptional factors and host selectivity needs to be taken care of to exploit it to its maximum potential (Zhang et al. 2017).

Several bioactive compounds have been identified with the aid of functional-based metagenomic screening. It facilitates the screening of large libraries due to application of high-throughput advanced screening methods and using robotic facilities in a relatively short period (Kennedy et al. 2010).

4.4.2.2 Sequence-Based Metagenomic Screening

Sequence-based metagenomic screening is a targeted metagenomic approach that involves screening metagenomic clones with the help of oligonucleotide primers and probes for desired genes. This technique aids in the discovery of novel sequences, and the target genes sequences coding for the protein can be sequenced by PCR-based sequencing and hybridization-based sequencing strategies (Madhavan et al. 2017; Nováková and Farkašovský 2013). PCR-based sequencing identifies conserved regions, thus making a phylogenetic evolutionary pathway for bioactive natural products and exploring the information about the complete genome (Lasken and McLean 2014). On the other hand, in hybridization-based sequencing, mRNA is extracted from the biological source and then hybridized with the already available or selected mRNA library, which contains a range of mRNA transcripts. Based on the hybridization signal intensities, the expression levels of the target transcript with known functions are determined (Suenaga 2012).

Sequence-based screening mostly depends on the high-fidelity amplification of relatively large DNA fragments. In contrast, genetic profiling is upgraded by inserting required base sequences and amplifying desired genes with the help of cDNA and poly (A) RNA libraries. High-fidelity polymerases like Q5 and Phusion play a critical role in reducing mutation rates and a high degree of accuracy in amplifying the desired genes during PCR (Hwang et al. 2017). Another critical step in sequence-based screening is employing appropriate DNA assembly methods. For the same, multiple DNA fragments are linked or merged in an end-to-end manner before insertion into host cells to get a larger sequence read at once. DNA assembly can be achieved by restriction enzyme fragmentation, homology-based sequencing, and then the linking of oligonucleotides. Lastly, the selection of a host for gene expression to produce the natural products. The proper host for the gene expression is very important, which aids in large-scale synthesis and expression of the potential bioproducts (Mahapatra et al. 2020). The sequence-based approach has helped in the identification of many novel natural products like chitinase (Hjort et al. 2010), phosphatase, glycerol dehydratase (enriched cultures) (Knietsch et al. 2003) etc., thus extending the possibilities of finding bioactive products with potential activities and efficiency (Madhavan et al. 2017).

4.4.2.3 High-Throughput Metagenomic Sequencing

High-throughput sequencing approaches involve screening large environmental samples for the particular natural products of interest. Both sequence-based and functional-based approaches include high-throughput-based approaches to screen, replicate the pathway and further redundancy rate is greatly reduced compared to classical culture-based approaches (Trindade et al. 2015).

Next-generation sequencing (NGS) led to parallel sequencing of massive DNA sequences and can generate millions of sequencing reads. Different types of NGS platforms like Illumina, Ion torrent, Ion proton, 454 genome sequencer, and SOLiD are being promoted for the sequencing strategies (Madhavan et al. 2017). High-throughput methods for functional-based screening, such as Substrate-induced gene expression (SIGEX), which involves FACS and metabolite-regulated expression (METREX) approaches, support screening of bioactive compounds from environmental samples (Malik et al. 2008).

4.5 Conclusion and Prospects

Metagenomic approaches offer opportunities to explore the genetic and metabolic diversity for the synthesis of targeted desired products having various industrial applications in therapeutical, agricultural, cosmetics, food, and biotechnological industries. With the advancement of high-throughput techniques, this process is made feasible; less time-consuming and larger samples can be easily processed (Staley and Sadowsky 2016). Therefore, applying metagenomic techniques for mining natural products from marine microorganisms would help recover novel, highly potential natural products and provide a better understanding of the biodiversity makeup. Although, there is considerable progress in the characterization of marine microorganisms for the production of bioactive compounds, further improvement of the metagenomic approaches and bioinformatic tools for screening and analyzing such huge amounts of data is in demand. Development of such new technologies would provide abundant opportunities to expand the understanding of the genetic composition of microorganisms and their expression to produce natural products which could be exploited for public welfare.

Acknowledgments Devika N. Nagar is grateful to Birla Institute of Technology and Science, K K Birla Goa Campus for the Institute Fellowship. Judith M. Braganca thanks Department of Science and Technology and Waste Management, Government of Goa, India for the Funding.

References

Abbes M, Baati H, Guermazi S et al (2013) Biological properties of carotenoids extracted from Halobacterium halobium isolated from a Tunisian solar saltern. BMC Complement Altern Med 13:255. https://doi.org/10.1186/1472-6882-13-255

Abdel-Mageed WM, Milne BF, Wagner M et al (2010) Dermacozines, a new phenazine family from deep-sea dermacocci isolated from a Mariana trench sediment. Org Biomol Chem 8:2352–2362. https://doi.org/10.1039/C001445A

Agrawal S, Adholeya A, Barrow CJ, Deshmukh SK (2018) In-vitro evaluation of marine derived fungi against Cutibacterium acnes. Anaerobe 49:5–13. https://doi.org/10.1016/J.ANAEROBE.2017.10.010

Al-Dhabi NA, Mohammed Ghilan AK, Esmail GA et al (2019) Bioactivity assessment of the Saudi Arabian marine Streptomyces sp. Al-Dhabi-90, metabolic profiling and its in vitro inhibitory property against multidrug resistant and extended-spectrum beta-lactamase clinical bacterial pathogens. J Infect Public Health 12:549–556. https://doi.org/10.1016/j.jiph.2019.01.065

Ameen F, AlNadhari S, Al-Homaidan AA (2021) Marine microorganisms as an untapped source of bioactive compounds. Saudi J Biol Sci 28:224–231. https://doi.org/10.1016/j.sjbs.2020.09.052

An FL, Liu WH, Zhang Y et al (2019) Spirocurvulaide, a novel spirobicyclic polyketide from a marine fungus Curvularia sp. IFB-Z10. Phytochem Lett 29:12–16. https://doi.org/10.1016/J.PHYTOL.2018.07.011

Arumugam M, Mitra A, Jaisankar P et al (2010) Isolation of an unusual metabolite 2-allyloxyphenol from a marine actinobacterium, its biological activities and applications. Appl Microbiol Biotechnol 86:109–117. https://doi.org/10.1007/s00253-009-2311-2

Bae SY, Liao L, Park SH et al (2020) Antitumor activity of asperphenin a, a lipopeptidyl benzophenone from marine-derived aspergillus sp. fungus, by inhibiting tubulin polymerization in colon cancer cells. Mar Drugs 18:110. https://doi.org/10.3390/md18020110

Barone R, de Santi C, Palma Esposito F et al (2014) Marine metagenomics, a valuable tool for enzymes and bioactive compounds discovery. Front Mar Sci 1:38. https://doi.org/10.3389/fmars.2014.00038

Bérdy J (2005) Bioactive microbial metabolites. J Antibiot 58:1–26. https://doi.org/10.1038/ja.2005.1

Bhattacharya A, Pletschke BI (2014) Review of the enzymatic machinery of Halothermothrix orenii with special reference to industrial applications. Enzym Microb Technol 55:159–169. https://doi.org/10.1016/J.ENZMICTEC.2013.10.011

Biddle JF, Fitz-Gibbon S, Schuster SC et al (2008) Metagenomic signatures of the Peru margin subseafloor biosphere show a genetically distinct environment. Proc Natl Acad Sci 105:10583–10588. https://doi.org/10.1073/pnas.0709942105

Bowers A, West N, Taunton J et al (2008) Total synthesis and biological mode of action of largazole: a potent class I histone deacetylase inhibitor. J Am Chem Soc 130:11219–11222. https://doi.org/10.1021/ja8033763

Brady SF (2007) Construction of soil environmental DNA cosmid libraries and screening for clones that produce biologically active small molecules. Nat Protoc 2:1297–1305. https://doi.org/10.1038/nprot.2007.195

Bugni TS, Ireland CM (2004) Marine-derived fungi: a chemically and biologically diverse group of microorganisms. Nat Prod Rep 21:143–163. https://doi.org/10.1039/B301926H

Chen L, Wei Y, Shi M et al (2019) An archaeal chitinase with a secondary capacity for catalyzing cellulose and its biotechnological applications in shell and straw degradation. Front Microbiol 10:1253. https://doi.org/10.3389/fmicb.2019.01253

Claverie M, McReynolds C, Petitpas A et al (2020) Marine-derived polymeric materials and biomimetics: an overview. Polymers (Basel) 12:1002. https://doi.org/10.3390/polym12051002

Corinaldesi C, Barone G, Marcellini F et al (2017) Marine microbial-derived molecules and their potential use in cosmeceutical and cosmetic products. Mar Drugs 15:118. https://doi.org/10.3390/md15040118

Corral P, Gutiérrez MC, Castillo AM et al (2013) Natronococcus roseus sp. nov., a haloalkaliphilic archaeon from a hypersaline lake. Int J Syst Evol Microbiol 63:104–108. https://doi.org/10.1099/ijs.0.036558-0

Daniel R (2005) The metagenomics of soil. Nat Rev Microbiol 3:470–478. https://doi.org/10.1038/nrmicro1160

Das S, Mangwani N (2015) Ocean acidification and marine microorganisms: responses and consequences. Oceanologia 57:349–361. https://doi.org/10.1016/j.oceano.2015.07.003

Dasari VRRK, Muthyala MKK, Nikku MY, Donthireddy SRR (2012) Novel Pyridinium compound from marine actinomycete, Amycolatopsis alba var. nov. DVR D4 showing antimicrobial and cytotoxic activities in vitro. Microbiol Res 167:346–351. https://doi.org/10.1016/J.MICRES.2011.12.003

de Vera CR, Díaz Crespín G, Hernández Daranas A et al (2018) Marine microalgae: promising source for new bioactive compounds. Mar Drugs 16:317. https://doi.org/10.3390/md16090317

Donia M, Hamann MT (2003) Marine natural products and their potential applications as anti-infective agents. Lancet Infect Dis 3:338–348. https://doi.org/10.1016/S1473-3099(03)00655-8

Egorova K, Antranikian G (2005) Industrial relevance of thermophilic archaea. Curr Opin Microbiol 8:649–655. https://doi.org/10.1016/j.mib.2005.10.015

Feling RH, Buchanan GO, Mincer TJ et al (2003) Salinosporamide a: a highly cytotoxic proteasome inhibitor from a novel microbial source, a marine bacterium of the new genus Salinospora. Angew Chem Int Ed 42:355–357. https://doi.org/10.1002/anie.200390115
Fenical W, Jensen PR (1993) Marine microorganisms: a new biomedical resource. In: Attaway DH, Zaborsky OR (eds) Pharmaceutical and bioactive natural products. Springer, Boston, MA, pp 419–457. https://doi.org/10.1007/978-1-4899-2391-2_12
Ferrer M, Beloqui A, Timmis KN, Golyshin PN (2008) Metagenomics for mining new genetic resources of microbial communities. J Mol Microbiol Biotechnol 16:109–123
Fimbres-Olivarria D, Carvajal-Millan E, Lopez-Elias JA et al (2018) Chemical characterization and antioxidant activity of sulfated polysaccharides from Navicula sp. Food Hydrocoll 75:229–236. https://doi.org/10.1016/J.FOODHYD.2017.08.002
Fortman JL, Sherman DH (2005) Utilizing the power of microbial genetics to bridge the gap between the promise and the application of marine natural products. Chembiochem 6:960–978. https://doi.org/10.1002/cbic.200400428
Galaviz-Silva L, Iracheta-Villarreal JM, Molina-Garza ZJ (2018) Bacillus and Virgibacillus strains isolated from three Mexican coasts antagonize Staphylococcus aureus and vibrio parahaemolyticus. FEMS Microbiol Lett 365:fny202. https://doi.org/10.1093/femsle/fny202
Gao X, Lu Y, Xing Y et al (2012) A novel anticancer and antifungus phenazine derivative from a marine actinomycete BM-17. Microbiol Res 167:616–622. https://doi.org/10.1016/J.MICRES.2012.02.008
Gao C-H, Tian X-P, Qi S-H et al (2010) Antibacterial and antilarval compounds from marine gorgonian-associated bacterium bacillus amyloliquefaciens SCSIO 00856. J Antibiot 63:191–193. https://doi.org/10.1038/ja.2010.7
Giani M, Garbayo I, Vílchez C, Martínez-Espinosa RM (2019) Haloarchaeal carotenoids: healthy novel compounds from extreme environments. Mar Drugs 17:524. https://doi.org/10.3390/md17090524
Helaly SE, Pesic A, Fiedler H-P, Süssmuth RD (2011) Elaiomycins B and C: Alkylhydrazide antibiotics from Streptomyces sp. BK 190. Org Lett 13:1052–1055. https://doi.org/10.1021/ol1031014
Hjort K, Bergström M, Adesina MF et al (2010) Chitinase genes revealed and compared in bacterial isolates, DNA extracts and a metagenomic library from a phytopathogen-suppressive soil. FEMS Microbiol Ecol 71:197–207. https://doi.org/10.1111/j.1574-6941.2009.00801.x
Hou J, Cui HL (2018) In vitro antioxidant, Antihemolytic, and anticancer activity of the carotenoids from halophilic archaea. Curr Microbiol 75:266–271. https://doi.org/10.1007/s00284-017-1374-z
Hu X, Cheng B, Du D et al (2019) Isolation and identification of a marine actinomycete strain and its control efficacy against citrus green and blue moulds. Biotechnol Biotechnol Equip 33:719–729. https://doi.org/10.1080/13102818.2019.1613175
Hu GP, Yuan J, Sun L et al (2011) Statistical research on marine natural products based on data obtained between 1985 and 2008. Mar Drugs 9:514–525. https://doi.org/10.3390/md9040514
Hwang J, Park SY, Park M et al (2017) Seasonal dynamics and metagenomic characterization of marine viruses in Goseong Bay, Korea. PLoS One 12:e0169841. https://doi.org/10.1371/journal.pone.0169841
Imhoff JF, Labes A, Wiese J (2011) Bio-mining the microbial treasures of the ocean: new natural products. Biotechnol Adv 29:468–482. https://doi.org/10.1016/j.biotechadv.2011.03.001
Ishihara K, Watanabe R, Uchida H et al (2017) Novel glycosylated mycosporine-like amino acid, 13-O-(β-galactosyl)-porphyra-334, from the edible cyanobacterium Nostoc sphaericum-protective activity on human keratinocytes from UV light. J Photochem Photobiol B Biol 172:102–108. https://doi.org/10.1016/J.JPHOTOBIOL.2017.05.019
Jenab A, Roghanian R, Emtiazi G (2020) Bacterial natural compounds with anti-inflammatory and immunomodulatory properties (mini review). Drug Des Devel Ther 14:3787–3801. https://doi.org/10.2147/DDDT.S261283

Kennedy J, Flemer B, Jackson SA et al (2010) Marine metagenomics: new tools for the study and exploitation of marine microbial metabolism. Mar Drugs 8:608–628. https://doi.org/10.3390/md8030608

Kennedy J, Marchesi JR, Dobson ADW (2008) Marine metagenomics: strategies for the discovery of novel enzymes with biotechnological applications from marine environments. Microb Cell Factories 7:1–8. https://doi.org/10.1186/1475-2859-7-27

Kenshole E, Herisse M, Michael M, Pidot SJ (2021) Natural product discovery through microbial genome mining. Curr Opin Chem Biol 60:47–54. https://doi.org/10.1016/j.cbpa.2020.07.010

Kerstin E, Degnes KF, Michael K et al (2010) Production of a new thiopeptide antibiotic, TP-1161, by a marine Nocardiopsis species. Appl Environ Microbiol 76:4969–4976. https://doi.org/10.1128/AEM.00741-10

Knietsch A, Bowien S, Whited G et al (2003) Identification and characterization of coenzyme B12-dependent glycerol dehydratase- and diol dehydratase-encoding genes from metagenomic DNA libraries derived from enrichment cultures. Appl Environ Microbiol 69:3048–3060. https://doi.org/10.1128/AEM.69.6.3048-3060.2003

König GM, Kehraus S, Seibert SF et al (2006) Natural products from marine organisms and their associated microbes. Chembiochem 7:229–238. https://doi.org/10.1002/cbic.200500087

Lasken RS, McLean JS (2014) Recent advances in genomic DNA sequencing of microbial species from single cells. Nat Rev Genet 15:577–584. https://doi.org/10.1038/nrg3785

Lauritano C, Martín J, de La Cruz M et al (2018) First identification of marine diatoms with anti-tuberculosis activity. Sci Rep 8:2284. https://doi.org/10.1038/s41598-018-20611-x

Leal MC, Puga J, Serôdio J et al (2012) Trends in the discovery of new marine natural products from invertebrates over the last two decades—where and what are we bioprospecting? PLoS One 7:e30580. https://doi.org/10.1371/journal.pone.0030580

Li H, Cao K, Cong P et al (2018a) Structure characterization and antitumor activity of the extracellular polysaccharide from the marine fungus Hansfordia sinuosae. Carbohydr Polym 190:87–94. https://doi.org/10.1016/J.CARBPOL.2018.02.077

Li J, Wang Y, Hao X et al (2019) Broad-spectrum antiviral natural products from the marine-derived Penicillium sp. IMB17-046. Molecules 24:2821. https://doi.org/10.3390/molecules24152821

Li Q, Zhu R, Yi W et al (2018b) Peniciphenalenins A–F from the culture of a marine-associated fungus Penicillium sp. ZZ901. Phytochemistry 152:53–60. https://doi.org/10.1016/J.PHYTOCHEM.2018.04.021

Liles MR, Williamson LL, Rodbumrer J et al (2008) Recovery, purification, and cloning of high-molecular-weight DNA from soil microorganisms. Appl Environ Microbiol 74:3302–3305. https://doi.org/10.1128/AEM.02630-07

Liu Z, Qiu P, Liu H et al (2019) Identification of anti-inflammatory polyketides from the coral-derived fungus Penicillium sclerotiorin: in vitro approaches and molecular-modeling. Bioorg Chem 88:102973. https://doi.org/10.1016/J.BIOORG.2019.102973

Lu J, Ma Y, Liang J et al (2012) Aureolic acids from a marine-derived Streptomyces sp. WBF16. Microbiol Res 167:590–595. https://doi.org/10.1016/J.MICRES.2012.06.001

Luo M, Ming Y, Wang L et al (2018a) Local delivery of deep marine fungus-derived equisetin from polyvinylpyrrolidone (PVP) nanofibers for anti-MRSA activity. Chem Eng J 350:157–163. https://doi.org/10.1016/J.CEJ.2018.05.159

Luo X, Yang J, Chen F et al (2018b) Structurally diverse polyketides from the mangrove-derived fungus diaporthe sp. SCSIO 41011 with their anti-influenza a virus activities. Front Chem 6:282. https://doi.org/10.3389/fchem.2018.00282

Ma M, Ge H, Yi W et al (2020) Bioactive drimane sesquiterpenoids and isocoumarins from the marine-derived fungus Penicillium minioluteum ZZ1657. Tetrahedron Lett 61:151504. https://doi.org/10.1016/j.tetlet.2019.151504

Madhavan A, Sindhu R, Parameswaran B et al (2017) Metagenome analysis: a powerful tool for enzyme bioprospecting. Appl Biochem Biotechnol 183:636–651. https://doi.org/10.1007/s12010-017-2568-3

Mahapatra GP, Raman S, Nayak S et al (2020) Metagenomics approaches in discovery and development of new bioactive compounds from marine Actinomycetes. Curr Microbiol 77: 645–656. https://doi.org/10.1007/s00284-019-01698-5
Malik S, Beer M, Megharaj M, Naidu R (2008) The use of molecular techniques to characterize the microbial communities in contaminated soil and water. Environ Int 34:265–276. https://doi.org/10.1016/j.envint.2007.09.001
Manivasagan P, Venkatesan J, Sivakumar K, Kim SK (2014) Pharmaceutically active secondary metabolites of marine actinobacteria. Microbiol Res 169:262–278. https://doi.org/10.1016/j.micres.2013.07.014
Mao X, Hong Y, Shao Z et al (2010) A novel cold-active and alkali-stable β-glucosidase gene isolated from the marine bacterium Martelella mediterranea. Appl Biochem Biotechnol 162: 2136–2148. https://doi.org/10.1007/s12010-010-8988-y
Martínez KA, Lauritano C, Druka D et al (2019) Amphidinol 22, a new cytotoxic and antifungal Amphidinol from the dinoflagellate Amphidinium carterae. Mar Drugs 17:385. https://doi.org/10.3390/md17070385
Mayer AMS, Glaser KB, Cuevas C et al (2010) The odyssey of marine pharmaceuticals: a current pipeline perspective. Trends Pharmacol Sci 31:255–265. https://doi.org/10.1016/j.tips.2010.02.005
Mesbah NM, Wiegel J (2014) Halophilic alkali- and thermostable amylase from a novel polyextremophilic Amphibacillus sp. NM-Ra2. Int J Biol Macromol 70:222–229. https://doi.org/10.1016/J.IJBIOMAC.2014.06.053
Mondol MAM, Shin HJ, Islam MT (2013) Diversity of secondary metabolites from marine bacillus species: chemistry and biological activity. Mar Drugs 11:2846–2872. https://doi.org/10.3390/md11082846
Moreno-Garcia L, Adjallé K, Barnabé S, Raghavan GSV (2017) Microalgae biomass production for a biorefinery system: recent advances and the way towards sustainability. Renew Sust Energ Rev 76:493–506. https://doi.org/10.1016/j.rser.2017.03.024
Nazir A (2016) Review on metagenomics and its applications. Imp J Intersd Res 2:1
Nováková J, Farkašovský M (2013) Bioprospecting microbial metagenome for natural products. Biologia (Poland) 68:1079–1086. https://doi.org/10.2478/s11756-013-0246-7
Olano C, Méndez C, Salas JA (2009) Antitumor compounds from marine actinomycetes. Mar Drugs 7:210–248. https://doi.org/10.3390/md702021
Oren A, Gunde-Cimerman N (2007) Mycosporines and mycosporine-like amino acids: UV protectants or multipurpose secondary metabolites? FEMS Microbiol Lett 269:1–10. https://doi.org/10.1111/j.1574-6968.2007.00650.x
Pang X, Lin X, Tian Y et al (2018) Three new polyketides from the marine sponge-derived fungus Trichoderma sp. SCSIO41004. Nat Prod Res 32:105–111. https://doi.org/10.1080/14786419.2017.1338286
Paul SI, Majumdar BC, Ehsan R et al (2021) Bioprospecting potential of marine microbial natural bioactive compounds. J Appl Biotechnol Rep 8:96–108. https://doi.org/10.30491/jabr.2020.233148.1232
Petruk G, Roxo M, de Lise F et al (2019) The marine gram-negative bacterium Novosphingobium sp. PP1Y as a potential source of novel metabolites with antioxidant activity. Biotechnol Lett 41:273–281. https://doi.org/10.1007/s10529-018-02636-4
Pietrow O, Panek A, Synowiecki J (2013) Extracellular proteolytic activity of Deinococcus geothermalis. Afr J Biotechnol 12:4020–4027
Poli A, Finore I, Romano I et al (2017) Microbial diversity in extreme marine habitats and their biomolecules. Microorganisms 5(2):25. https://doi.org/10.3390/microorganisms5020025
Poli A, Vizzini A, Prigione V, Varese GC (2018) Basidiomycota isolated from the Mediterranean Sea—phylogeny and putative ecological roles. Fungal Ecol 36:51–62. https://doi.org/10.1016/j.funeco.2018.09.002

Qin Y, Huang Z, Liu Z (2014) A novel cold-active and salt-tolerant α-amylase from marine bacterium Zunongwangia profunda: molecular cloning, heterologous expression and biochemical characterization. Extremophiles 18:271–281. https://doi.org/10.1007/s00792-013-0614-9

Rateb ME, Ebel R (2011) Secondary metabolites of fungi from marine habitats. Nat Prod Rep 28: 290–344. https://doi.org/10.1039/C0NP00061B

Rath CM, Janto B, Earl J et al (2011) Meta-omic characterization of the marine invertebrate microbial consortium that produces the chemotherapeutic natural product ET-743. ACS Chem Biol 6:1244–1256. https://doi.org/10.1021/cb200244t

Ravikumar S, Gnanadesigan M, Saravanan A et al (2012) Antagonistic properties of seagrass associated Streptomyces sp. RAUACT-1: a source for anthraquinone rich compound. Asian Pac J Trop Med 5:887–890. https://doi.org/10.1016/S1995-7645(12)60165-5

Riesenfeld CS, Schloss PD, Handelsman J (2004) Metagenomics: genomic analysis of microbial communities. Annu Rev Genet 38:525–552. https://doi.org/10.1146/annurev.genet.38.072902.091216

Rodrigo-Baños M, Garbayo I, Vílchez C et al (2015) Carotenoids from Haloarchaea and their potential in biotechnology. Mar Drugs 13:5508–5532. https://doi.org/10.3390/md13095508

Rotter A, Barbier M, Bertoni F et al (2021) The essentials of marine biotechnology. Front Mar Sci 8:158. https://doi.org/10.3389/fmars.2021.629629

Sato S, Iwata F, Yamada S et al (2011) Usabamycins A–C: new anthramycin-type analogues from a marine-derived actinomycete. Bioorg Med Chem Lett 21:7099–7101. https://doi.org/10.1016/J.BMCL.2011.09.086

Sayed KA (2000) Natural products as antiviral agents. Stud Nat Prod Chem 24:473–572. https://doi.org/10.1016/S1572-5995(00)80051-4

Schneider K, Keller S, Wolter FE et al (2008) Proximicins a, B, and C—antitumor furan analogues of netropsin from the marine actinomycete Verrucosispora induce upregulation of p53 and the cyclin kinase inhibitor p21. Angew Chem Int Ed 47:3258–3261. https://doi.org/10.1002/anie.200705295

Seth-Pasricha M, Bidle KA, Bidle KD (2013) Specificity of archaeal caspase activity in the extreme halophile Haloferax volcanii. Environ Microbiol Rep 5:263–271. https://doi.org/10.1111/1758-2229.12010

Shin D, Byun WS, Moon K et al (2018) Coculture of marine Streptomyces sp. with bacillus sp. produces a new piperazic acid-bearing cyclic peptide. Front Chem 6:498. https://doi.org/10.3389/fchem.2018.00498

Shin HJ, Mondol MAM, Yu TK et al (2010) An angiogenesis inhibitor isolated from a marine-derived actinomycete, Nocardiopsis sp. 03N67. Phytochem Lett 3:194–197. https://doi.org/10.1016/J.PHYTOL.2010.07.005

Shivale N, Marar T, Samant M, Harmalkar MN (2018) Screening of antioxidant activity of marine bacteria isolated from marine soil obtained from north-West coastal region of India. Int J Biol 7: 1. https://doi.org/10.31032/IJBPAS/2018/7.3.4382

Siddharth S, Vittal RR (2018) Evaluation of antimicrobial, enzyme inhibitory, antioxidant and cytotoxic activities of partially purified volatile metabolites of marine Streptomyces sp.S2A. Microorganisms 6:72. https://doi.org/10.3390/microorganisms6030072

Simon C, Daniel R (2009) Achievements and new knowledge unraveled by metagenomic approaches. Appl Microbiol Biotechnol 85:265–276. https://doi.org/10.1007/s00253-009-2233-z

Singh OV, Gabani P (2011) Extremophiles: radiation resistance microbial reserves and therapeutic implications. J Appl Microbiol 110:851–861. https://doi.org/10.1111/j.1365-2672.2011.04971.x

Skjånes K, Rebours C, Lindblad P (2013) Potential for green microalgae to produce hydrogen, pharmaceuticals and other high value products in a combined process. Crit Rev Biotechnol 33: 172–215

Sogin ML, Morrison HG, Huber JA et al (2006) Microbial diversity in the deep sea and the underexplored "rare biosphere". Proc Natl Acad Sci 103:12115–12120. https://doi.org/10.1073/pnas.0605127103
Solanki R, Khanna M, Lal R (2008) Bioactive compounds from marine actinomycetes. Indian J Microbiol 48(4):410–431. https://doi.org/10.1007/s12088-008-0052-z
Sran KS, Bisht B, Mayilraj S, Roy Choudhury A (2019) Structural characterization and antioxidant potential of a novel anionic exopolysaccharide produced by marine microbacterium aurantiacum FSW-25. Int J Biol Macromol 131:343–352. https://doi.org/10.1016/J.IJBIOMAC.2019.03.016
Staley C, Sadowsky MJ (2016) Application of metagenomics to assess microbial communities in water and other environmental matrices. J Mar Biol Assoc U K 96:121–129. https://doi.org/10.1017/S0025315415001496
Steele HL, Jaeger KE, Daniel R, Streit WR (2008) Advances in recovery of novel biocatalysts from metagenomes. J Mol Microbiol Biotechnol 16:25–37. https://doi.org/10.1159/000142892
Subramani R, Aalbersberg W (2013) Culturable rare Actinomycetes: diversity, isolation and marine natural product discovery. Appl Microbiol Biotechnol 97:9291–9321. https://doi.org/10.1007/s00253-013-5229-7
Suenaga H (2012) Targeted metagenomics: a high-resolution metagenomics approach for specific gene clusters in complex microbial communities. Environ Microbiol 14:13–22. https://doi.org/10.1111/j.1462-2920.2011.02438.x
Suganya T, Varman M, Masjuki HH, Renganathan S (2016) Macroalgae and microalgae as a potential source for commercial applications along with biofuels production: a biorefinery approach. Renew Sust Energ Rev 55:909–941. https://doi.org/10.1016/j.rser.2015.11.026
Taori K, Paul VJ, Luesch H (2008) Structure and activity of largazole, a potent antiproliferative agent from the Floridian marine cyanobacterium Symploca sp. J Am Chem Soc 130:1806–1807. https://doi.org/10.1021/ja7110064
Tisthammer KH, Cobian GM, Amend AS (2016) Global biogeography of marine fungi is shaped by the environment. Fungal Ecol 19:39–46. https://doi.org/10.1016/j.funeco.2015.09.003
Trindade M, van Zyl LJ, Navarro-Fernández J, Elrazak AA (2015) Targeted metagenomics as a tool to tap into marine natural product diversity for the discovery and production of drug candidates. Front Microbiol 6:890. https://doi.org/10.3389/fmicb.2015.00890
Tringe SG, von Mering C, Kobayashi A et al (2005) Comparative metagenomics of microbial communities. Science 308:554–557. https://doi.org/10.1126/science.1107851
Uchiyama T, Abe T, Ikemura T, Watanabe K (2005) Substrate-induced gene-expression screening of environmental metagenome libraries for isolation of catabolic genes. Nat Biotechnol 23:88–93. https://doi.org/10.1038/nbt1048
Uchiyama T, Watanabe K (2008) Substrate-induced gene expression (SIGEX) screening of metagenome libraries. Nat Protoc 3:1202–1212. https://doi.org/10.1038/nprot.2008.96
Uzyol K, Sariyar Akbulut B, Denizci A, Kazan D (2012) Thermostable α-amylase from moderately halophilic Halomonas sp. AAD21. Turk J Biol 36:327–338. https://doi.org/10.3906/biy-1106-7
Vala AK, Sachaniya B, Dudhagara D et al (2018) Characterization of L-asparaginase from marine-derived aspergillus Niger AKV-MKBU, its antiproliferative activity and bench scale production using industrial waste. Int J Biol Macromol 108:41–46. https://doi.org/10.1016/J.IJBIOMAC.2017.11.114
Vollmers J, Wiegand S, Kaster AK (2017) Comparing and evaluating metagenome assembly tools from a microbiologist's perspective—not only size matters! PLoS One 12(1):e169622. https://doi.org/10.1371/journal.pone.0169662
Wang C, Fan Q, Zhang X et al (2018a) Isolation, characterization, and pharmaceutical applications of an exopolysaccharide from aerococcus uriaeequi. Mar Drugs 16:337. https://doi.org/10.3390/md16090337
Wang L, Li C, Yu G et al (2019) Dicitrinones E and F, citrinin dimers from the marine derived fungus Penicillium citrinum HDN-152-088. Tetrahedron Lett 60:151182. https://doi.org/10.1016/J.TETLET.2019.151182

Wang W, Liao Y, Chen R et al (2018b) Chlorinated azaphilone pigments with antimicrobial and cytotoxic activities isolated from the deep sea derived fungus Chaetomium sp. NA-S01-R1. Mar Drugs 16:61. https://doi.org/10.3390/md16020061

Waters AL, Hill RT, Place AR, Hamann MT (2010) The expanding role of marine microbes in pharmaceutical development. Curr Opin Biotechnol 21:780–786. https://doi.org/10.1016/j.copbio.2010.09.013

Williams PG (2009) Panning for chemical gold: marine bacteria as a source of new therapeutics. Trends Biotechnol 27:45–52. https://doi.org/10.1016/j.tibtech.2008.10.005

Xu P, Ding L, Wei J et al (2020) A new aquatic pathogen inhibitor produced by the marine fungus aspergillus sp. LS116. Aquaculture 520:734670. https://doi.org/10.1016/j.aquaculture.2019.734670

Yang SQ, Li XM, Li X et al (2018) Two new diketomorpholine derivatives and a new highly conjugated ergostane-type steroid from the marine algal-derived endophytic fungus aspergillus alabamensis EN-547. Mar Drugs 16:114. https://doi.org/10.3390/md16040114

Yu Z, Lang G, Kajahn I et al (2008) Scopularides a and B, cyclodepsipeptides from a marine sponge-derived fungus, Scopulariopsis brevicaulis. J Nat Prod 71:1052–1054. https://doi.org/10.1021/np070580e

Zhang J, Li B, Qin Y et al (2020) A new abyssomicin polyketide with anti-influenza a virus activity from a marine-derived Verrucosispora sp. MS100137. Appl Microbiol Biotechnol 104:1533–1543. https://doi.org/10.1007/s00253-019-10217-2

Zhang YM, Liu BL, Zheng XH et al (2017) Anandins a and B, two rare steroidal alkaloids from a marine Streptomyces anandii H41-59. Mar Drugs 15:355. https://doi.org/10.3390/md15110355

Zhang D, Shu C, Lian X, Zhang Z (2018) New antibacterial bagremycins F and G from the marine-derived streptomyces sp. ZZ745. Mar Drugs 16:330. https://doi.org/10.3390/md16090330

Zhao Y, Liu D, Proksch P et al (2018a) Truncateols O-V, further isoprenylated cyclohexanols from the sponge-associated fungus Truncatella angustata with antiviral activities. Phytochemistry 155:61–68. https://doi.org/10.1016/J.PHYTOCHEM.2018.07.017

Zhao DL, Wang D, Tian XY et al (2018b) Anti-phytopathogenic and cytotoxic activities of crude extracts and secondary metabolites of marine-derived fungi. Mar Drugs 16:36. https://doi.org/10.3390/md16010036

Marine Derived Pharmaceuticals in Biomedical Research: Current Developments and Future Prospects

5

Y. Santhosh Kumar and G. V. Swarnalatha

Abstract

Medical advancements are allowing people to live longer and healthier lives. We must still make significant progress to treat terminal diseases, overcome resistance to current prognoses, and uncover new medication sources. Due to this, pharmaceutical companies are constantly looking for new ways to develop safe and effective medicines. Because of this diversity, the oceans are the world's most important biosphere. Marine organisms must produce many secondary metabolites to survive in the harsh ocean environment. Though their presence may provide an evolutionary advantage in the long run, secondary metabolites aren't required to function for primary or secondary metabolisms. Some substances in the ocean could be used to develop new medications. Because marine animal metabolites have biological activity, they can be used in human therapy. Because of their intriguing biological and pharmacological potential, marine-derived pharmaceuticals have recently received a lot of attention. This chapter summarizes studies on pharmaceutical medicines obtained from marine species and their potential therapeutic effects in biomedical research.

Keywords

Anticancer drugs · Marine pharmacology · Angiogenesis · Antiviral drugs · Drug delivery

Y. S. Kumar
Faculty of Allied Health Sciences, Chettinad Hospital and Research Institute, Chettinad Academy of Research and Education (Deemed to be University), Chennai, Tamil Nadu, India

G. V. Swarnalatha (✉)
Department of Biochemistry, Rayalaseema University, Kurnool, India

P. Veera Bramhachari, C. V. Berde (eds.), *Marine Bioactive Molecules for Biomedical and Pharmacotherapeutic Applications*,
https://doi.org/10.1007/978-981-99-6770-4_5

5.1 Introduction

Compounds that provide marine plants and animals their active pharmacological properties are the focus of marine pharmacology, a branch of pharmaceutical science. It is possible to extract these compounds from their remains. Oceans are our largest and most diverse habitat at over 70% of Earth's total area. There is a broad range of biological and chemically diverse species to be found there, making them a valuable resource in and of itself. In recent years, huge quantities of medications, active compounds, and other metabolites originating from marine species have been identified. This tendency is predicted to reverse soon, although most medications are still derived from natural or synthetic sources. Approximately 50,000 aquatic components have been identified, and more than 1500 new marine chemicals have been discovered yearly since 2008. Structure, complexity, and variation are all common characteristics that set them apart (Malve 2016; Kiuru et al. 2014). After Faulkner's departure in 2002, the Natural Products Report was edited by Blunt and his colleagues. Blunt et al. 2016 annual reports on novel marine natural products continued. In 1600 BC, the Phoenicians were the first to extract the tyrian purple dye from marine mollusks to create it. In terms of marine products, this is the oldest. For a long time, most research on marine natural products has been on metabolites collected from fish and marine algae.

Examples of marine biopolymers include the vitamins A and D present in fish liver oil, agar, carrageenan, and polyunsaturated fatty acids like eicosapentenoic acid and docosahexenoic acid. The extraction of spongeothymidine and spongeouridine from the Caribbean sponge Tethyacrypta in the 1950s marked the start of the early development of marine drugs (Bergmann and Feeney 1951; Newman and Cragg 2004; Sagar et al. 2010). *Acremoniumchrysogenum*, an antibiotic fungus discovered in 1940 in Mediterranean Sea water near Sardinia, was the source of the cephalosporin antibiotic class (Konig 1992). Because the fungus produces cephalosporin C, we have cephalosporins today. In 1969, Weinheimer and Spraggins published a study showing that the gorgonian Plexaurahomomalla has many prostaglandins (Weinheimer and Spraggins 1969). There has been a significant increase in patent requests for marine natural products since the mid-1980s (Bongiorni 1996). The vast majority of substances come from marine crustaceans. On the other hand, marine microorganisms are gaining more and more interest (Jensen and Fenical 2000; Schweder et al. 2005; Tarman et al. 2013).

This chapter emphasizes the existing approved treatments made from aquatic animals and other chemicals. Biopolymers, vitamins, minerals, and UV-protecting marine compounds are some products derived from marine sources (mycosporine acids). Nutraceuticals originating from marine sources have been thoroughly examined by Suleriaet al. Suleria et al. 2015, and Martins et al. 2014 are an excellent source for information on marine-derived cosmetics, cosmeceuticals, and other products.

Marine life accounts for more than 80% of the planet's total diversity of plants and animals. There are a variety of marine organisms that make up the foundation of bioactive mixes. These include wipes, tunicates, corals, nudibranchs, ocean bunnies (mollusks), prawns (shells), and marine bacteria. Marine invertebrates have evolved a defensive system that involves releasing toxic materials to defend themselves from predators. The study of secondary metabolites has been made possible to help these molecules. Researchers have discovered a connection between the technique they employ to fend off invaders and enzymes and receptors that play a similar role in human disease. New human diseases, which are inescapable, have forced the field of marine pharmacology to step in as a potential research venue for innovative treatments.

5.2 History

People have reaped the benefits of marine medicine in numerous ways since the dawn of time. In the 1940s, a fungus called *acremonium chrysogenum* that was isolated from the Mediterranean Sea near Sardinia produced cephalosporin C, as evidenced by the use of Bathini fish for asthma treatment, oyster powder for GIT disturbance treatment, pearl oxide for tuberculosis treatment, and dried sea horses for respiratory and sexual function. The cephalosporin class of antibiotics was born as a result of this. The discovery of spongothymidine and spongouridine from the Tethyacrypta sponge in the 1950s marked the beginning of real marine drug development (Werner Bergmann). Arabinose nucleosides were extracted from marine sponges collected off the coast of Florida in 1969 and 1976, respectively, to generate cytarabine (Ara-C) and vidarabine (Ara-A). Shimomura and two other American scientists shared the 2008 Nobel Prize in Chemistry for their work on the anticancer properties of a jellyfish protein.

5.3 Marine Pharmacology in India

Natural products are a vital source of inspiration for the creation of new medications. Biodiversity in the ocean has great promise as a future supply of medicines due to the emergence of antibiotic resistance in hazardous bacteria, such as tuberculosis and malaria resistance. Since they transmit and manage life, peptides are vital to the biological communication pathway. Many marine animals include bioactive peptides, proven products that have piqued the interest of pharmaceutical companies worldwide. There are several maritime habitats and species to be found throughout India's 8000-kilometer coastline. This country can utilize its resources better to study clinical issues. Innovative medications have come out of Indian institutions like Goa's national institute of Oceanology, Central Drug Research Institute in Lucknow, Institution: Bose College in Kolkata, Bhubaneswar's regional research Laboratory and Mumbai, the Central Institute of Fisheries Education (CIFE). Many other Indian educational institutions, academic institutions, and pharmaceutical companies have noticed this topic's importance.

5.4 Marine Drug Development

The process's initial phase is collecting and classifying marine species. It was common practice to gather creatures in locations that were easily accessible, along beaches or from the surface of the sea. Recently developed underwater life support systems have allowed scientists to collect data from previously unexplored areas. Many cases necessitate genetic testing to be identified. The organisms that have been collected and identified should, if at all possible, be cultured. It is necessary to have a significant amount of cleaned and dried material for further inquiry in this case. The procedures for dealing with marine species are, for the most part, the same as those for dealing with organisms that live on land: drug candidate pharmacodynamics, pharmacokinetics, and safety parameter research; target identification; supply assurance; possible derivatization and generation of substance libraries; isolation of compounds with potential biological activity or unique structures (Kiuru et al. 2014; Suleria et al. 2015). Alejandro M. S. Mayer of Midwestern University in Illinois manages the internet resource "The Global Marine Pharmaceuticals Pipeline," which identifies marine-derived pharmaceuticals and summarizes information about licensed drugs and drugs currently engaging in clinical studies. Clinical trials for a total of twenty marine-derived drugs are now underway.

5.4.1 Biomedical Applications of Marine-Derived Pharmaceuticals

As a result of the many beneficial properties of marine polymers, several researchers have begun to focus on marine polymers/IL solutions in their efforts to design biomaterials that would accelerate the healing process of wounds. As a consequence of the utilization of the bio macromolecule combination of silk fibroin/chitosan in (Bmim)(OAc) or chitosan/cellulose/keratin in (Bmim)(Cl), hDFb adhesion and proliferation were improved, in addition to mechanical strength, bactericidal activity, and control of the structure's growth. These results were achieved by using the methods mentioned above. It was suggested by Roger RD and colleagues that chitin and alginate should be co-dissolved in (Bmim)(OAc) to form fibers that may be employed in wound dressings and that these fibers should then be extruded into a coagulation bath (Silva et al. 2012). These fibers were used to treat a full-thickness dermal wound model (rat model, histological study), and they were left in situ for up to 14 days before being removed. The model was performed on rats. It was found those 7 days after treatment, the hyperplastic epidermis continued to cover the dermal fibrosis in the chitin and calcium alginate-coated lesion sites, demonstrating typical wound healing with re-epithelialization. (CEL/CHT) composite films were produced to generate a moist microenvironment that promotes healthy tissue regeneration to heal wounds. This was done to treat wounds (Harkins et al. 2014). They claim that the antibacterial, hemostatic, and biocompatible composites were made using (Bmim)(Cl) as a single solvent. They also claim that it is safe to use on fibroblasts and may be used to absorb entire blood that has been treated to prevent clotting. The composites absorbed the same quantity of blood as standard wound

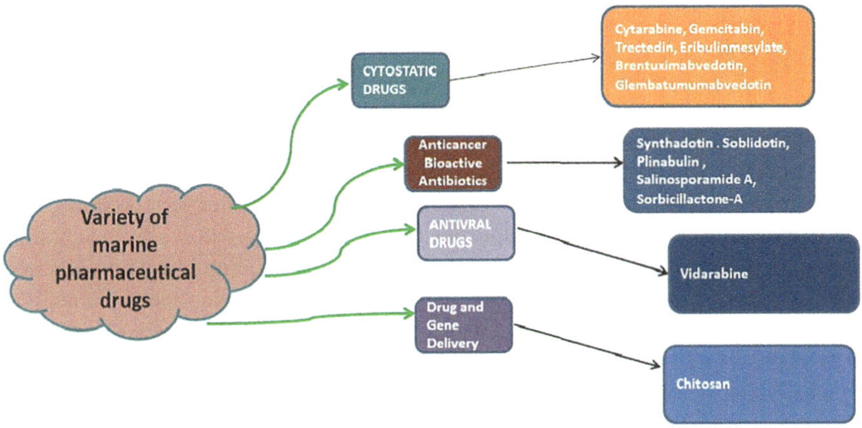

Fig. 5.1 Marine drugs and compounds used in biomedical research

dressings, demonstrating that they are functionally equivalent (Shamshina et al. 2014). Figure 5.1 summarizes the marine drugs with biomedical applications.

5.5 Marine Compounds as Cytostatic Drugs

Many times, cancers fall somewhere in the middle of the global mortality rankings. Novel anticancer drugs are urgently required with a high activity level but few side effects. This requirement can be met in part by marine-derived natural compounds. The discovery of the nucleosides sphongothymidine and spongouridine in the Caribbean sponge *Tethyacrypta* led to the creation of marine cytostatics. These nucleosides were starting point for marine cytostatic nucleosides (Bergmann and Feeney 1951; Bergmann and Burke 1955; Newman and Cragg 2004). *Tectytethyacrypta* is the more widespread name for this sponge, which was previously known as *Cryptotethyacrypta*. Based on this finding, Cytarabine, a synthetic counterpart of cytosine arabinoside, was developed as a strong antileukemic drug and was approved by the Food and Drug Administration in 1969. This pyrimidine nucleoside serves as a metaphor for a prodrug. Deoxycytidine competes for the physiological substrate with cytosine arabinoside triphosphate, which undergoes an intracellular change. This strategy also inhibits DNA synthesis because DNA polymerase is inhibited. Acute myeloid leukemia, as well as other forms of leukemia and non-lymphoma, can be treated with it, according to Löwenberg 2013. According to the type of cancer and the stage of treatment, Hodgkin's dosage is different. Digestive issues are the most common side effect. Additional side effects include pneumonia, anorexia, and disorientation. It is recommended that intranasal administration of liposomal cytarabine manufactured by Mundipharma in Limburg, Germany, for treating lymphomatous meningitis. Under the brand name Gemzar, the medication known as gemcitabine (difluoro-cytarabine) is offered for sale by

Lilly Deutschland GmbH in Giessen, Germany. Malignancies of the bronchial tract, bladder, pancreas, breast, and ovaries can all be treated with this medication (Mayer et al. 2010; Anjum et al. 2016; Mutschler et al. 2016).

The ascidian *Ecteinascidia turbinata* was the source from which tabectedin was isolated. Coral-dwelling species in the Mediterranean create this molecule in self-defense against the numerous bacteria they encounter. Based on the newly collected data and genetic research, we postulate that trabectedin is produced by microorganisms associated with tunicates (Giddings and Newman 2013; Schofield et al. 2015). Trabectedin is a member of the tetrahydroisoquinoline alkaloid family from a chemical perspective. One of the three tetrahydroquinoline moieties is a 10-member heterocyclic ring that contains a cysteine residue. The other eight rings each contain one of the other seven chiral centers. It is possible to create enough cyanosafracin B for long-term use by semi-synthesizing it from a byproduct of the bacteria *Pseudomonas fluorescens*. This would allow for sufficient production of cyanosafracin B. (Cuevas and Francesch 2009). Trectedin may be the first anticancer medicine that targets both cancer cells and the microenvironment surrounding the tumor. For instance, the cytostatic action of the compound is closely connected to the fact that it binds to certain triplets in the minor groove of DNA. It accomplishes this by inhibiting transcription factors from binding to DNA, which stops transcription from taking place in the first place. An inter-strand cross-linking lesion, also known as an adduct, is indistinguishable from newly formed adducts. This type of lesion restricts transcription by preserving the integrity of double-stranded DNA.

The RNA polymerase II elongating enzyme also directly interacts with this protein (Pol II). It is possible to affect the tumor's microenvironment by reducing the number of macrophages linked with the tumor and by controlling the production of cytokines and substances that promote Angiogenesis. It is reasonable to anticipate that the activities of trabectedin will synergistically affect the tumor cells and the environment around them. The anticancer drug trabectedin is less effective in NER-deficient cells than in NER-sufficient cells (D'incalci et al. 2014; López-Guerrero et al. 2015). This is not the case with other anticancer drugs that interact with DNA, such as platinum-based chemicals and similar substances. Trabectedin was granted permission by the EU in 2007 and approval by the FDA in 2015. Patients with soft tissue sarcoma who are resistant to anthracycline and ifosfamide treatment and those who are not candidates for treatment with these medications are now eligible for treatment with this drug, as it has been approved for use by the FDA. To qualify for approval, patients must have recurrent platin-sensitive ovarian cancer and receive pegylated liposomal doxorubicin in combination with the medication.

In an open-phase III multicenter research study, 672 women with recurrent ovarian cancer who were given trabectedin and pegylated liposomal doxorubicin had a significantly longer progression-free survival (6 weeks) than those who were given doxorubicin treatment alone. The National Cancer Institute conducted the study. Patients who were given trabectedin and pegylated liposomal doxorubicin throughout the experiment were considered participants. A similar tolerability profile was observed even after treatment with doxorubicin (Lindequist 2016). According to recent studies, a patient with metastatic leiomyosarcoma treated with first-line trabectedin experienced a response that lasted for an extended period (Maruzzo et al. 2016). If the initial chemotherapy treatment doesn't work, trabectedin can be administered as an adjunct to the second-line treatment being executed. The standard dose for treating soft tissue is 1.5 mg/m^2, administered intravenously once every 3 weeks (Mutschler et al. 2016). The hepatic CYP3A4 enzyme is the most critical in the process of the medication's metabolism (Brandon et al. 2006). There is a possibility of drug interactions with other compounds that are metabolized by this enzyme. Only 0.7% of individuals have rhabdomyolysis, a side effect of trabectedin that is considered significant despite its rarity. In cases of significant myelosuppression and liver dysfunction, particularly during the first three cycles of treatment, close patient monitoring is required, and therapeutic modifications must be implemented in these cases. In addition, close patient monitoring is necessary in cases with a significant risk of infection (Grosso et al. 2012). Marine sponges were the first to discover halichondrin B.

The *Halichondriaokadai* and the *Lysodendoryx* Sp. are two examples of these marine sponges. Eribulinmesylate is the synthetic analogue of the naturally occurring compound called halichondrin B. Halaven is the brand name under which eribulinmesylate is sold by Eisai Europe Ltd., which is located in Hatfield, Hertfordshire, in the United Kingdom. On a biological level, it is classified as a polyketide, although it is not a polyketide that is formed from taxanes. One particularly noteworthy illustration of the supply issue is halichondrin B. Only 300 mg of halichondrin B was extracted from the *Lissodendoryxn.*sp.1 deep sea sponge, an exceptionally rare species. Because of its extremely low yield, Halichondrin B could not be isolated economically (Hart et al. 2000; Radjasa et al. 2011). We now know that dinoflagellates that live in sponges are the most likely source for the production of halichondrin B. This was a discovery that was made quite recently. This fact had not been brought to our attention at any point. The issue with the availability of halichondrin was addressed by the use of chemical synthesis (Aicher et al. 1992). Women with locally advanced or metastatic breast cancer who have progressed after undergoing at least two rounds of chemotherapy are eligible for these treatments (after the first round since 2014).

The Food and Medicine Administration (FDA) and the European Union (EU) blessed the drug in 2010. An inhibitor is a substance that stops the aggregation of tubulin and prevents the creation of microtubules. An inhibitor also stops the synthesis of microtubules. This is accomplished by manipulating the dynamics of the microtubules in a convoluted fashion, ultimately leading to the outcomes described above. Failure of the mitotic spindles causes the cell cycle to become stuck in the G2/M phase, ultimately leading to apoptosis (Smith et al. 2010). Eribulin mesylate is given by intravenous administration on day 1 and day 8 of a therapy cycle that lasts for 3 weeks. It is recommended that a dosage of 1.23 μg of eribulin and 1.4 mg of eribulin mesylate be applied to each square meter of skin. Patients with locally advanced or metastatic breast cancer who had previously undergone two rounds of chemotherapy and were given eribulin had a median survival time of 13.2 months, whereas patients in the control group had a survival time of 10.6 months (hazard ratio = 0.805; 95 percent confidence range = 0.677–0.958; $p = 0.014$). Asthenia and neutropenia are two of the most often experienced adverse reactions to chemotherapy (Morris 2010; Pean et al. 2012).

Brentuximabvedotin (Adcetris, Takeda Pharma A/S, Taastrup, Denmark), a medicine marketed under Adcetris, combines a highly cytotoxic chemical with an antibody targeting tumors. An anti-CD30 antibody found on tumor cells, such as those in Hodgkin lymphoma or systemic anaplastic large cell lymphoma, is used with the deadly chemical monomethylauristatin (MMAE). A valine-citrulline dipeptide connects these two components, which enzymes can break down (Doronina et al. 2006; Ansell 2011). It is possible to synthesize olastatin 10 from marine creatures, and this synthetic version is called auristatin. First discovered in the mollusk *Dolabella auricularia*, which lives in the Indian Ocean, dolastatins are a category of cytotoxic peptides. In the beginning, just a very small amount was discovered. This new genus and species, *Caldora* and *penicillata*, are the true makers of dolastatins, and we know this since they were recently renamed Symploca and penicillata (Luesch et al. 2001; Engene et al. 2015). Dolastatin 10 is a linear-structured depsipeptide. Because of its high level of toxicity, it cannot be delivered by itself; however, the attached antibody allows it to be targeted exclusively to tumor cells that bear the particular antigen.

Intravascular delivery of brentuximabvedotin results in clathrin-mediated endocytosis of the antibody, which then moves to the lysosome, where the linker peptide is broken and MMAE released into the cytosol (Sutherland et al. 2006). It inhibits microtubule polymerization by binding to tubulin and causing cell cycle G2/M arrest and death. CD30-negative lymphoma cells can be affected by MMAE if it is allowed to permeate into their microenvironment (Fromm et al. 2012).

Patients with systemic anaplastic large cell lymphoma (ALCL) who have failed at least one prior multidrug chemotherapy regimen, as well as those with Hodgkin lymphoma who have relapsed after autologous stem cell transplantation (ASCT) after two multidrug regimens in Hodgkin lymphoma patients who are not candidates for ASCT, or after two multidrug regimens in Hodgkin lymphoma patients who are, the medicine should be administered three times per week through intravenous infusion at a dose of 1.8 mg per kilogram of body weight (Mutschler et al. 2016). More than 100 patients with recurrent Hodgkin lymphoma who had previously received autologous stem cell transplantation and high-dose chemotherapy were enrolled in a noncomparative, multicenter phase II research.

The study looked into the efficacy of the treatment. During this research, 75% (102 total) of the patients showed an objective response rate (Younes et al. 2012). In the same kind of clinical research, 86% of 58 patients with recurrent ALCL who had previously been treated with at least one other therapy demonstrated an overall objective response to the treatment (Pro et al. 2012). After receiving treatment, most patients diagnosed with systemic ALCL and Hodgkin lymphoma noticed a reduction in the size of their tumors. According to the forecasts, the 89% and 52% survival rates over a year were respectively achieved. However, peripheral neuropathy, the most serious adverse effect, can be treated. It has been demonstrated that the anticancer agent brintuximabvedotin is effective when used with various other medications (Deng et al. 2013; Garnock-Jones 2013). Currently in phase II of clinical testing is the antibody-drug combination known as glembatumumabvedotin, which is being explored as a potential treatment for breast cancer (Newman and Cragg 2004).

5.5.1 Anticancer Bioactive Antibiotics Derived from Marine Sources

Toxins were initially developed to eliminate competing bacteria; however, it has since been shown that they can have a wide variety of effects on the physiology of the body and may even lead to the discovery of new cancer treatments. These molecules usually aim for specific signal transduction pathways, which can be found in a diverse range of animal species (Adjei 2000). Anthracycline, actinomycin, and aureolic acid are some antibiotics that are effective against tumors. Antitumor antibiotics are considered some of the most promising drugs in cancer chemotherapy. Peptolides, dactinomycin, and other members of these families have been shown to block many glioma metabolic enzymes, including glycolysis, glutaminolysis, and lipogenesis. These findings suggest that the anti-glioma mechanism of actinomycin D may target multiple tumor metabolic regulators (Bhanot et al. 2011; Cragg et al. 1997; Lu et al. 2012; Zhang et al. 2016).

Anthracyclines and other anticancer medications are often employed in therapeutic settings nowadays. Anthracyclines are mostly known for inhibiting topoisomerase II, the primary mechanism behind their anticancer effect (Tewey et al. 1984). In addition to being cytotoxic to HeLa cells, the naturally occurring fermentation product known as geldanamycin, which is formed from the benzoquinone known as ansamycin, inhibits the heat shock protein known as HSP90 (Cardenas et al. 1998; Lin et al. 2017). A randomized phase III study combined pegylated liposomal doxorubicin and trabectedin to treat ovarian cancer in living patients (Monk et al. 2010). The data were evaluated alongside the effects of doxorubicin taken by it. The action mechanism of the trabectedin complex was helped along by the combination of three tetrahydroisoquinoline rings during its development. When the chemical structure was found, it was hypothesized that it might impede major DNA-transcription interaction due to its ability to attach to DNA and directly engage transcription factors covalently. This was because it could do both things (for example, SP-1). Paclitaxel, an inhibitor of the protein kinase C (PKC), has its effects modified by bryostatin, an additional chemotherapy medicine. The bryozoan species known as Bugulaneritina was the origin of one of the most well-known anticancer agents, known as Bryostatin 1. Patients suffering from lymphoblastic leukemia who have experienced Bcl-2 ubiquitination and Bcl-2 destruction at the hands of the proteasome may benefit from this medication (Carter and Keam 2003).

Protein kinase C is a robust activator, and bryostatins are potent activators of protein kinase C to regulate how cells become stimulated, develop, and differentiate (PKC). In addition, didemnin B and aplidine have been associated with the arrest of the cell cycle, the reduction of protein synthesis, and antiangiogenic characteristics. There are several alternative paths for the kahalalide F molecule, and each of these pathways includes a unique interaction with proteins found in the cell membrane.

Due to its delayed neuromuscular toxicity and likely anti-proliferative impact, apliidine can inhibit cell proliferation (Sekar et al. 2015). Clinical trials in phases I and II are currently being carried out to examine the effectiveness of this activity. Squalamine, neovastat, and LAF389 have all been investigated to see whether or not they can inhibit the process of Angiogenesis (Wall et al. 2000). Squalamine and neovastat are now being investigated in phases II and III. Because of this, it was clear that both Squalamine and LAF389 were aiming at the phospholipid bilayer because of their effect on sodium-proton exchangers and sodium-hydrogen antiporters. Of the two candidates, LAF389 was the only one participating in phase I clinical studies. Neovastat, on the other hand, prevents the attachment of the VEGF ligand to its receptor (Tugues et al. 2015). Plinabulin (NPI-2358) is currently being tested in phase II clinical studies due to its effectiveness against human tumor cell lines resistant to many drugs. Plinabulin, a potent and selective VDA, was discovered to have been produced from an extract of a marine fungus. Plinabulin was discovered to be present in an extract of a marine fungus.

Studies on Synthadotin (ILX-651) and Soblidotin (TZT 1027), two anticancer drugs derived from marine bacterium Tasidotin and Synthadotin, are currently in phase II clinical trials, and both show promise. Synthadotin (ILX-651) and Soblidotin (TZT 1027) are anticancer drugs derived from the marine bacterium Tasidotin and Syn (Simmons et al. 2005). It has been discovered that a marine bacteria known as *Salinisporatropica* produces Salinosporamide A, a proteasome inhibitor with a prolonged half-life. In phase I clinical studies, it is more effective against a wider range of blood cancers and a variety of models of solid tumors, while also generating less cytotoxicity in normal cells. These findings result from the fact that it has generated less cytotoxicity in normal cells (Feling et al. 2003). Researchers have discovered that the alkaloid Sorbicillactone-A possesses antileukemic characteristics. It was discovered that *Ircinia fasciculate* and *Penicillium chrysogenum* were connected to the discovery in some way. NSC 630176, or depsipeptide, is a bicyclic peptide produced from *Chromobacterium violaceum*. This peptide was responsible for reducing C-MYC oncogene expression (NSC). At the same time as it blocked a histone deacetylase, it brought an immediate halt to the G0-G1 phase of the cell cycle. Figure 5.2 presents the mechanism that underlies the anticancer effect of the compound.

5.6 Marine Compounds as Antiviral Drugs

In the 1950s, researchers began investigating the medicinal potential of sponge nucleosides, which led to the development of vidarabine. Adenine arabinoside triphosphate, or AATP, is rapidly converted into vidarabine within the cell. This active form can inhibit the DNA polymerase and DNA synthesis processes of herpes, vaccinia, and varicella zoster viruses. In the European Union, eye drops are used to treat acute keratoconjunctivitis as well as recurrent epithelial keratitis, both of which are caused by herpes simplex types 1 and 2. (EU). In June 2001, it was taken off the market in the United States because there was a narrow therapeutic

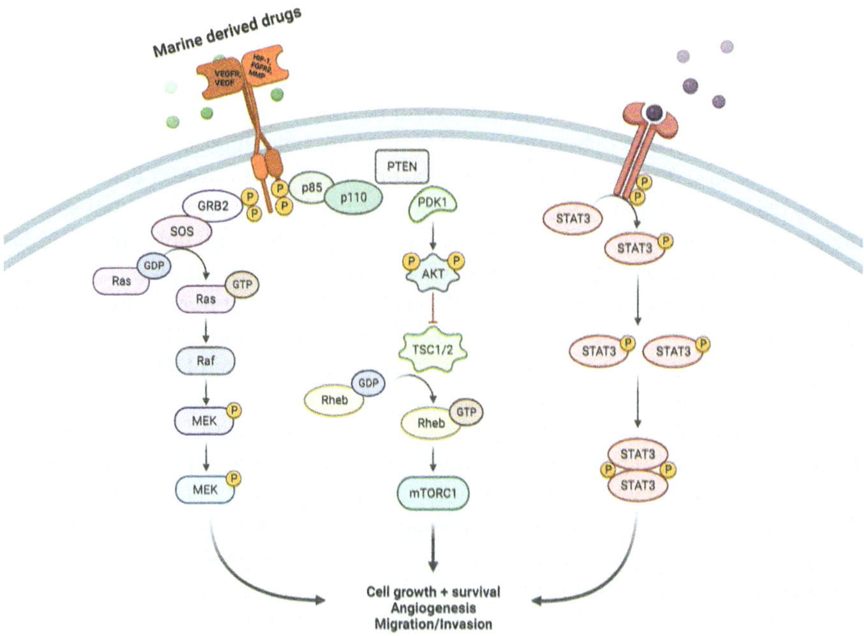

Fig. 5.2 Anticancer drugs are produced from marine sources and their mechanism of action by inhibition of angiogenesis through VEGFR/VEGF pathway

window (Mayer et al. 2010; Sagar et al. 2010; Martins et al. 2014; Anjum et al. 2016).

5.6.1 Drug and Gene Delivery

Ionic liquids have garnered the interest of many researchers working in the pharmaceutical industry. This is primarily because ionic liquids can be utilized in the pharmaceutical industry as biomolecule-stabilizing agents, solvents, or carriers for drugs that are difficult to dissolve, such as APIs-IL (Gomes et al. 2019). Research has been done into creating stimuli-responsive biomaterials based on chitosan and several other ILs (Dias et al. 2013; Hua et al. 2011). In 2011, Hua and his colleagues developed a novel method that could be used to intravenously distribute hydrophobic medications in a manner that was sensitive to the provided inputs. The Schiff reaction with IL-1-butyl-3-methylimidazolium chloride, which was utilized to mix the medications with chitosan, was one method that may be employed to accomplish this goal. The standard IL approach was not used in developing chitosan biomaterials; rather, bio-ILs, which are biocompatible with ILs, were used (Chen et al. 2012). To investigate the release of sodium phosphate dexamethasone from chitosan-based biomaterials, ammonium-based bio-ILs such as choline chloride and choline dihydrogen phosphate were utilized. The journal *Drugs* 2020, volume 18, issue 346, cited 22 out of the 30 investigations that were conducted. On the basis of the various ionic interactions between chitosan and dexamethasone, it was hypothesized that these two substances could be used as electrically regulated drug release systems for iontophoretic applications. Although chitosan and dexamethasone engage in a diverse set of ionic interactions, it was discovered that this is, in reality, the case. Increasing the efficiency of gene transfer has been accomplished by the application of a variety of techniques, including grafting with polyethyleneimine in (Bmim) and the production of O-alkylated chitosan in (Bmim)(Cl) (Chen et al. 2015; Maneewattanapinyo et al. 2019). It has been shown that noncytotoxic vectors based on chitosan can traverse the membranes of living cells. Because chitosan contains positively charged amine groups, it can transfer plasmid DNA (pDNA) into cells by endocytosis and membrane destability (Carvalho et al. 2014).

Chitosan-based vectors that do not cause cytotoxicity have also been demonstrated. This improvement in solubility is likely attributable to the singular properties of ILs, which make it possible for the selective alkylation of hydroxyl groups of chitosan while allowing it to keep its amino groups in organic solvents. This process makes it possible for chitosan to dissolve more easily. Because of this, a medicine combining lidocaine and diclofenac was loaded into a transdermal patch made of gelatin and polyvinyl alcohol. The patch was used to deliver the medication (Lourenço et al. 2011). The high release values of lidocaine and diclofenac were both kept under control by the gelatin/poly(vinyl alcohol) patch, which also displayed good stability throughout the study period of 3 months when incubated at 4° C or under circumstances of ambient temperature. Because of the patch's ability to regulate the high drug release values of lidocaine and diclofenac, it was possible to exert control over these two substances. The technique has succeeded in increasing low water-soluble medicines' physicochemical and biological characteristics. These pharmaceuticals may benefit from the technique.

5.6.2 Future Perspectives and Conclusions

Therapeutics obtained from marine sources offer an innovative and extensive source of chemical diversity, which can contribute to the design and development of new therapeutic drugs that can be useful. Regrettably, these secondary metabolites are typically only present in extremely minute amounts, and the natural stocks are insufficient to support the creation of readily available medications. As a result, developing methods that can extract vast amounts of secondary metabolites is the most essential task that can be undertaken at this time. According to the currently available data, the marine ecosystem may be used as a tool to find novel biological targets that can be treated with therapeutic strategy. Patients at a higher risk for severe toxicities can be identified with a better knowledge of the molecular determinants of the treatment response. To successfully use this demanding new tool in clinical care, researchers, the pharmaceutical marine biotech industry, and government regulatory bodies need to communicate proactively.

References

Adjei AA (2000) Signal transduction pathway targets for anticancer drug discovery. Curr Pharm Des 6:361–378

Aicher TD, Buszek KR, Fang FG, Forsyth CJ, Jung SH, Kishi Y, Matelich MC, Scola PM, Spero DM, Yoon SK (1992) Total synthesis of halichondrin B and norhalichondrin B. J Am Chem Soc 114(8):3162–3164

Anjum K, Abbas SQ, Shah SA, Akhter N, Batool S, Ul Hassan SS (2016) Marine sponges as a drug treasure. Biomol Ther 24(4):347

Ansell SM (2011) Brentuximabvedotin: delivering an antimitotic drug to activated lymphoma cells. Expert Opin Investig Drugs 20(1):99–105

Bergmann W, Burke DC (1955) Contributions to the study of marine products. XXXIX. The nucleosides of sponges. III. 1 Spongothymidine and spongouridine2. J Org Chem 20(11): 1501–1507

Bergmann W, Feeney RJ (1951) Contributions to the study of marine products. XXXII. The nucleosides of sponges. I. J Org Chem 16(6):981–987

Bhanot A, Sharma R, Noolvi MN (2011) Natural sources as potential anticancer agents: a review. Int J Phytomedicine 3:9–26

Blunt JW, Copp BR, Keyzers RA, Munroa MH, Prinsepd MR (2016) Natural product reports. Nat Prod Rep 33:382–431

Bongiorni L (1996) Marine natural products for industrial application. Chem Ind 1996:54–58

Brandon EF, Sparidans RW, Guijt KJ, Löwenthal S, Meijerman I, Beijnen JH, Schellens JH (2006) In vitro characterization of the human biotransformation and CYP reaction phenotype of ET-743 (Yondelis®, Trabectidin®), a novel marine anticancer drug. Investig New Drugs 24(1):3–14

Cardenas ME, Sanfridson A, Cutler NS, Heitman J (1998) Signal-transduction cascades as targets for therapeutic intervention by natural products. Trends Biotechnol 16:427–433

Carter NJ, Keam SJ (2003) Trabectedin. Drugs 2007, 67, 2257–2276. 224. Kortmansky, J.; Schwartz, G.K. Bryostatin-1: a novel PKC inhibitor in clinical development. Cancer Investig 21:924–936

Carvalho T, Augusto V, Rocha Â, Lourenço NMT, Correia NT, Barreiros S, Vidinha P, Cabrita EJ, Dionísio M (2014) Ion jelly conductive properties using dicyanamide-based ionic liquids. J Phys Chem B 118:9445–9459

Chen H, Cui S, Zhao Y, Wang B, Zhang S, Peng X (2012) O-alkylation of chitosan for gene delivery by using ionic liquid in an in- situ reactor. Engineering 4:114–117

Chen H, Cui S, Zhao Y, Zhang C, Zhang S, Peng X (2015) Grafting chitosan with polyethylenimine in an ionic liquid for efficient gene delivery. PLoS One 10:e0121817

Cragg GM, Newman DJ, Weiss RB (1997) Coral reefs, forests, and thermal vents: the worldwide exploration of nature for novel antitumor agents. Semin Oncol 24:156–163

Cuevas C, Francesch A (2009) Development of Yondelis®(trabectedin, ET-743). A semisynthetic process solves the supply problem. Nat Prod Rep 26(3):322–337

D'incalci M, Badri N, Galmarini CM, Allavena P (2014) Trabectedin, a drug acting on both cancer cells and the tumour microenvironment. Br J Cancer 111(4):646–650

Deng C, Pan B, O'Connor OA (2013) Brentuximabvedotin. Clin Cancer Res 19(1):22–27

Dias AMA, Cortez AR, Barsan MM, Santos JB, Brett CMA, de Sousa HC (2013) Development of greener multi-responsive chitosan biomaterials doped with biocompatible ammonium ionic liquids. ACS Sustain Chem Eng 1:1480–1492

Doronina SO, Mendelsohn BA, Bovee TD, Cerveny CG, Alley SC, Meyer DL, Oflazoglu E, Toki BE, Sanderson RJ, Zabinski RF, Wahl AF (2006) Enhanced activity of monomethylauristatin F through monoclonal antibody delivery: effects of linker technology on efficacy and toxicity. Bioconjug Chem 17(1):114–124

Engene N, Tronholm A, Salvador-Reyes LA, Luesch H, Paul VJ (2015) Caldorapenicillata gen. Nov., comb. nov.(cyanobacteria), a pantropical marine species with biomedical relevance. J Phycol 51(4):670–681

Feling RH, Buchanan GO, Mincer TJ, Kauffman CA, Jensen PR, Fenical W (2003) Salinosporamide a: a highly cytotoxic proteasome inhibitor from a novel microbial source, a marine bacterium of the new genus Salinospora. Angew Chem Int Ed 42:355–357

Fromm JR, McEarchern JA, Kennedy D, Thomas A, Shustov AR, Gopal AK (2012) Clinical binding properties, internalization kinetics, and clinicopathologic activity of brentuximabvedotin: an antibody-drug conjugate for CD30-positive lymphoid neoplasms. Clin Lymphoma Myeloma Leuk 12(4):280–283

Garnock-Jones KP (2013) Brentuximabvedotin: a review of its use in patients with Hodgkin lymphoma and systemic anaplastic large cell lymphoma following previous treatment failure. Drugs 73(4):371–381

Giddings LA, Newman DJ (2013) Microbial natural products: molecular blueprints for antitumor drugs. J Ind Microbiol Biotechnol 40(11):1181–1210

Gomes JM, Silva SS, Reis RL (2019) Biocompatible ionic liquids: fundamental behaviours and applications. Chem Soc Rev 48:4317–4335

Grosso F, D'Incalci M, Cartoafa M, Nieto A, Fernández-Teruel C, Alfaro V, Lardelli P, Roy E, Gómez J, Kahatt C, Soto-Matos A (2012) A comprehensive safety analysis confirms rhabdomyolysis as an uncommon adverse reaction in patients treated with trabectedin. Cancer Chemother Pharmacol 69(6):1557–1565

Harkins AL, Duri S, Kloth LC, Tran CD (2014) Chitosan-cellulose composite for wound dressing material.Part 2.Antimicrobial activity, blood absorption ability, and biocompatibility. J Biomed Mater Res Part B Appl Biomater 102:1199–1206

Hart JB, Lill RE, Hickford SJ, Blunt JW, Munro MH (2000) The halichondrins: chemistry, biology, supply and delivery. In: Drugs from the sea. Karger, Basel, pp 134–153

Hua D, Jiang J, Kuang L, Jiang J, Zheng W, Liang H (2011) Smart chitosan-based stimuli-responsive nanocarriers for the controlled delivery of hydrophobic pharmaceuticals. Macromolecules 44:1298–1302

Jensen PR, Fenical W (2000) Marine microorganisms and drug discovery: current status and future potential. In: Drugs from the sea. Karger, Basel, pp 6–29

Kiuru P, D'Auria MV, Muller CD, Tammela P, Vuorela H, Yli-Kauhaluoma J (2014) Exploring marine resources for bioactive compounds. Plantamedica 80(14):1234–1246

Konig GM (1992) Meeresorganismen als quelle pharmazeutisch bedeutsamer naturstoffe. Dtsch Apoth Ztg 132(14):673–683

Lin H-N, Wang K-L, Wu Z-H, Tian R-M, Liu G-Z, Xu Y (2017) Biological and chemical diversity of bacteria associated with a marine flatworm. Mar Drugs 15:281

Lindequist U (2016) Marine-derived pharmaceuticals–challenges and opportunities. Biomol Ther 24(6):561

López-Guerrero JA, Romero I, Poveda A (2015) Trabectedin therapy as an emerging treatment strategy for recurrent platinum-sensitive ovarian cancer. Chin J Cancer 34(1):41–49

Lourenço NMT, Österreicher J, Vidinha P, Barreiros S, Afonso CAM, Cabral JMS, Fonseca LP (2011) Effect of gelatin–ionic liquid functional polymers on glucose oxidase and horseradish peroxidase kinetics. React Funct Polym 71:489–495

Löwenberg B (2013) Sense and nonsense of high-dose cytarabine for acute myeloid leukemia, blood. J Am Soc Hematol 121(1):26–28

Lu J, Ma Y, Liang J, Xing Y, Xi T, Lu Y (2012) Aureolic acids from a marine-derived Streptomyces sp. WBF16. Microbiol Res 167:590–595

Luesch H, Moore RE, Paul VJ, Mooberry SL, Corbett TH (2001) Isolation of dolastatin 10 from the marine cyanobacteriumSymploca species VP642 and total stereochemistry and biological evaluation of its analogue symplostatin 1. J Nat Prod 64(7):907–910

Malve H (2016) Exploring the ocean for new drug developments: marine pharmacology. J Pharm Bioallied Sci 8(2):83

Maneewattanapinyo P, Yeesamun A, Watthana F, Panrat K, Pichayakorn W, Suksaeree J (2019) Controlled release of lidocaine–diclofenac ionic liquid drug from freeze-thawed gelatin/poly (vinyl alcohol) transdermal patches. AAPS PharmSciTech 20:322

Martins A, Vieira H, Gaspar H, Santos S (2014) Marketed marine natural products in the pharmaceutical and cosmeceutical industries: tips for success. Mar Drugs 12(2):1066–1101

Maruzzo M, Brunello A, Diminutto A, Rastrelli M, Basso U (2016) Long-term response to first-line trabectedin in an elderly female patient with a metastatic leiomyosarcoma unfit for anthracycline. Anti-Cancer Drugs 27(3):264

Mayer AM, Glaser KB, Cuevas C, Jacobs RS, Kem W, Little RD, McIntosh JM, Newman DJ, Potts BC, Shuster DE (2010) The odyssey of marine pharmaceuticals: a current pipeline perspective. Trends Pharmacol Sci 31(6):255–265

Monk BJ, Herzog TJ, Kaye SB, Krasner CN, Vermorken JB, Muggia FM, Pujade-Lauraine E, Lisyanskaya AS, Makhson AN, Rolski J (2010) Trabectedin plus pegylated liposomal doxorubicin in recurrent ovarian cancer. J Clin Oncol 28:3107–3114

Morris PG (2010) Advances in therapy: eribulin improves survival for metastatic breast cancer. Anti-Cancer Drugs 21(10):885–889

Mutschler E, Geisslinger G, Menzel S, Ruth P, Schmidtko A (2016) Pharmakologiekompakt: allgemeine und klinischePharmakologie. WissenschaftlicheVerlagsgesellschaft Stuttgart, Toxikologie

Newman DJ, Cragg GM (2004) Marine natural products and related compounds in clinical and advanced preclinical trials. J Nat Prod 67(8):1216–1238

Pean E, Klaar S, Berglund EG, Salmonson T, Borregaard J, Hofland KF, Ersbøll J, Abadie E, Giuliani R, Pignatti F (2012) The European medicines agency review of eribulin for the treatment of patients with locally advanced or metastatic breast cancer: summary of the scientific assessment of the committee for medicinal products for human use. Clin Cancer Res 18(17): 4491–4497

Pro B, Advani R, Brice P, Bartlett NL, Rosenblatt JD, Illidge T, Matous J, Ramchandren R, Fanale M, Connors JM, Yang Y (2012) Brentuximabvedotin (SGN-35) in patients with relapsed or refractory systemic anaplastic large-cell lymphoma: results of a phase II study. J Clin Oncol 30(18):2190–2196

Radjasa OK, Vaske YM, Navarro G, Vervoort HC, Tenney K, Linington RG, Crews P (2011) Highlights of marine invertebrate-derived biosynthetic products: their biomedical potential and possible production by microbial associants. Bioorg Med Chem 19(22):6658–6674

Sagar S, Kaur M, Minneman KP (2010) Antiviral lead compounds from marine sponges. Mar Drugs 8(10):2619–2638

Schofield MM, Jain S, Porat D, Dick GJ, Sherman DH (2015) Identification and analysis of the bacterial endosymbiont specialized for production of the chemotherapeutic natural product ET-743. Environ Microbiol 17(10):3964–3975

Schweder T, Lindequist U, Lalk M (2005 Jan) Screening for new metabolites from marine microorganisms. Mar Biotechnol I(1):1–48

Sekar M, Poomalai S, Gunasekaran M, Mani P, Krishnamurthy A (2015) Bioactive compounds from marine yeast inhibits lung cancer. J Appl Pharm Sci 5:7–15

Shamshina JL, Gurau G, Block LE, Hansen LK, Dingee C, Walters A, Rogers RD (2014) Chitin–calcium alginate composite fibers for wound care dressings spun from ionic liquid solution. J Mater Chem B 2:3924–3936

Silva SS, Santos TC, Cerqueira MT, Marques AP, Reys LL, Silva TH, Caridade SG, Mano JF, Reis RL (2012) The use of ionic liquids in the processing of chitosan/silk hydrogels for biomedical applications. Green Chem 14:1463–1470

Simmons TL, Andrianasolo E, McPhail K, Flatt P, Gerwick WH (2005) Marine natural products as anticancer drugs. Mol Cancer Ther 4:333–342

Smith JA, Wilson L, Azarenko O, Zhu X, Lewis BM, Littlefield BA, Jordan MA (2010) Eribulin binds at microtubule ends to a single site on tubulin to suppress dynamic instability. Biochemistry 49(6):1331–1337

Suleria HA, Osborne S, Masci P, Gobe G (2015) Marine-based nutraceuticals: an innovative trend in the food and supplement industries. Mar Drugs 13(10):6336–6351

Sutherland MS, Sanderson RJ, Gordon KA, Andreyka J, Cerveny CG, Yu C, Lewis TS, Meyer DL, Zabinski RF, Doronina SO, Senter PD (2006) Lysosomal trafficking and cysteine protease metabolism confer target-specific cytotoxicity by peptide-linked anti-CD30-auristatin conjugates. J Biol Chem 281(15):10540–10547

Tarman K, Lindequist U, Mundt S (2013) Metabolites of marine microorganisms and their pharmacological activities. Marine Microbiol 14:393–416

Tewey KM, Chen GL, Nelson EM, Liu L-F (1984) Intercalative antitumor drugs interfere with the breakage-Reunion reaction of mammalian DNA topoisomerase II. J Biol Chem 259:9182–9187

Tugues S, Burkhard SH, Ohs IA, Vrohlings M, Nussbaum K, Vom Berg J, Kulig P, Becher B. New insights into IL-12-mediated tumor suppression. Cell Death & Differentiation. 2015 Feb;22 (2):237–46.

Wall NR, Mohammad RM, Reddy KB, Al-Katib AM (2000) Bryostatin 1 induces ubiquitination and proteasome degradation of Bcl-2 in the human acute lymphoblastic leukemia cell line, Reh. Int J Mol Med 5:165–236

Weinheimer AJ, Spraggins RL (1969) The occurrence of two new prostaglandin derivatives (15-epi-PGA2 and its acetate, methyl ester) in the gorgonian Plexaurahomomalla chemistry of coelenterates. XV. Tetrahedron Lett 10:5185–5188

Younes A, Gopal AK, Smith SE, Ansell SM, Rosenblatt JD, Savage KJ, Ramchandren R, Bartlett NL, Cheson BD, De Vos S, Forero-Torres A (2012) Results of a pivotal phase II study of brentuximabvedotin for patients with relapsed or refractory Hodgkin's lymphoma. J Clin Oncol 30(18):2183

Zhang X, Ye X, Chai W, Lian X-Y, Zhang Z (2016) New metabolites and bioactive actinomycins from marine-derived Streptomyces sp. ZZ338. Mar Drugs 14:181

Marine Polysaccharides: Prospects for Nanostructures Preparation and Their Exploitation in Cancer Therapy

6

Devanabanda Mallaiah

Abstract

Cancer is the most deadly non-infectious disease in the world and still challenging the scientific world. Although many drugs have been used to treat cancer at earlier stages, death is inevitable for advanced-stage cancers. Multiple systemic side effects, low bioavailability, and drug resistance are the major obstacles in conventional chemotherapeutics for cancers. Recently, different types of nanomaterials have emerged as a new strategy for treating various cancers. Nanostructures prepared from polymers from marine sources have received considerable interest due to their large availability, low price, less toxicity, and biodegradability. The chapter covers the polymers derived from marine sources, nanostructures preparation from polymers, and their exploitation in cancer therapy.

Keywords

Marine polymers · Marine organisms · Nanomaterials · Nanostructures · Cancer therapy

6.1 Introduction

Different organisms inhabiting the marine environment, such as animals, plants, fungi, and algae, are rich resources for marine polysaccharides (MP) with various physicochemical and biological properties. The most studied MP includes agarose, alginates, chitosan, carrageenan, fucoidan, hyaluronic acid, and Ulva. The MP is

D. Mallaiah (✉)
Department of Biotechnology and Bioinformatics, Yogi Vemana University, Kadapa, India

P. Veera Bramhachari, C. V. Berde (eds.), *Marine Bioactive Molecules for Biomedical and Pharmacotherapeutic Applications*,
https://doi.org/10.1007/978-981-99-6770-4_6

preferred over synthetic polymers because of polysaccharides virtues like non-toxicity, biodegradability, availability, low price, and ease of chemical modifications (Lee et al. 2017). Some MP showed important inhibitory effects against cancer cells, pathogenic microorganisms, inflammation, and oxidative stress. An array of nanostructures (NS), prepared from MP, respond to external signals such as temperature and pH. The drug delivery systems (DDS) constructed from MP are also used against serious diseases, including advanced therapies like gene delivery and regenerative medicine. The interest in the diverse applications of MP is explained by their excellent biological activities and simple extraction and purification methods (Cardoso et al. 2016). The MP is easily modified or fabricated into various NS like microparticles, nanoparticles, nanofibers, beads, scaffolds, hydrogels, and sponge forms (Manivasagan and Oh 2016).

Compared with terrestrial polysaccharides, the MP isolated from marine organisms such as crabs, seaweeds, squids, and sharks showed absolute benefits regarding ease of preparation and biodiversity. The MP and their derivatives are excellent biomass for preparing nano-based DDS (Sun et al. 2021). The NS based on MP has been used in biomedical applications such as bone diseases, diabetes and cardiovascular diseases (Wan et al. 2021). The NS from MP also explored with special attention to pharmaceuticals delivery, tissue engineering, cosmetics, biosensors, water treatment, and the food industry (Mondal et al. 2020). In addition to various applications, NS prepared from MP has been exploited in cancer treatment to improve drug solubility in aqueous solutions, stability, and drug deposition in tumor tissue (Oliveira et al. 2021). In this book chapter, we describe the MP structures, the preparation of nanoparticles from MP, and their applications in various cancer treatments.

6.2 Marine Polysaccharides Structures

Carbohydrates are divided into three classes, 1. Monosaccharides, 2. Oligosaccharides, 3. Polysaccharides. Polysaccharides are high molecular weight and complex structures formed by multiple monosaccharide units. Glycosidic bonds link the monosaccharides into linear and branched chains. The linear chains are formed by either α-1,4 (starch) or β-1,4 (cellulose) glycosidic linkages; the linking point in branched form is α, 1–6. The polysaccharides made from one type of monosaccharides are called homopolysaccharides, and more than one type of monosaccharides are called heteropolysaccharides. The MP are major heteropolysaccharides present in different marine organisms. Based on their source, MP are classified into marine animal, plant, and microbial MP (Shen et al. 2021). Based on their charge, MP is also divided into neutral, polyanionic, and polycationic (Cardoso et al. 2016). Different types of active MP were isolated from various marine animals, plants (seaweeds), and microorganisms (Manivasagan and Kim 2014). The MP names, composition, and linkage details were summarized (Table 6.1).

Table 6.1 Composition and linkages of marine polysaccharides

Marine polysaccharide	Composition	Linkage
Chitosan	N-acetyl glucosamine and glucosamine	GlcNAc(β1→4)GlcA
Hyaluronic acid	N-acetyl glucosamine and glucuronic acid	4)GlcA(β1→3)GlcNAc (β1
Fucoidan	Fucose residues	(α1→4) Fuc
Alginates	Guluronic acid and Mannuronic acid	Gul (α1→β) Mau

GlcNAc N-acetyl glucosamine, *GlcA* glucosamine, *Fuc* fucose, *Gul* guluronic acid, *Mau* mannuronic acid

Multiple functional groups (carboxyl, hydroxyl and amine) are present on the MP backbone, allowing chemical, biochemical, and enzymatic modification. The drug molecules are grafted to the multifunctional groups of MP through reactions. Polysaccharide derivatives are produced by various reactions such as esterification, etherification, and amidation (Dash et al. 2011). Compared to land polysaccharides, MP showed biocompatibility, non-toxicity, biodegradability, availability, and environmental friendliness (Jing et al. 2021). Although many advantages of MP are there, some MP still have ambiguity in structure and metabolism in the body, limited clinical applications and some others have low solubility in common solvents (Sun et al. 2021).

6.3 Nanoparticles Synthesis from Different Marine Polysaccharides

Different extraction and purification protocols were used to isolate MP from marine organisms. Various biomedical applications will again exploit the nanostructures fabricated from MP by various methods (Fig. 6.1).

Nanoparticles (NP) based on MP are synthesized by two methods: 1. self-assembly, 2. covalent cross-linking. The self-assembled NP of MP is formed by polyelectrolyte complexation, hydrophobic interaction, and ionic interactions. The NP is formed by many covalent cross-linking methods, such as chemical cross-linking and radiation cross-linking (Liu et al. 2020; Li et al. 2015; Zhao and Mitomo 2009).

6.4 Nanostructures from Marine Polysaccharides and Anti-cancer Activity

6.4.1 Chitosan Nanoparticles

Chitin is a polymer of N- acetyl-glucosamine present in fungi cell walls, crustacean exoskeleton, and some insects, algae, and mollusks. The deacetylation of chitin is more than 50% in aqueous alkaline solutions at 60–80 °C referred to as chitosan.

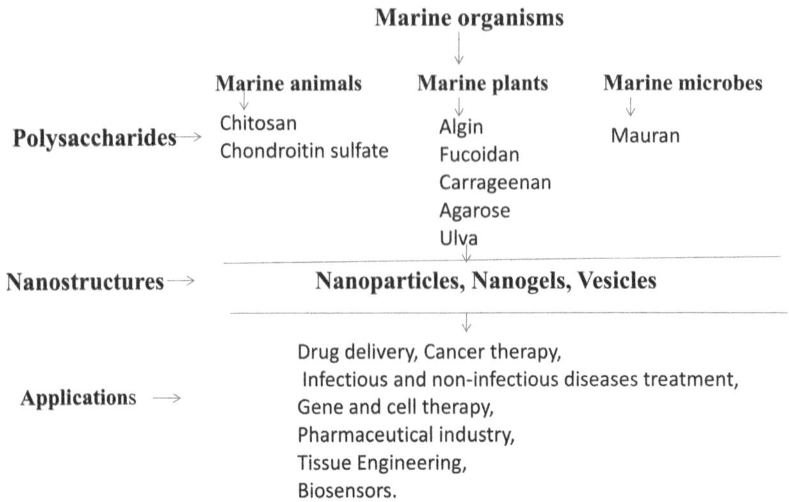

Fig. 6.1 Different marine polysaccharides, sources, nanostructures, and their applications

Chitosan is a random copolymer of N-acetyl-glucosamine and glucosamine (Claverie et al. 2020).Chitosan nanoparticles are prepared from the traditional ionotropic gelation technique. In the ionic gelation method, positively charged chitosan and negatively charged counterions such as tripolyphosphate are cross-linked by inter- and intra-molecular interactions. The ionic gelation technique has been improved by methodological modifications such as electrospraying and spinning disc processing (Desai 2016). The term ionic gelation is preferred when chitosan gelation is produced by small anionic molecules such as phosphate, citrate, and sulfate, while polyelectrolyte complexation is used when anionic macromolecules are employed. Many other methods, such as emulsification and desolvation, are used to prepare chitosan nanoparticles (Grenha 2012).

Safe and efficient nanoparticles were loaded with steroid D-homo lactone and showed strong anticancer activity against breast cancer cells (MDA-MB-231) and PC-3 cancer cells (Kuzminac et al. 2022). Dual-responsive chitosan nanoparticles loaded with cinnamaldehyde and doxorubicin were fabricated to improve in vivo drug delivery and multidrug resistance in breast cancer (Chen et al. 2022). Mebendazole-loaded/folic acid-targeted chitosan nanoparticles were prepared, showing their high anticancer therapeutic efficiency against triple-negative breast cancer and their biocompatible nature (Kefayat et al. 2022). The oral formulations of berberine-loaded chitosan nanoparticles were prepared, demonstrating their protective effect against urethane-induced lung cancer (Mahmoud et al. 2022). The PEGylated chitosan nanoparticles loaded with ascorbic acid and oxaliplatin were fabricated and showed a higher apoptotic effect against breast cancer cells than non-PEGylated chitosan nanoparticles (Fahmy et al. 2022).

The cisplatin-encapsulated chitosan nanoparticles and cisplatin-encapsulated chitosan nanoparticles with surface-linked rituximab formulations prepared and showed cisplatin-loaded chitosan nanoparticles are more successful formulations and also their combating effect against breast cancer cells (Sultan et al. 2022). Folate-functionalized chitosan nanoparticles incorporated with cytarabine showed their targeted delivery and improved cytotoxic effect on MCF-7 human breast cancer cell lines (Geethakumari et al. 2022). Dual targeting (folate and EGFR) chitosan nanoparticles loaded with docetaxel were prepared and showed enhanced cytotoxicity toward lung cancer cells (Vikas et al. 2021). Chitosan nanoparticles loaded with TGF-β1 were prepared and showed their inhibitory role against cervical cancer cells by down-regulation of microRNA-155 and Tim −3 (Yan et al. 2021). chitosan nanoformulations with ellagic acid and tween-80 were prepared, and reported that a combination of ellagic acid, tween-80 and chitosan nanoparticles showed enhanced antitumor effect towards breast cancer (Kaur et al. 2021). Hyaluronic acid-coated chitosan nanoparticles loaded with horseradish peroxidase and indole-3- acetic acid showed a potential anticancer effect against bladder carcinoma cell lines (Pereira et al. 2021). Strigol1/albumin/chitosan nanoparticles were prepared, demonstrating their enhanced apoptotic, antiproliferative effects against HepG2 cancer cell lines (Al-Malki et al. 2021).

The drug doxorubicin-loaded PEGylated chitosan nanoparticles labeled with monoclonal antibodies fabricated and proved their most cytotoxic effect towards breast cancer cell lines (Helmi et al. 2021). The chemotherapeutic drugs (oxaliplatin and resveratrol) were loaded separately in NO carboxy methyl chitosan nanoparticles. They demonstrated their combined high anticancer effect on colorectal cancer than free drugs or either type of nanoparticles alone (Wang et al. 2021). The gemcitabine encapsulated fucoidan/chitosan nanoparticles functionalized with ErbB-2 antibody and demonstrated their high antitumor effect towards breast cancer cells (Oliveira et al. 2021). The codelivery of STAT3 and PD-L1 by hyaluronate-TAT trimethyl/thiolated chitosan nanoparticles restrained tumor growth in tumor-bearing mice (Bastaki et al. 2021). The boswellic acid-loaded chitosan nanoparticles showed an enhanced antiproliferative effect on A549 lung cancer cells (Solanki et al. 2020). The chitosan nanoparticles encapsulated with IR780 and 5-Aminolevulinic acid reagents developed and showed their photothermally enhanced photodynamic therapy against colon cancer (Chen et al. 2020). The hydrophobically modified chitosan nanoparticles were prepared by encapsulating protocatechuic acid and showed anticancer effects against A549 human lung cancer cells (Yee Kuen et al. 2020). The glycol chitosan nanoparticles are conjugated with indocyanine green or cyanine 5.5 and used as theranostic nanoparticles in image-guided lung cancer surgery in small and large animal models (On et al. 2020). The chitosan hydrochloride/carboxymethyl chitosan nanoparticles are loaded with anti-β catenin siRNA and exploited to inhibit colorectal cancer growth (Yan et al. 2020).

The alginate conjugated- trimethyl chitosan nanoparticles loaded with S1P/sphingosine-1-phosphate receptor 1 and IL-6/glycoprotein 130 siRNA synthesized and demonstrated their tumor suppressive effect (Rostami et al. 2020). The cancer cell membrane-coated chitosan nanoparticles were used to co-deliver Ca^{2+} channel siRNA and doxorubicin and showed an inhibitory effect on multidrug resistance of cancer (Zhao et al. 2020). The pectin/chitosan nanoparticles encapsulated with essential oil developed and showed their anticancer effect against breast cancer cell lines (MCF-7) (Attallah et al. 2020).

The drug (mertansine or cabazitaxel) loaded photosensitizer-chitosan nanoparticles to show their chemo-photodynamic effect on breast cancer cells (Pandya et al. 2020). The hyaluronic acid-conjugated chitosan nanoparticles were prepared and showed their pH-sensitive controlled release of doxorubicin and celecoxib drugs in nonsmall lung cancer (Lee et al. 2020). The novel α-santalol functionalized chitosan nanoparticles were used to inhibit polo-like kinase1in triple-negative breast cancer cell lines (Zhang et al. 2020). The hyaluronate-N,N,N-trimethyl chitosan nanoparticles encapsulated with IL-6 and STAT3 siRNA showed their effect in reducing cancer cell progression (Masjedi et al. 2020). Chitosan nanoparticles loaded with 5-fluorouracil demonstrated their efficient therapeutic anticancer effect on skin cancer lines (Patel and Yadav 2020).

The β-and γ-cyclodextrins/curcumin-loaded chitosan nanoparticles were prepared and demonstrated their efficient anti-tumor activity against A549 cancer cell lines (Alizadeh and Malakzadeh 2020). The placental CSA-binding peptide VAR2CSA was coated on N-2-hydroxypropyl trimethyl ammonium chloride chitosan nanoparticles and demonstrated their anticancer activity (Zhao et al. 2020). The polycaprolactone nanofibers loaded in chitosan nanoparticles were used to deliver phytochemicals such as resveratrol and ferulic acid and showed their anticancer potential (Balan et al. 2020). The doxorubicin-loaded alginate/chitosan nanoparticles were prepared by emulsion process and assessed in vitro cytotoxic effect against breast cancer cells (Rosch et al. 2019). Quercetin-loaded chitosan nanoparticles were prepared and showed their use in sustained and specific drug delivery towards colorectal cancer (Rashedi et al. 2019). The cyanine 3 labeled siRNA loaded in hyaluronic acid modified chitosan nanoparticles and used to demonstrate their anti-proliferation effect in nonsmall cell lung cancer (Zhang et al. 2019). The anti-HER2 monoclonal antibody conjugated, doxorubicin-loaded chitosan nanoparticles developed and demonstrated their *vitro cytotoxic effect on* breast cancer cell lines (Naruphontjirakul and Viravaidya-Pasuwat 2019). The chlorin e6 conjugated, doxorubicin encapsulated chitosan nanoparticles fabricated and demonstrated their significant antitumor effect against breast cancer cells (Bhatta et al. 2019).

Chitosan nanoparticles, in combination with reversine and X-ray irradiation, were used to improve the sensitization of breast cancer cells to X-ray irradiation (Piña Olmos et al. 2019). The hyaluronic acid-linked chitosan nanoparticles loaded with raloxifene developed and demonstrated their selective inhibitory and apoptotic effect against A549 cancer cell lines (Almutairi et al. 2019). The chitosan nanoparticles loaded with phloretin hydrophobic phytochemical fabricated and demonstrated their enhanced mitochondrial-mediated apoptosis in oral cancer cell lines of humans (Mariadoss et al. 2019). The chitosan nanoparticles increased cytotoxic effect and decreased cell viability against colon cancer cell lines in the presence of near-infrared laser than in the absence (Abuelmakarem et al. 2019). The 3, 3′ Diindolylmethane encapsulated chitosan nanoparticles showed an inhibitory effect in 7, 12-Dimethylbenz (a) anthracene-induced rat mammary cancer model (Isabella and Mirunalini 2019).

The chitosan extracted from crayfish was used to prepare nanoparticles and demonstrated their significant inhibitory action against various breast cancer cell lines rather than chitosan alone (Taher et al. 2019). In one study, docetaxel-encapsulated D-α-tocopherol polyethylene glycol 1000 succinate linked chitosan nanoparticles fabricated with and without Trastuzumab. The study observed enhanced cellular uptake and cytotoxic effects in breast cancer cell lines with targeted and non-targeted chitosan nanoparticles than conventional Docel™ formulation (Kumar Mehata et al. 2019). The anti-Mucin1 aptamer-linked chitosan nanoparticles loaded with insulin-like growth factor receptor 1 siRNA and docetaxel developed and showed their significant anticancer effect towards metastatic breast cancer cells (Jafari et al. 2019).

6.4.2 Alginate Nanoparticles

Alginates are anionic marine linear polysaccharides isolated from brown algae. Alginates are homopolymer or alternating copolymer blocks of (Lee et al. 2017; Sun et al. 2021) linked β-D-mannuronic acid (M) and α-L-guluronic acid (G) residues. An in vitro and in vivo study fabricated 3D printed tablets with oxaliplatin-loaded alginate nanoparticles showed good tumor targeting, anticancer activity and safety profile in colon cancer treatment (Mirdamadian et al. 2022). The fluorescein-labeled wheat germ agglutinin-conjugated on docetaxel-loaded alginate nanoparticles demonstrated its promising drug delivery agent for colon cancer (Chiu and Lim 2021). The chitosan-β-cyclodextrin-TPP-Folic acid/alginate nanoparticles were prepared and showed effective cytotoxic effects against breast cancer cell lines (Afzali et al. 2021). The curcumin diethyl di succinate loaded chitosan/alginate nanoparticles were prepared and showed more stability, bioavailability, and cytotoxic effect against HepG2 cancer cells compared with free drug (Sorasitthiyanukarn et al. 2021).

Garcinia mangostana L. extract was loaded in catechol-alginate nanoparticles and showed anticancer activity against urothelial carcinoma cell lines (Sahatsapan et al. 2020). The inhibitory immune checkpoints (A2AR and CTLA-4) were suppressed using Si RNA-loaded polyethylene glycol-chitosan-alginate nanoparticles and demonstrated synergistic enhanced antitumor T-cell responses (Ghasemi-Chaleshtari et al. 2020). Erythrocyte, camouflaged arsenic trioxide-loaded sodium alginate nanoparticles, developed and demonstrated antitumor effects towards NB4 and 7721 cells (Lian et al. 2019). The paclitaxel encapsulated egg white/sodium alginate nanoparticles prepared and showed an anticancer effect against CT26 colorectal cancer cells (Zhang et al. 2019). The curcumin prodrug diglutaric acid loaded in chitosan/alginate nanoparticles showed better antitumor activity towards various cancer cell lines (Sorasitthiyanukarn et al. 2018).

6.4.3 Fucoidan Nanoparticles

Fucoidan is a marine sulfated polysaccharide extracted from brown seaweeds. Fucoidans are polymer chains of alternative linked (Lee et al. 2017; Manivasagan and Oh 2016; Sun et al. 2021) α-L-fucopyranosyl residues. The structure and composition mainly depend on the types of algae sources. Some algae sources are rich in fucose and sulfate, whereas others contain residues such as mannose, galactose, glucose and xylose (Luthuli et al. 2019). Like other marine polysaccharides, fucoidans are also used to prepare nanoparticles and exploited in various biomedical applications, including cancer treatment.

Fucoidan nanoparticles were loaded with quinacrine and used to attenuate the growth and metastasis of pancreatic cancer. Drug-encapsulated fucoidan nanoparticles reduced the tumor volume, inhibited metastasis and increased survival rate compared to drug solution alone (Etman et al. 2021). The pro-oxidant piperlongumine encapsulated in chitosan-fucoidan nanoparticles showed efficient toxic effects against prostate cancer cells through intracellular reactive oxygen species generation by piperlongumine (Choi et al. 2019). Fucoidan nanoparticles showed higher cytotoxic effects against osteosarcoma than native fucoidan. Fucoidan nanoparticles also showed more permeation in transport studies, higher bioactivity and bioavailability than native fucoidan (Kimura et al. 2013). Fucoidan nanoparticles prepared from the emulsion method enhanced the inherent therapeutic effect of fucoidan against breast cancer in mouse models. Fucoidan nanoparticles significantly inhibited the progression of tumors and invasion of cancer cells (Chiang et al. 2021). Nanoparticles prepared from poly-allylamine hydrochloride and fucoidan self-assembly and loaded with methotrexate found promising anticancer effects against cancer cell lines (Wang et al. 2019). Multi-stimuli-responsive nanoparticles prepared from polycationic protamine and fucoidan and used to improve drug delivery, P-selectin-mediated endocytosis and enhanced anticancer activity against breast cancer cell lines (Lu et al. 2017).

6.4.4 Hyaluronic Acid Nanoparticles

Hyaluronic acid was used to prepare nanoparticles due to its biocompatibility, biodegradability, and active targeting capability. The CD44, which is over-expressing on cancer cells, is a hyaluronic acid receptor. The nanoparticles formed by conjugating photosensitizer, BFVPBT to HA through redox responsive disulfide linkage and used to prove simultaneously in two-photon imaging and photodynamic therapy (Huang et al. 2019). pH-responsive nano micelles were prepared from hyaluronic acid (HA), dasatinib, and olaparib drugs. They successfully showed enhanced hyaluronic nanoparticles' efficiency compared to free drugs against triple-negative breast cancer (Zeng et al. 2022). Hyaluronic nanoparticles were prepared by incorporating drugs such as doxorubicin and camptothecin and showed synergistic potency against leukemia and T-cell lymphoma cell lines (Krishnan et al. 2022). Hyaluronic nanoparticles conjugated with α1- acid glycoprotein are designed to suppress metastasis and drug resistance in breast cancer cells (Omar et al. 2022). Nanoparticles fabricated from a boron dipyrromethene and HA self-assemble was used to show synergistic photodynamic and photothermal therapeutic effect against cancer (Chen et al. 2021).

A nano-platform synthesized for responsive and cascade release of nitric oxide, reactive oxygen species and reactive nitrogen species used for gas/ photothermal/ photodynamic/immunotherapy of cancers (Ding et al. 2021). Theranostic cross-linked hyaluronic acid nanoparticles were prepared by loading gadolinium-diethylenetriamine penta-acetic acid for MRI imaging and irinotecan as a chemo-therapeutic agent for glioma cells (Costagliola di Polidoro et al. 2021). Nanoparticles synthesized from catechol-modified chitosan and hyaluronic acids were used in the local delivery of the doxorubicin drug to oral cancer cells (Pornpitchanarong et al. 2020). Nanocomplexes were synthesized by conjugating perfluorocarbons and pyropheophorbide with hyaluronic acid and used to enhance the photodynamic therapy of ocular choroidal melanoma (Li et al. 2020). When compared, cross-linked multifunctional hyaluronic acid nanoparticles demonstrated enhanced antitumor and anti-migration effects in CD44–over expressed 4 T1-Luc breast cancer cells than free docetaxel drug (Fang et al. 2020).

The curcumin loaded in phenyl butyric acid and hyaluronic acid nano-assembled structure demonstrated efficient tumor growth suppression and apoptosis-inducing effects in A549-tumor-bearing mice (Lee et al. 2019). The pH-responsive hyaluronic acid nanoparticles loaded with sorafenib and cisplatin are used to treat hepatocellular carcinoma (Zhang et al. 2019). An enzyme and pH-responsive hyaluronic nanoparticles prepared using hyaluronic acid, chlorine e6, and doxorubicin showed more therapeutic effect of nanocomplex against A549 cell lines than free drug (Ren et al. 2019). A modified hyaluronic acid was loaded with doxorubicin and prepared pH-responsive nanoparticles. The results showed that drug-loaded nanoparticles exhibited higher dose-dependent antitumor activity than free drugs (Tian et al. 2019).

6.5 Conclusions

Different marine polysaccharides-based nanostructures, such as nanoparticles, have been fabricated by various methods and applied in biomedical fields, including cancer treatment. Many in vitro and in vivo studies demonstrated that nanoparticles from marine polysaccharides are efficient in drug delivery and therapy against different types of cancers. In addition, more pre-clinical and clinical studies are yet mandated to develop commercial marine polysaccharides-based nanoproducts against cancers.

References

Abuelmakarem HS, Sliem MA, El-Azab J, Farghaly MMA, Ahmed WA (2019) Toward highly efficient cancer imaging and therapy using the environment-friendly chitosan nanoparticles and NIR laser. Biosensors (Basel) 9(1):28

Afzali E, Eslaminejad T, Yazdi Rouholamini SE, Shahrokhi-Farjah M, Ansari M (2021) Cytotoxicity effects of curcumin loaded on chitosan alginate Nanospheres on the KMBC-10 spheroids cell line. Int J Nanomedicine 16:579–589

Alizadeh N, Malakzadeh S (2020) Antioxidant, antibacterial and anticancer activities of β-and γ-CDs/curcumin loaded in chitosan nanoparticles. Int J Biol Macromol 147:778–791

Al-Malki AL, Bakkar A, Huwait EA, Barbour EK, Abulnaja KO, Kumosani TA, Moselhy SS (2021) Strigol1/albumin/chitosan nanoparticles decrease cell viability, induce apoptosis and alter metabolomics profile in HepG2 cancer cell line. Biomed Pharmacother 142:111960

Almutairi FM, Abd-Rabou AA, Mohamed MS (2019) Raloxifene-encapsulated hyaluronic acid-decorated chitosan nanoparticles selectively induces apoptosis in lung cancer cells. Bioorg Med Chem 27(8):1629–1638

Attallah OA, Shetta A, Elshishiny F, Mamdouh W (2020) Essential oil loaded pectin/chitosan nanoparticles preparation and optimization *via* box-Behnken design against MCF-7 breast cancer cell lines. RSC Adv 10(15):8703–8708

Balan P, Indrakumar J, Murali P, Korrapati PS (2020) Bi-faceted delivery of phytochemicals through chitosan nanoparticles impregnated nanofibers for cancer therapeutics. Int J Biol Macromol 142:201–211

Bastaki S, Aravindhan S, Ahmadpour Saheb N, Afsari Kashani M, Evgenievich Dorofeev A, Karoon Kiani F, Jahandideh H, Beigi Dargani F, Aksoun M, Nikkhoo A, Masjedi A, Mahmoodpoor A, Ahmadi M, Dolati S, Namvar Aghdash S, Jadidi-Niaragh F (2021) Codelivery of STAT3 and PD-L1 siRNA by hyaluronate-TAT trimethyl/thiolated chitosan nanoparticles suppresses cancer progression in tumor-bearing mice. Life Sci 266:118847

Bhatta A, Krishnamoorthy G, Marimuthu N, Dihingia A, Manna P, Biswal HT, Das M, Krishnamoorthy G (2019) Chlorin e6 decorated doxorubicin encapsulated chitosan nanoparticles for photo-controlled cancer drug delivery. Int J Biol Macromol 136:951–961

Cardoso MJ, Costa RR, Mano JF (2016) Marine origin polysaccharides in drug delivery systems. Mar Drugs 14(2):34

Chen B, Cao J, Zhang K, Zhang YN, Lu J, Zubair Iqbal M, Zhang Q, Kong X (2021) Synergistic photodynamic and photothermal therapy of BODIPY-conjugated hyaluronic acid nanoparticles. J Biomater Sci Polym Ed 32(15):2028–2045

Chen Q, Jia C, Xu Y, Jiang Z, Hu T, Li C, Cheng X (2022) Dual-pH responsive chitosan nanoparticles for improving in vivo drugs delivery and chemoresistance in breast cancer. Carbohydr Polym 290:119518

Chen G, Zhao Y, Xu Y, Zhu C, Liu T, Wang K (2020) Chitosan nanoparticles for oral photothermally enhanced photodynamic therapy of colon cancer. Int J Pharm 589:119763

Chiang CS, Huang BJ, Chen JY, Chieng WW, Lim SH, Lee W, Shyu WC, Jeng LB (2021) Fucoidan-based nanoparticles with inherently therapeutic efficacy for cancer treatment. Pharmaceutics. 13(12):1986

Chiu HI, Lim V (2021) Wheat germ agglutinin-conjugated disulfide cross-linked alginate nanoparticles as a docetaxel carrier for colon cancer therapy. Int J Nanomedicine 16:2995–3020

Choi DG, Venkatesan J, Shim MS (2019) Selective anticancer therapy using pro-oxidant drug-loaded chitosan-Fucoidan nanoparticles. Int J Mol Sci 20(13):3220

Claverie M, McReynolds C, Petitpas A, Thomas M, Fernandes SCM (2020) Marine-derived polymeric materials and biomimetics: an overview. Polymers (Basel). 12(5):1002

Costagliola di Polidoro A, Zambito G, Haeck J, Mezzanotte L, Lamfers M, Netti PA, Torino E (2021) Theranostic Design of angiopep-2 conjugated hyaluronic acid nanoparticles (Thera-ANG-cHANPs) for dual targeting and boosted imaging of glioma cells. Cancers (Basel) 13(3):503

Dash M, Chiellini F, Ottenbrite RM, Chiellini E (2011) Chitosan-a versatile semi synthetic polymer in biomedical applications. Prog Polym Sci 36:981–1014

Desai KG (2016) Chitosan nanoparticles prepared by ionotropic gelation: an overview of recent advances. Crit Rev Ther Drug Carrier Syst 33(2):107–158

Ding M, Shao K, Wu L, Jiang Y, Cheng B, Wang L, Shi J, Kong X (2021) A NO/ROS/RNS cascaded-releasing nano-platform for gas/PDT/PTT/immunotherapy of tumors. Biomater Sci 9(17):5824–5840

Etman SM, Mehanna RA, Bary AA, Elnaggar YSR, Abdallah OY (2021) Undaria pinnatifida fucoidan nanoparticles loaded with quinacrine attenuate growth and metastasis of pancreatic cancer. Int J Biol Macromol 170:284–297

Fahmy SA, Ramzy A, Mandour AA, Nasr S, Abdelnaser A, Bakowsky U, Azzazy HME (2022) PEGylated chitosan nanoparticles encapsulating ascorbic acid and oxaliplatin exhibit dramatic apoptotic effects against breast cancer cells. Pharmaceutics. 14(2):407

Fang H, Zhao X, Gu X, Sun H, Cheng R, Zhong Z, Deng C (2020) CD44-targeted multifunctional nanomedicines based on a single-component hyaluronic acid conjugate with all-natural precursors: construction and treatment of metastatic breast tumors *in vivo*. Biomacromolecules 21(1):104–113

Geethakumari D, Bhaskaran Sathyabhama A, Raji Sathyan K, Mohandas D, Somasekharan JV, Thavarool PS (2022) Folate functionalized chitosan nanoparticles as targeted delivery systems for improved anticancer efficiency of cytarabine in MCF-7 human breast cancer cell lines. Int J Biol Macromol 199:150–161

Ghasemi-Chaleshtari M, Kiaie SH, Irandoust M, Karami H, Nabi Afjadi M, Ghani S, Aghaei Vanda N, Ghaderi Sede MJ, Ahmadi A, Masjedi A, Hassannia H, Atyabi F, Hojjat-Farsangi M, Namdar A, Ghalamfarsa G, Jadidi-Niaragh F (2020) Concomitant blockade of A2AR and CTLA-4 by siRNA-loaded polyethylene glycol-chitosan-alginate nanoparticles synergistically enhances antitumor T-cell responses. J Cell Physiol 235(12):10068–10080

Grenha A (2012) Chitosan nanoparticles: a survey of preparation methods. J Drug Target 20(4): 291–300

Helmi O, Elshishiny F, Mamdouh W (2021) Targeted doxorubicin delivery and release within breast cancer environment using PEGylated chitosan nanoparticles labeled with monoclonal antibodies. Int J Biol Macromol 184:325–338

Huang YQ, Sun LJ, Zhang R, Hu J, Liu XF, Jiang RC, Fan QL, Wang LH, Huang W (2019) Hyaluronic acid nanoparticles based on a conjugated oligomer photosensitizer: target-specific two-photon imaging, redox-sensitive drug delivery, and synergistic chemo-photodynamic therapy. ACS Appl Bio Mater 2(6):2421–2434

Isabella S, Mirunalini S (2019) 3, 3'-Diindolylmethane-encapsulated chitosan nanoparticles accelerate molecular events during chemical carcinogen-induced mammary cancer in Sprague Dawley rats. Breast Cancer 26(4):499–509

Jafari R, Majidi Zolbanin N, Majidi J, Atyabi F, Yousefi M, Jadidi-Niaragh F, Aghebati-Maleki L, Shanehbandi D, Soltani Zangbar MS, Rafatpanah H (2019) Anti-Mucin1 aptamer-conjugated chitosan nanoparticles for targeted co-delivery of docetaxel and IGF-1R siRNA to SKBR3 metastatic breast cancer cells. Iran Biomed J 23(1):21–33

Jing X, Sun YD, Ma X, Hu H (2021) Marine polysaccharides: green and recyclable resources as wound dressings, vol 5. Mater Chem Front, p 5595

Kaur H, Ghosh S, Kumar P, Basu B, Nagpal K (2021) Ellagic acid-loaded, tween 80-coated, chitosan nanoparticles as a promising therapeutic approach against breast cancer: in-vitro and in-vivo study. Life Sci 284:119927

Kefayat A, Hosseini M, Ghahremani F, Jolfaie NA, Rafienia M (2022) Biodegradable and biocompatible subcutaneous implants consisted of pH-sensitive mebendazole-loaded/folic acid-targeted chitosan nanoparticles for murine triple-negative breast cancer treatment. J Nanobiotechnol 20(1):169

Kimura R, Rokkaku T, Takeda S, Senba M, Mori N (2013) Cytotoxic effects of fucoidan nanoparticles against osteosarcoma. Mar Drugs 11(11):4267–4278

Krishnan V, Dharamdasani V, Bakre S, Dhole V, Wu D, Budnik B, Mitragotri S (2022) Hyaluronic acid nanoparticles for immunogenic chemotherapy of leukemia and T-cell lymphoma. Pharmaceutics. 14(2):466

Kumar Mehata A, Bharti S, Singh P, Viswanadh MK, Kumari L, Agrawal P, Singh S, Koch B, Muthu MS (2019) Trastuzumab decorated TPGS-g-chitosan nanoparticles for targeted breast cancer therapy. Colloids Surf B Biointerfaces 173:366–377

Kuzminac IZ, Ćelić AS, Bekić SS, Kojić V, Savić MP, Ignjatović NL (2022) Hormone receptor binding, selectivity and cytotoxicity of steroid D-homo lactone loaded chitosan nanoparticles for the treatment of breast and prostate cancer cells. Colloids Surf B Biointerfaces 216:112597

Lee R, Choi YJ, Jeong MS, Park YI, Motoyama K, Kim MW, Kwon SH, Choi JH (2020) Hyaluronic acid-decorated glycol chitosan nanoparticles for pH-sensitive controlled release of doxorubicin and celecoxib in nonsmall cell lung cancer. Bioconjug Chem 31(3):923–932

Lee SY, Hong EH, Jeong JY, Cho J, Seo JH, Ko HJ, Cho HJ (2019) Esterase-sensitive cleavable histone deacetylase inhibitor-coupled hyaluronic acid nanoparticles for boosting anticancer activities against lung adenocarcinoma. Biomater Sci 7(11):4624–4635

Lee YE, Kim H, Seo C, Park T, Lee KB, Yoo SY, Hong SC, Kim JT, Lee J (2017) Marine polysaccharides: therapeutic efficacy and biomedical applications. Arch Pharm Res 40(9): 1006–1020

Li MQ, Tang ZH, Zhang DW, Sun H, Liu HY, Zhang Y, Zhang YY, Chen XS (2015) Doxorubicin-loaded polysaccharide nanoparticles suppress the growth of murine colorectal carcinoma and inhibit the metastasis of murine mammary carcinoma in rodent models. Biomaterials 51:161–172

Li J, Xue Y, Tian J, Liu Z, Zhuang A, Gu P, Zhou H, Zhang W, Fan X (2020) Fluorinated-functionalized hyaluronic acid nanoparticles for enhanced photodynamic therapy of ocular choroidal melanoma by ameliorating hypoxia. Carbohydr Polym 237:116119

Lian Y, Wang X, Guo P, Li Y, Raza F, Su J, Qiu M (2019) Erythrocyte membrane-coated arsenic trioxide-loaded sodium alginate nanoparticles for tumor therapy. Pharmaceutics 12(1):21

Liu M, Li H, Wang X, Jing L, Jiang P, Li Y (2020) Experimental study of the vascular normalization window for tumors treated with apatinib and the efficacy of sequential chemotherapy with apatinib in lung cancer-bearing mice and patients. Cancer Med 9:2660–2673

Lu KY, Li R, Hsu CH, Lin CW, Chou SC, Tsai ML, Mi FL (2017) Development of a new type of multifunctional fucoidan-based nanoparticles for anticancer drug delivery. Carbohydr Polym 165:410–420. https://doi.org/10.1016/j.carbpol.2017.02.065. Epub 2017 Feb 20

Luthuli S, Wu S, Cheng Y, Zheng X, Wu M, Tong H (2019) Therapeutic effects of Fucoidan: a review on recent studies. Mar Drugs 17:487

Mahmoud MA, El-Bana MA, Morsy SM, Badawy EA, Farrag AE, Badawy AM, Abdel-Wahhab MA, El-Dosoky MA (2022) Synthesis and characterization of berberine-loaded chitosan nanoparticles for the protection of urethane-induced lung cancer. Int J Pharm 618:121652

Manivasagan P, Kim SK (2014) Extracellular polysaccharides produced by marine bacteria. Adv Food Nutr Res 72:79–94

Manivasagan P, Oh J (2016) Marine polysaccharide-based nanomaterials as a novel source of nanobiotechnological applications. Int J Biol Macromol 82:315–327

Mariadoss AVA, Vinayagam R, Senthilkumar V, Paulpandi M, Murugan K, Xu B et al (2019) Phloretin loaded chitosan nanoparticles augments the pH-dependent mitochondrial-mediated intrinsic apoptosis in human oral cancer cells. Int J Biol Macromol 130:997–1008

Masjedi A, Ahmadi A, Atyabi F, Farhadi S, Irandoust M, Khazaei-Poul Y, Ghasemi Chaleshtari M, Edalati Fathabad M, Baghaei M, Haghnavaz N, Baradaran B, Hojjat-Farsangi M, Ghalamfarsa G, Sabz G, Hasanzadeh S, Jadidi-Niaragh F (2020) Silencing of IL-6 and STAT3 by siRNA loaded hyaluronate-N,N,N-trimethyl chitosan nanoparticles potently reduces cancer cell progression. Int J Biol Macromol 149:487–500

Mirdamadian SZ, Varshosaz J, Minaiyan M, Taheri A (2022) 3D printed tablets containing oxaliplatin loaded alginate nanoparticles for colon cancer targeted delivery. An in vitro/in vivo study. Int J Biol Macromol 205:90–109

Mondal S, Manivasagan P, Oh J (2020) Encyclopedia of marine biotechnology. In: Marine polysaccharide-based nanomaterials. Wiley, Hoboken, NJ, pp 1231–1248

Naruphontjirakul P, Viravaidya-Pasuwat K (2019) Development of anti-HER2-targeted doxorubicin-core-shell chitosan nanoparticles for the treatment of human breast cancer. Int J Nanomedicine 14:4105–4121

Oliveira C, Gonçalves CS, Martins EP, Neves NM, Reis RL, Costa BM, Silva TH, Martins A (2021) Fucoidan/chitosan nanoparticles functionalized with anti-ErbB-2 target breast cancer cells and impair tumor growth in vivo. Int J Pharm 600:120548

Oliveira C, Reis RL, Martins A, Silva TH (2021) Marine-derived polymeric nanostructures for cancer treatment. Nanomedicine (Lond) 16(22):1931–1935

Omar H, Fardous R, Alhindi YM, Aodah AH, Alyami M, Alsuabeyl MS, Alghamdi WM, Alhasan AH, Almalik A (2022) α_1-acid glycoprotein-decorated hyaluronic acid nanoparticles for suppressing metastasis and overcoming drug resistance breast cancer. Biomedicine 10(2):414

On KC, Rho J, Yoon HY, Chang H, Yhee JY, Yoon JS, Jeong SY, Kim HK, Kim K (2020) Tumor-targeting glycol chitosan nanoparticles for image-guided surgery of rabbit Orthotopic VX2 lung cancer. Pharmaceutics. 12(7):621

Pandya AD, Øverbye A, Sahariah P, Gaware VS, Høgset H, Masson M, Høgset A, Mælandsmo GM, Skotland T, Sandvig K, Iversen TG (2020) Drug-loaded photosensitizer-chitosan nanoparticles for combinatorial chemo- and photodynamic-therapy of cancer. Biomacromolecules 21(4):1489–1498

Patel G, Yadav BKN (2020) Study of 5-fluorouracil loaded chitosan nanoparticles for treatment of skin cancer. Recent Pat Nanotechnol 14(3):210–224

Pereira FM, Melo MN, Santos ÁKM, Oliveira KV, Diz FM, Ligabue RA, Morrone FB, Severino P, Fricks AT (2021) Hyaluronic acid-coated chitosan nanoparticles as carrier for the enzyme/prodrug complex based on horseradish peroxidase/indole-3-acetic acid: characterization and potential therapeutic for bladder cancer cells. Enzym Microb Technol 150:109889

Piña Olmos S, Díaz Torres R, Elbakrawy E, Hughes L, Mckenna J, Hill MA, Kadhim M, Ramírez Noguera P, Bolanos-Garcia VM (2019) Combinatorial use of chitosan nanoparticles, reversine, and ionising radiation on breast cancer cells associated with mitosis deregulation. Biomol Ther 9(5):186

Pornpitchanarong C, Rojanarata T, Opanasopit P, Ngawhirunpat T, Patrojanasophon P (2020) Catechol-modified chitosan/hyaluronic acid nanoparticles as a new avenue for local delivery of doxorubicin to oral cancer cells. Colloids Surf B Biointerfaces 196:111279

Rashedi J, Ghorbani Haghjo A, Mesgari Abbasi M, Dastranj Tabrizi A, Yaqoubi S, Sanajou D, Ashrafi Jigheh Z, Namvaran A, Mohammadi A, Mohammadi Khoshraj J, Baradaran B (2019) Antitumor effect of quercetin loaded chitosan nanoparticles on induced colon cancer in Wistar rats. Adv Pharm Bull 9(3):409–415

Ren Q, Liang Z, Jiang X, Gong P, Zhou L, Sun Z, Xiang J, Xu Z, Peng X, Li S, Li W, Cai L, Tang J (2019) Enzyme and pH dual-responsive hyaluronic acid nanoparticles mediated combination of photodynamic therapy and chemotherapy. Int J Biol Macromol 130:845–852

Rosch JG, Winter H, DuRoss AN, Sahay G, Sun C (2019) Inverse-micelle synthesis of doxorubicin-loaded alginate/chitosan nanoparticles and *in vitro* assessment of breast cancer cytotoxicity. Colloid Interface Sci Commun 28:69–74

Rostami N, Nikkhoo A, Khazaei-Poul Y, Farhadi S, Sadat Haeri M, Moghadaszadeh Ardebili S, Aghaei Vanda N, Atyabi F, Namdar A, Baghaei M, Haghnavaz N, Kazemi T, Yousefi M, Ghalamfarsa G, Sabz G, Jadidi-Niaragh F (2020) Coinhibition of S1PR1 and GP130 by siRNA-loaded alginate-conjugated trimethyl chitosan nanoparticles robustly blocks development of cancer cells. J Cell Physiol 235(12):9702–9717

Sahatsapan N, Ngawhirunpat T, Rojanarata T, Opanasopit P, Patrojanasophon P (2020) Catechol-functionalized alginate nanoparticles as Mucoadhesive carriers for Intravesical chemotherapy. AAPS PharmSciTech 21(6):212

Shen S, Chen X, Shen Z, Chen H (2021) Marine polysaccharides for wound dressings application: an overview. Pharmaceutics. 13(10):1666

Solanki N, Mehta M, Chellappan DK, Gupta G, Hansbro NG, Tambuwala MM, Aa Aljabali A, Paudel KR, Liu G, Satija S, Hansbro PM, Dua K, Dureja H (2020) Antiproliferative effects of boswellic acid-loaded chitosan nanoparticles on human lung cancer cell line A549. Future Med Chem 12(22):2019–2034

Sorasitthiyanukarn FN, Muangnoi C, Ratnatilaka Na Bhuket P, Rojsitthisak P, Rojsitthisak P (2018) Chitosan/alginate nanoparticles as a promising approach for oral delivery of curcumin diglutaric acid for cancer treatment. Mater Sci Eng C Mater Biol Appl 93:178–190

Sorasitthiyanukarn FN, Muangnoi C, Rojsitthisak P, Rojsitthisak P (2021) Chitosan-alginate nanoparticles as effective oral carriers to improve the stability, bioavailability, and cytotoxicity of curcumin diethyl disuccinate. Carbohydr Polym 256:117426

Sultan MH, Moni SS, Madkhali OA, Bakkari MA, Alshahrani S, Alqahtani SS, Alhakamy NA, Mohan S, Ghazwani M, Bukhary HA, Almoshari Y, Salawi A, Alshamrani M (2022) Characterization of cisplatin-loaded chitosan nanoparticles and rituximab-linked surfaces as target-specific injectable nano-formulations for combating cancer. Sci Rep 12(1):468

Sun Y, Ma X, Hu H (2021) Marine polysaccharides as a versatile biomass for nano drug delivery systems construction. Mar Drugs 19(6):345

Taher FA, Ibrahim SA, El-Aziz AA, Abou El-Nour MF, El-Sheikh MA, El-Husseiny N, Mohamed MM (2019) Antiproliferative effect of chitosan nanoparticles (extracted from crayfish Procambarus clarkii, Crustacea: Cambaridae) against MDA-MB-231 and SK-BR-3 human breast cancer cell lines. Int J Biol Macromol 126:478–487

Tian G, Sun X, Bai J, Dong J, Zhang B, Gao Z, Wu J (2019) Doxorubicin-loaded dual-functional hyaluronic acid nanoparticles: preparation, characterization and antitumor efficacy in vitro and in vivo. Mol Med Rep 19(1):133–142

Vikas VMK, Mehata AK, Sharma V, Priya V, Varshney N, Mahto SK, Muthu MS (2021) Bioadhesive chitosan nanoparticles: dual targeting and pharmacokinetic aspects for advanced lung cancer treatment. Carbohydr Polym 274:118617

Wan MC, Qin W, Lei C, Li QH, Meng M, Fang M, Song W, Chen JH, Tay F, Niu LN (2021) Biomaterials from the sea: future building blocks for biomedical applications. Bioact Mater 6(12):4255–4285

Wang P, Kankala RK, Chen B, Long R, Cai D, Liu Y, Wang S (2019) Poly-allylamine hydrochloride and fucoidan-based self-assembled polyelectrolyte complex nanoparticles for cancer therapeutics. J Biomed Mater Res A 107(2):339–347

Wang Y, Ma J, Qiu T, Tang M, Zhang X, Dong W (2021) In vitro and in vivo combinatorial anticancer effects of oxaliplatin- and resveratrol-loaded N,O-carboxymethyl chitosan nanoparticles against colorectal cancer. Eur J Pharm Sci 163:105864

Yan L, Gao S, Shui S, Liu S, Qu H, Liu C, Zheng L (2020) Small interfering RNA-loaded chitosan hydrochloride/carboxymethyl chitosan nanoparticles for ultrasound-triggered release to hamper colorectal cancer growth in vitro. Int J Biol Macromol 162:1303–1310

Yan M, Zhang Y, Chang S (2021) Chitosan nanoparticles loaded with TGF-β1 inhibit cervical cancer cell progression through Down-regulation of MicroRNA-155 and activation of Tim-3 pathway. J Biomed Nanotechnol 17(9):1850–1857

Yee Kuen C, Galen T, Fakurazi S, Othman SS, Masarudin MJ (2020) Increased cytotoxic efficacy of protocatechuic acid in A549 human lung cancer delivered via hydrophobically modified-chitosan nanoparticles as an anticancer modality. Polymers (Basel) 12(9):1951

Zeng X, Wang H, Zhang Y, Xu X, Yuan X, Li J (2022) pH-responsive hyaluronic acid nanoparticles for enhanced triple negative breast cancer therapy. Int J Nanomedicine 17: 1437–1457

Zhang W, Cai J, Wu B, Shen Z (2019) pH-responsive hyaluronic acid nanoparticles coloaded with sorafenib and cisplatin for treatment of hepatocellular carcinoma. J Biomater Appl 34(2): 219–228

Zhang J, Huang J, Huang K, Zhang J, Li Z, Zhao T, Wu J (2019) Egg white coated alginate nanoparticles with electron sprayer for potential anticancer application. Int J Pharm 564:188–196

Zhang J, Wang Y, Li J, Zhao W, Yang Z, Feng Y (2020) α-Santalol functionalized chitosan nanoparticles as efficient inhibitors of polo-like kinase in triple negative breast cancer. RSC Adv 10(9):5487–5501

Zhang W, Xu W, Lan Y, He X, Liu K, Liang Y (2019) Antitumor effect of hyaluronic-acid-modified chitosan nanoparticles loaded with siRNA for targeted therapy for non small cell lung cancer. Int J Nanomedicine 14:5287–5301

Zhao K, Cheng G, Zhang B, Li D, Han J, Chen J, Wang B, Li M, Xiao T, Le A, Sun B, Jin Z, Zhang J, Fan X (2020) Targeting delivery of partial VAR2CSA peptide guided N-2-Hydroxypropyl trimethyl ammonium chloride chitosan nanoparticles for multiple cancer types. Mater Sci Eng C Mater Biol Appl 106:110171

Zhao Z, Ji M, Wang Q, He N, Li Y (2020) Ca^{2+} signaling modulation using cancer cell membrane coated chitosan nanoparticles to combat multidrug resistance of cancer. Carbohydr Polym 238:116073

Zhao L, Mitomo H (2009) Hydrogels of dihydroxypropyl chitosan cross-linked with irradiation at paste-like condition. Carbohydr Polym 76:314–319

Marine Microbial Cell Mediated Nanomaterials Synthesis: Prospectus, Current Development and Challenges

7

Abullais Ansari, Laxmi Jadhav, Ashwini Sarje, Smita Patil, and Ashok Bankar

Abstract

Nanoparticles (NPs) have a wide range of applications in various sectors. The NPs synthesis is classified into two basic approaches: top-down and bottom-up. These approaches can be further classified into chemical, physical and biological methods. Biological methods are simple, inexpensive, and eco-friendly. Several microorganisms, such as bacteria, cyanobacteria, actinomycetes, yeasts, fungi, and algae, have synthesized NPs. The marine environment is well known for its microbial diversity. Such microbes can survive in hostile environments such as extreme pH, lack of nutrients, and high concentrations of metals and salts. Marine microorganisms are well-known for their ability to produce various types of metallic NPs like CO_3O_4, Fe, Au, Ag, Cu, Co, and Ti. Biomolecules produced by marine microbes play a crucial role as capping and stabilizing agents for NPs. The NPs synthesized by marine microorganisms are studied for their antimicrobial, antifungal, and anticancer activities. These NPs are effective against several multidrug-resistant microorganisms and exhibit a synergistic effect when combined with antibiotics. Anticancer activity of these NPs is studied against varieties of cancer cell lines such as human cervical cancer cell line (HeLa), lung cancer cell line (A549), and histolytic lymphoma cancer cell line (V937). Bio-inspired metal NPs are synthesized from various microbial sources with diverse shapes, sizes, and optical and dielectric properties. These NPs are used in medical bio-imaging due to their optical properties, such as high transparency, plasmon

A. Ansari · L. Jadhav · A. Sarje · S. Patil · A. Bankar (✉)
MES Abasaheb Garware College Affiliated to Savitribai Phule Pune University, Pune, Maharashtra, India

Camp Education Society's Dr. Arvind B. Telang Senior College of Arts, Science and Commerce, Pune, Maharashtra, India

P. Veera Bramhachari, C. V. Berde (eds.), *Marine Bioactive Molecules for Biomedical and Pharmacotherapeutic Applications*,
https://doi.org/10.1007/978-981-99-6770-4_7

resonance, high/low refractive index, and new photoluminescence properties. Since recent advances in nanotechnology, significant advances in metallic nanomaterial-based biosensors have been observed. Nanotechnology has helped minimize biosensors nanosized biosensors that can easily detect and interact with small molecules, from single-cell to double-stranded DNA. In this chapter, we have emphasized the mechanism of synthesis, mode of action and applications of NPs in various sectors.

Keywords

Marine · Biomolecules · Nanomaterials · Green synthesis and medicine

7.1 Introduction

The microbial biodiversity of marine ecosystems has high significance in bio-nanotechnology. Several metal NPs viz. Ca, Pd, Si, Ag, Au, and Fe are synthesized using various microorganisms. These microorganisms include bacteria, yeasts, fungi, actinomycetes, cyanobacteria, and algae. It has been observed that marine microorganisms have great potential to produce inorganic nanomaterials by reducing the available inorganic elements available at the bottom of the sea. It has been reported that marine microorganisms can synthesize a variety of nanostructures, such as mineral crystals and metallic NPs of desired size, shape, and composition (Asmathunisha and Kathiresan 2013).

Several microorganisms have evolved unique mechanisms to survive in extreme environmental conditions such as extreme pH, lack of nutrients, and high concentrations of metals and salts. The marine environment is a well-known example of a habitat with high salinity and a rich microbial flora that adapts well to different metal ions. Metal ions can be easily converted into NPs of desired shape and size due to the unique inherent mechanisms of marine microbes (Baker et al. 2013). Marine microbes like bacteria, yeasts, fungi, and microalgae have been synthesized for NPs synthesis. Nanomaterials are produced via intracellular and extracellular pathways reported earlier (Patil and Kim 2018). The chemical reaction between seawater and mineral salts forms metal-rich effluent, leading to extreme environmental conditions (Colmer and Hinkle 1947). Therefore, marine microorganisms can be a promising tool for synthesizing NPs (Pugazhenthiran et al. 2009).

Nanotechnology has several applications in different fields of science and technology. Nanomaterials have been used in pharmaceuticals due to their several advantages and are particularly interesting to many researchers (Adibkia et al. 2007). Over the last 20 years, NPs have been reported in drug delivery applications to enhance therapeutic effects (Sabzevari et al. 2013). The mode of action of NPs on human pathogens is not yet fully understood and needs to be understood in more detail (Besinis et al. 2014). This chapter emphasizes the importance of marine microbial sources for NPs synthesis and its medicinal importance.

7.2 Methods of Nanomaterial Synthesis

There are many methods reported for synthesizing metallic NPs. These methods are highly valued for obtaining nanomaterials of the desired size, shape, and chemical composition. Therefore, nanotechnology has evolved into an excellent field of study. Extensions of several dimensional (1D, 2D, 3D) mesostructured NPs combined with many organic molecules have received special attention in the study. Nanomaterial synthesis methods are classified as (1) top-down approach and (2) bottom-up approach. In a top-down approach, particles are converted to nanomaterials using various physical and chemical techniques such as chemical etching, milling, sputtering, laser ablation, and grinding. In a bottom-up approach, nanomaterials are created by incorporating atoms, and molecules to molecules, or clusters to clusters. Such nanomaterials are produced through multiple physical, chemical, and biological processes (Fig. 7.1) (Marchiol 2012; Khandel and Shahi 2016; Goutam et al. 2020).

These methods are of high importance because of their excellent properties that are accountable for the structural shape and size of NPs. Figure 7.2 illustrates a "top-down" and "bottom-up" approach used in nanomaterials production. In a top-down approach, nanomaterials production occurs after dissolving bulk material into small particles using a variety of techniques such as lithography, chemical etching, pulsed laser, mechanical, thermal dehydration, sputtering, ablation, and photoreduction (Chang et al. 2013; Hou et al. 2019; Bukka et al. 2019).

The top-down approach has some advantages, such as being cost-effective and useful for obtaining nanomaterials with desired shape and size (Mijatovic et al.

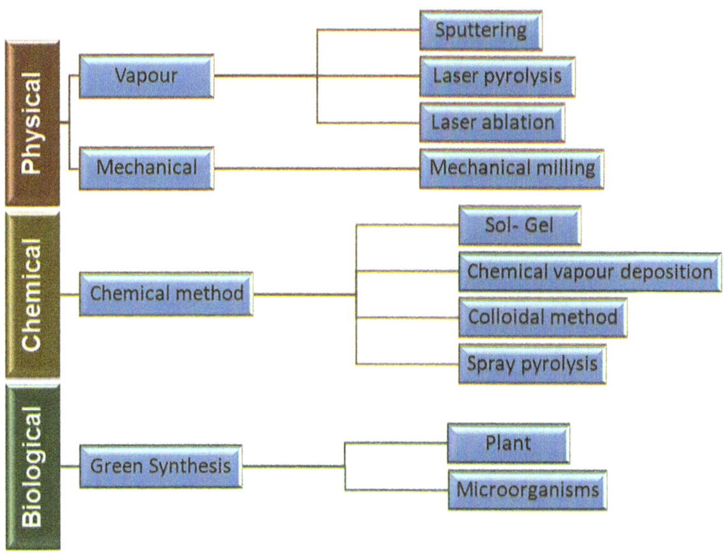

Fig. 7.1 Different methods of nanomaterials synthesis (Marchiol 2012; Khandel and Shahi 2016; Goutam et al. 2020)

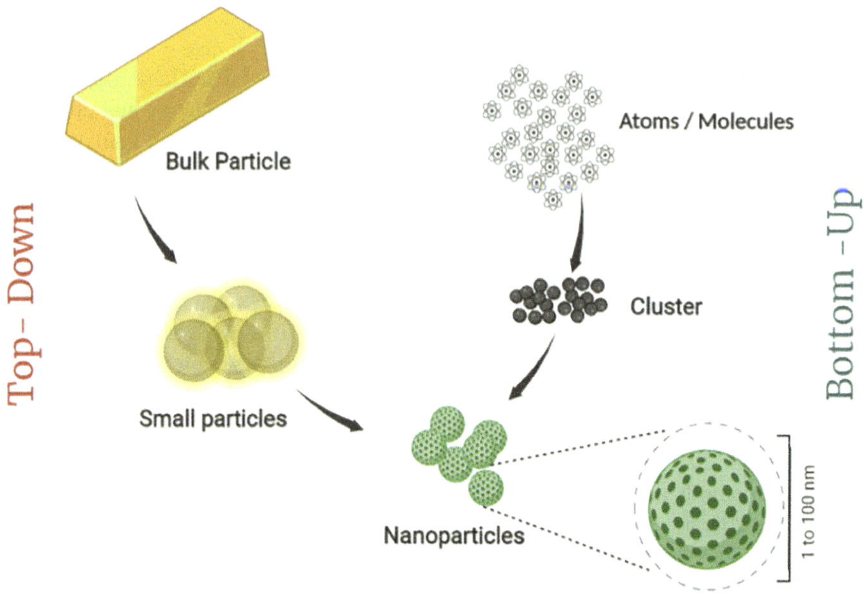

Fig. 7.2 Diagrammatic representation of top-down and bottom-up approaches (Mijatovic et al. 2005; Aryal et al. 2019)

2005). This technique also offers nanomaterials that are uniform in size and shape. Therefore, this approach provides high-quality yields for monodisperse nanomaterials. This approach proved suitable for scaling up to mass production (Aryal et al. 2019). The imperfect surface structure is a major disadvantage of the top-down technique.

A bottom-up approach uses mainly wet chemical methods such as microemulsion, coprecipitation, pyrolysis, radiation-induced, chemical vapor deposition, solid gel chemistry, and electrodeposition processes (Mazari et al. 2021). This approach is also recognized as self-assembly because nanomaterials are made using small components such as atoms, molecules, and particles (Thiruvengadathan et al. 2013).

The process used in this approach has several drawbacks, including high cost, high use, and consumption of toxic and harmful chemicals used. In many chemical processes, chemicals like sodium citrate, ethylene glycol, hydrazine, and sodium borohydride are employed as reducing agents for nanomaterial synthesis. These reducing agents were found to be inadequate due to undesired particle size and size distribution regulation. Therefore, capping agents like polyvinyl alcohol, polyethylene glycol, polyvinyl pyrrolidone, and ascorbic acid are needed (Mukherji et al. 2018). Other chemicals like acetyl trimethyl ammonium bromide, sodium dodecyl sulfate, phospholipids, and gemini surfactants are commonly used to avoid agglomeration of NPs (Talib and Hui-Fen-Wu 2016). Thus, the absorption of reducing and stabilizing agents on the surface of NPs can decrease the effectiveness of synthesized nanomaterials by reducing the surface area and lowering the product reactivity (Kango et al. 2013). Therefore, most of these processes are expensive and harmful and can limit their applications on a large scale (Hu and Zhu 2015).

Biological methods are simple, inexpensive and eco-friendly. Many researchers monitor the difference between the green and traditional wet-chemistry techniques for synthesizing nanomaterials (Sharma et al. 2019). The phytotoxicity and cytotoxicity of bio-inspired nanomaterials are significantly reduced compared to wet chemical NPs. Bio-inspired NPs are considered safe and useful in medicines (Shankar et al. 2021). Therefore, nanomaterials synthesis using biological methods has proven to be more significant. Biological agents like plants, bacteria, yeasts, fungi, and algae are useful reducing agents for NPs synthesis (Nagajyothi et al. 2014; Sharma et al. 2019; Jan et al. 2021).

7.3 A Role of Biomolecules Obtained from Marine Microbes for Nanomaterials Synthesis

Various kinds of marine microorganisms have been employed for nanomaterials synthesis. However, this section described the importance of different biomolecules obtained from marine microbial sources for synthesizing green NPs.

7.3.1 Marine Bacterial Biomolecules Source for Green NPs Synthesis

The marine environment is known for its diverse microorganisms that can survive in hostile environments. Marine bacteria produce biomolecules or metabolites such as alkaloids, steroids, terpenoids, peptides, and polyketides. These biomolecules can be reducing agents for NPs synthesis. Therefore, marine bacteria have been given special consideration due to their numerous uses in nanotechnology, biotechnology, and pharmacology (Stincone and Brandelli 2020; Andryukov et al. 2019; Jadhav et al. 2022). A new strain, the *Marinomonas* sp. ef1 from psychrophilic marine habitats, is used to synthesize AgNPs. In this method, the reaction mixture of *Marinomonas* sp. ef1 and 1 mM of $AgNO_3$ was incubated at 22 °C for 24 h. Small-sized AgNPs (50 nm) showed potential antimicrobial activity against 12 frequently associated nosocomial pathogens, including *S. aureus* and two *Candida* strains. Genome sequencing studies have discovered that enzymes associated with the nitro-FMN-reductase superfamily may play a crucial role in the synthesis of NPs (John et al. 2020a, b).

The AgNPs are synthesized by *Nocardiopsis dassonvillei*-DS013, a new strain of actinomycetes found in coastal marine sediments. Synthesized circular NPs were found to be 30–80 nm in size. These synthesized NPs have demonstrated antimicrobial activity against several clinical isolates such as *Enterococcus* sp., *E. coli*, *Klebsiella* sp., *Pseudomonas* sp., *Shigella* sp., *Proteus* sp., *Streptococcus* sp., and *Bacillus subtilis* (Dhanaraj et al. 2020). Similarly, the marine strain of *Pseudomonas* sp. also synthesized AgNPs. These NPs showed higher antimicrobial activity than chemically assisted AgNPs against nosocomial pathogens like *E. coli, S. aureus* and *C. albicans* (John et al. 2020a, b). The exopolysaccharides, including poly-γ-glutamic acid γ-PGA-APA, EPS B3-15, and EPS T14, were obtained from marine bacterial isolates such as *B. licheniformis strain* B3-15, *B. horneckiae strain* APA, and *B. licheniformis strain* T14. The Au and AgNPs have been synthesized using exopolysaccharides as reducing and stabilizing agents. These nanomaterials also displayed superior antimicrobial activity against several human pathogens, such as *S. aureus, C. albicans, P. aeruginosa*, and *E. coli* (Scala et al. 2019). AgNPs nanoparticles were also synthesized by using a cell-free extract of *Streptomyces* sp. Al-Dhabi-89. These NPs displayed strong antimicrobial activity towards multidrug-resistant hospital pathogens. The MIC was 15.6, 7.81, and 7.81 μg/ml towards multidrug-resistant *Proteus mirabilis, S. aureus,* and *Acinetobacter baumannii*, respectively (Al-Dhabi et al. 2018).

The AgNPs synthesized by the marine strain of *Bacillus* sp. KFU36 and *Streptomyces rochei* MHM13 have been used for wound healing and have shown promising anticancer activity against the tumor cell lines HepG2, MCF-7, CACO, A-549, HEP-2, PC-3, and HELA (Abd-Elnaby et al. 2016; Almalki and Khalifa 2020). Therefore, as shown in Table 7.1, some bacteria can produce a variety of medicinal nanomaterials.

7.3.2 Marine Fungal Biomolecules Source for Green NPs Synthesis

The fungi have been reported for secondary metabolite secretion, which contains several biomolecules such as proteins, enzymes, or other biomolecules that act as capping and stabilizing agents (Zhang et al. 2011). The marine fungus *Aspergillus brunneoviolaceus* has been reported for AgNPs synthesis. NPs showed excellent inhibitory activity against some clinical pathogens like *S. aureus, B. subtilis, E. coli, P. aeruginosa,* and *Salmonella* spp. (Mistry et al. 2021). Another marine strain, *A. sydowii,* has been used for AuNPs synthesis in the 8.7–15.6 nm (Vala 2015). Similarly, AuNPs are synthesized using the marine endophyte *Penicillium citrinum*. These NPs displayed superior antioxidant activity due to their ability to scavenge free radicals (Manjunath et al. 2017). The marine strains such as *P. fellutanum* and *P. polonicum* ARA 10 were used for AgNPs synthesis (Kathiresan et al. 2009; Neethu et al. 2018). Therefore, yeasts and fungi are commonly used for NPs synthesis as shown in Table 7.2.

7.3.3 Marine Algal Source for Green Nanoparticle Synthesis

Marine algae have been discovered as potential candidates for NPs synthesis. This NPs synthesis is based on dose-dependent and the type of algae used. Several biomolecules, such as peptides, polysaccharides, and pigments from marine algae,

Table 7.1 Marine bacterial cells mediated NPs and their applications

Marine bacteria	NPs	Mode of synthesis	Applications	References
Bacillus sp. KFU36	Ag	Extracellular	Anticancer	Almalki and Khalifa (2020)
Marinomonas sp. ef1	Ag	Extracellular	Antimicrobial	John et al. (2020a, b)
N. dassonvillei-DS013	Ag	Extracellular	Antibacterial	Dhanaraj et al. (2020)
Pseudomonas sp. ef1	Ag	Extracellular	Antimicrobial	John et al. (2020a, b)
Streptomyces sp. 192ANMG	Ag	Extracellular	Antibacterial	Hamed et al. (2020)
Streptomyces sp. 17ANMG	Ag	Extracellular	Antibacterial	Hamed et al. (2020)
B. horneckiae strain APA	Ag	Extracellular	Antimicrobial	Scala et al. (2019)
B. licheniformis strain B3-15	Ag, Au, Ag	Extracellular	Antimicrobial	Scala et al. (2019)
B. licheniformis strain T14	Ag, Au, AgCl	Extracellular	Antimicrobial	Scala et al. (2019)
Pseudomonas sp. H64	Ag	Extracellular	Antibacterial Antifungal	
Streptomyces sp. *Al-Dhabi*-89	Ag	Extracellular	Antibacterial	Al-Dhabi et al. (2018)
Marinobacter lipolyticus	Ag	Extracellular	Antifungal	Oves et al. (2017)
S. atrovirens	Ag	Extracellular	Anticancer MCF-7 cell line	Subbaiya et al. (2017)
E. coli VM1	Ag	Extracellular	Anticancer	Maharani et al. (2016)
S. rochei MHM13	Ag	Extracellular	Antibacterial	Abd-Elnaby et al. (2016)
Streptomyces sp. KC179795	Au	Extracellular	Cytotoxic	Manivasagan and Oh (2015)
Streptomyces sp. MBRC-82	Au	Extracellular	α-Amylase production	Manivasagan and Oh (2015)
Bacillus sp. MSh-1	SeNPs	Extracellular	Antioxidant	Forootanfar et al. (2014)
P. aeruginosa JQ989348	Ag	Extracellular	Antibacterial	Ramalingam et al. (2014)
S. marcescens	CdTe QDs	Extracellular	Bioimaging and biolabeling	Pawar et al. (2013)
B. pumilus	CdTe	Extracellular	Bioimaging and biolabeling	Pawar et al. (2013)
B. subtilis MSBN17	Ag	Extracellular	Antimicrobial	Sathiyanarayanan et al. (2013)
Nocardiopsis sp. MBRC-1	Ag	Extracellular	Antimicrobial and anticancer	Manivasagan et al. (2013)

Table 7.2 Fungi/yeast mediated nanoparticles and their applications

Fungi/yeast	NPs	Mode of synthesis	Application	References
A. brunneoviolaceus	Ag	Extracellular	Antibacterial Antioxidant	Mistry et al. (2021)
P. polonicum ARA 10/aqueous	Ag	Extracellular	Antibacterial	Neethu et al. (2018)
P. polonicum MF185681	Ag	Extracellular	Antibacterial	Neethu et al. (2018)
Trichoderma harzianum	CS	Extracellular	Antioxidant Antibacterial	Saravanakumar et al. (2018)
P. citrinum	Au	Extracellular	Antioxidant	Manjunath et al. (2017)
A. niger	Ag	Extracellular	Antimicrobial	Kathiresan et al. (2010)
C. glabrata	CdS	Extracellular and intracellular	Physiological	Dameron et al. (1989)
Yarrowia lipolytica NCIM3589	Au	Cell surface	–	Pimprikar et al. (2009)

are used to reduce metals. These biomolecules stabilize and capping agents for NPs (Kannan et al. 2013).

Cobalt oxide NPs synthesized by the marine red algae were tested due to their antibacterial, antioxidant, anticancer, anticoagulant, and thrombolytic properties (Ajarem et al. 2022). AuNPs synthesized by the marine strain *Gelidium pusillum* are used in medicine as a versatile anticancer cell agent by inducing apoptosis (Jeyarani et al. 2020). The AgNPs synthesized by the marine red algae *G. corneum* demonstrated excellent antimicrobial activity against *C. albicans* (Yılmaz Oztürk et al. 2020). Green algae *Chaetomorpha* sp. had been employed for the synthesis of Ag-AgNPs (Karimi and Samimi 2019).

The AuNPs produced by the marine algae *G. acerosa* displayed antibacterial activity against *E. Coli (*MTCC443*), Serratia marcescens (*MTCC97*), Klebsiella pneumonia* (MTCC432), and *B. subtilis* (MTCC441) (Senthilkumar et al. 2019). The AuNPs synthesized by the marine macroalgae *P. tetrastromatica* have been used as nanocatalysts for the organic dye degradation process (Princy and Gopinath 2018). Ruthenium NPs are synthesized by *Dictyota dichotoma* and display excellent anticancer properties (Ali et al. 2017). In addition, AgNPs synthesized from *Turbinaria ornata* demonstrated cytotoxic activity against the in-vitro retinoblastoma Y79 cell line (Remya et al. 2017). AgNPs produced by *Hypnea musciformis* displayed excellent larvicidal activity against the cabbage pest *Plutella xylostella* and the dengue vector *Aedes aegypti* (Roni et al. 2015). AuNPs have been synthesized from the new marine brown algae *Ecklonia cava* at 80 °C within 1 min. These nanomaterials have multiple uses in the medical field (Venkatesan et al. 2014). Therefore, as shown in Table 7.3, several algae have been used for metallic NPs synthesis.

Table 7.3 Algae/diatoms mediated nanoparticles and their applications

Algae/diatoms	NPs	Mode of synthesis	Application	References
Red algae	Co_3O_4	Extracellular	Antibacterial activity	Ajarem et al. (2022)
Chaetomorpha antennina	Fe	Extracellular	Nano-fertilizer, ameliorate drought stress in the plants	Sreelakshmi et al. (2021)
Acanthophora spicifera	Au	Extracellular	Antibacterial	Babu et al. (2020)
Gelidium corneum	Ag	Extracellular	Antimicrobial	Yılmaz Oztürk et al. (2020)
G. pusillum	Au	–	Cytotoxic	Jeyarani et al. (2020)
Chaetomorpha sp.	Ag-AgCl	–	Colorimetric sensor for mercury detection	Karimi and Samimi (2019)
Corallina elongate	Ag	–	Antibacterial	Hamouda et al. (2019)
Gracilaria verrucose	Au	–	Biocompatibility activity HEK-293 cells nontoxic till 100 μg/mL	Chellapandian et al. (2019)
Ulva lactuca	Ag	–	Antifungal	Sahayaraj et al. (2019)
U. prolifera	Fe_3O_4	–	Cytotoxic	Mashjoor et al. (2019)
Padina tetrastromatica	Au	–	Catalytic reduction against 4-nitrophenol	Princy and Gopinath (2018)
U. armoricana	Au	–	Catalytic reduction against 4-nitrophenol	Mukhoro et al. (2018)
Dictyota dichotoma	Ru	–	Anticancer	Ali et al. (2017)
T. ornate	Ag	–	Cytotoxic	Remya et al. (2017)
U. fasciata	Ag	–	Antimicrobial	Lakshmi et al. (2017)
Caulerpa racemosa	Ag	Extracellular	Degradation of methylene blue	Edison et al. (2016)
Sargassum tenerrimum and Turbinaria conoides	Au	–	Reduction of dyes—Rhodamine B and sulforhodamine 101	Ramakrishna et al. (2016)
Hypnea musciformis	Ag	–	Larvicidal	Roni et al. (2015)
U. lactuca	Ag	Extracellular	Antiplasmodial activity	Murugan et al. (2015)
Ecklonia cava	Au	–	Antimicrobial	Venkatesan et al. (2014)
S. plagiophyllum	Ag	–	Antibacterial activity	Dhas et al. (2014)

(continued)

Table 7.3 (continued)

Algae/diatoms	NPs	Mode of synthesis	Application	References
C. racemose	Ag	Extracellular	Antibacterial activity	Kathiraven et al. (2014)
Padina gymnospora	Au	–	–	Singh et al. (2013)
U. faciata	Ag	–	Antibacterial	El-Rafie et al. (2013)

7.4 Mode of Action of Metallic NPs

This section describes the mechanism of action of various NPs. The antimicrobial activity of AgNPs has been demonstrated against various clinical pathogens. As shown in Fig. 7.3, AgNPs are reported to exhibit antimicrobial activity by damaging the cell wall and cell membrane of pathogens.

NPs interact with microbial cells through various forces, including hydrophobic interactions, van der Waals, and electrostatic forces (Armentano et al. 2014; Li et al. 2015; Luan et al. 2016). In addition, NPs enter the cell and cause oxidative stress

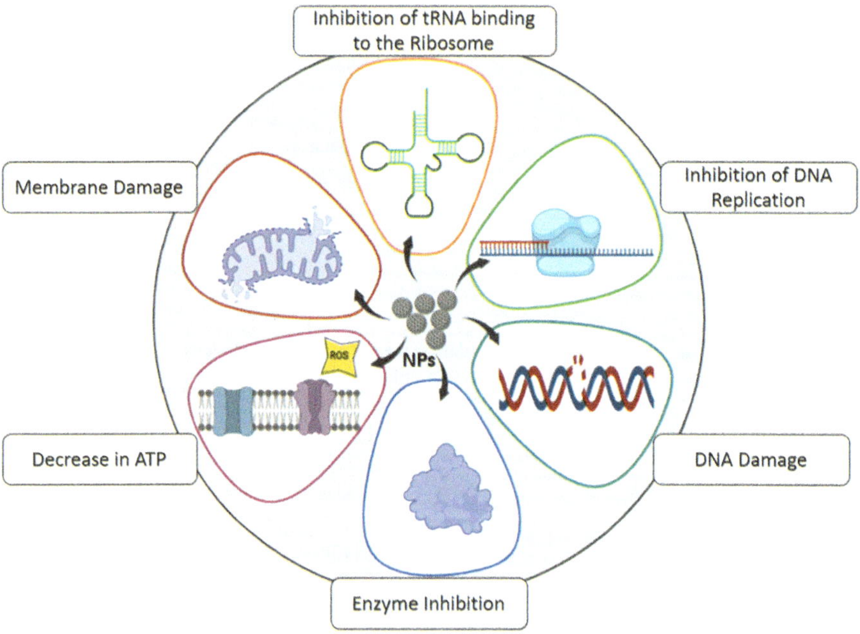

Fig. 7.3 Mode of actions of metallic NPs on microbial cells (Dakal et al. 2016; Gurunathan et al. 2012; Leung et al. 2014; Buzea et al. 2007; Pal et al. 2007; Egger et al. 2009; Yun et al. 2013; Iavicoli et al. 2013; Lima et al. 2013; Tiwari et al. 2011; Lolina and Narayanan 2013)

inside the cell. NPs destroy the organelles and alter signaling transduction pathways (Dakal et al. 2016). AgNPs have been reported to inhibit microbial cells by causing a release of ions, cavities induction, and damaging bacterial membrane. AgNPs can bind functional groups of enzymes, such as disulfide or sulfhydryl groups, that limit the metabolic processes. AgNPs bind to DNA and inhibit replication. The cell cycle is arrested by DNA damage in the G2/M phase (Buzea et al. 2007; Pal et al. 2007; Egger et al. 2009; Yun et al. 2013; Iavicoli et al. 2013). It has also been studied that TiO_2 NPs may increase cell membrane fluidity and interrupt cell integrity by lipid peroxidation (Allahverdiyev et al. 2011). AuNPs can reduce intracellular ATP levels by changing the membrane potential. It also prevents tRNA from binding to the ribosome (Lima et al. 2013; Lolina and Narayanan 2013). CuO NPs showed inhibitory activity against some vital bacterial enzymes (Ahamed et al. 2014). MgO and CaO NPs can easily damage cell membranes, resulting in leakage of intracellular contents and cell death (Leung et al. 2014; Vidic et al. 2013).

7.5 Biomedical Applications of Metallic NPs

7.5.1 Antimicrobial Activity of NPS

During the "golden age" of antibiotics, more than 20 classes of antibiotics were prepared. However, this has led to resistance in several bacteria and poses a major challenge to the pharmaceutical industry (Sánchez-López et al. 2020). With the advent of superbugs, scientists sought better antibacterial and antifungal agents. Metal-based NPs have shown more promising activity than traditional antibiotics. Biosynthesis of metal NPs is a simpler and more environmentally friendly option that can be regulated by optimizing physicochemical parameters such as pH, temperature, and metal concentration (Bankar et al. 2010a, b).

7.5.1.1 Antibacterial Activity of NPs

Several green sources, such as bacteria, fungi, and yeasts, have been used to synthesize metallic NPs. These NPs have shown promising antibacterial activity toward several bacterial pathogens (Sánchez-López et al. 2020). NPs dissolve faster in certain solutions than larger compounds. Therefore, more metal ions are released and show better antimicrobial activity (Karlsson et al. 2013). Ag and CuNPs are currently widely studied for antibacterial and antifungal activity. Ag has long been used as an antimicrobial agent. AgNPs exhibit superior antimicrobial activity compared to metal salts (Sánchez-López et al. 2020).

AgNPs affect the cellular metabolic activity of bacteria, disrupting cell membranes and causing various secondary effects such as DNA damage and forming reactive oxygen species (ROS) (Park et al. 2011). AgNPs synthesis by using marine yeast—*Candida* sp. *VITDKGB* shows antibacterial activity towards multidrug-resistant *S. aureus* and *K. pneumoniae* reported previously (Kumar et al. 2011).

The Cu, CU_2O, and CuO NPs are specifically used for antibacterial activity against *E. coli, V. cholerae, P. aeruginosa, S. typhi, S. aureus, Enterococcus faecalis*, and *S. faecalis* (Rajendran et al. 2017). Bacterial-mediated CuNPs in the range of 4.8 $\pm$ 1.6 nm displayed excellent antibacterial activity towards Gram-positive and Gram-negative bacteria (Ahmad et al. 2003). The CuONPs synthesized using Actinomycete VITBN4 exhibited antibacterial activity towards human and fish pathogens such as *S. aureus, B. cereus, P. mirabilis,* and *A. caviae, A. hydrophila, Edwardsiella tarda,* and *V. anguillarum*, respectively (Nabila and Kannabiran 2018). In addition to CuNPs and AgNPs, metal NPs such as Cd NPs also showed antibacterial activity towards some human pathogens such as *Serratia nematoda, E. coli, K. planticola*, and *Vibrio* sp. (Rajeshkumar et al. 2014). ZnO NPs synthesized using *Pseudomonas putida* exhibited in vitro antibacterial and antibiofilm activity against *P. oleovorans, E. faecalis, Pseudomonas otitidis*, and *A. baumannii* (Jayabalan et al. 2019).

7.5.1.2 Antifungal Activity of NPs

Metallic NPs showed excellent antifungal activity. Its molecular mechanism is poorly understood, but recent studies have attempted to assume the mechanisms involved. AgNPs can bind cell surfaces and penetrate the cells. In addition, they have the potential to bind compounds containing phosphorous such as DNA, subsequently inhibiting the respiratory chain and thus exhibiting previously reported antifungal activity (Reidy et al. 2013).

AgNPs synthesis by supernatant of *A. terreus* has been studied earlier (Li et al. 2012). It showed excellent antifungal activity towards four different strains of *Candida* species, including *C. albicans, C. kruzei, C. parapsilosis*, and *C. tropicalis*. Also, AgNPs demonstrated inhibitory action against filamentous fungi like *A. fumigatus* and *A. flavus* (Li et al. 2012). Novel *Nocardiopsis* sp. *MBRC-1* has been employed for the synthesis of AgNPs at room temperature. The obtained spherical NPs showed strong antifungal activity toward human pathogens such as *C. albicans, A. niger*, and *A. brasiliensis* (Manivasagan et al. 2013). Similarly, potential antifungal activity towards *C. kruzei, C. tropicalis*, and *C. albicans* was reported using AGNPs synthesized using *Brevundimonas* sp. MSK4 (Rajamanickam et al. 2013).

The synergistic action of fluconazole and AgNPs was studied on pathogenic fungi such as *P. herbarum, C. albicans, F. semitectum,* and *P. glomerata* (Gajbhiye et al. 2009). The fungus-mediated AGNPs displayed effective antifungal activity toward *F. oxysporum* and *C. albicans* (Musarrat et al. 2010). The CuNPs have been employed as fungicidal agents against several plant pathogens, such *as Fusarium* sp., *P. digitatum, Rhizoctonia solani,* and *Curvularia lunata* (Pariona et al. 2019). SeNPs synthesized by six different strains of *Trichoderma* sp. recently showed excellent antifungal activity against *S. graminicola* (Nandini et al. 2017). The Fe and NiNPs exhibited antifungal activity against the phytopathogenic fungi (Cruz-Luna et al. 2021).

7.5.2 Anticancer Activity of Metallic NPs

Due to the lack of efficient therapeutic cancer agents, researchers are now testing metal NPs as a possible anticancer agent. Various studies revealed that Cu, Au, and AgNPs can be effective anticancer agents. The AuNPs are inert, nontoxic, and promising anticancer agents because of their optical properties (Huang et al. 2007) and are currently under clinical trials for cancer diagnosis (Singh et al. 2018). The AuNPs synthesized by *Nocardiopsis* sp. *MBRC-48* exhibited strong cytotoxic and antioxidant activity against the human cervical cancer cell line (HeLa). The viability of the HeLa cell line was reported to decrease with the increase in concentrations of AuNPs (Manivasagan and Oh 2015). Similar cytotoxic activity has been observed against the HeLa cell line induced by AgNPs synthesized by *Nocardiopsis* sp. *MBRC-1* (Manivasagan et al. 2013). AgNPs synthesized by *E. coli VMI* showed anticancer activity against HeLa and the human lung cancer cell line (A549) (Maharani et al. 2016). The CuNPs exhibited cytotoxic activity towards the HeLa cells and the histolytic lymphoma cancer cell line (V937) (Jose et al. 2011).

7.5.3 Use of NPs in Medical Imaging

Bio-inspired metal NPs are synthesized from various microbial sources with diverse shapes, sizes, and optical and dielectric properties. These NPs are used in medical bioimaging because of their optical properties, such as high transparency, plasmon resonance, high/low refractive index, and new photoluminescence properties (Iskandar 2009). Nanophotonics is a unique field in which light interacts with particles much smaller than its wavelength, leading to new phenomena such as localized surface plasmon resonance (LSPR) and size-dependent semiconductor bandgap. These new properties can create novel sensitive medical bioimaging tools (Zhu et al. 2012).

Metallic NPs in the 30–150 nm size range are used for bioimaging (Debbage and Jaschke 2008). AuNPs in this size range are generally stable, easily distinguishable from the environment, and can bind strongly to the target without interfering with their activity. These properties of NPs make them the best contrast agent in bioimaging (Mei and Wu 2018). Based on the nanoscale, these AgNPs are used to develop various devices for real-time detection of DNA, RNA and proteins (Mei and Wu 2018).

7.5.4 Metallic NPs Based Biosensors

Since recent advances in nanotechnology, significant advances in metallic nanomaterial-based biosensors have been observed. The electrical, chemical, and electrochemical properties of nanomaterials such as metal NPs, graphene NPs, carbon nanotubes, and hybrid NPs have significantly improved biosensor performance (Claussen et al. 2012). Nanotechnology has helped minimize biosensors nanosized biosensors that can easily detect and interact with small molecules, from single-cell to double-stranded DNA (Gruner 2006; Whitesides 2003). Traditional biosensors consist of three main components: a) Biological recognition receptor: enzyme, antibody, and whole cell etc. b) Transducer: converts a biological response into a detectable optical or electrochemical signal; c) Signal display: to displays the result (Terry et al. 2005). Metal NPs are incorporated into the transducer for high sensitivity for nano-based biosensors. Precious metals like Ag, Au, and Pt have been immensely studied for applications in various nano biosensors (Malekzad et al. 2017).

Recently, as concerns about diabetes have increased, there is a high demand for better, more accurate glucose biosensors. In 2014, the amperometric glucose detector was developed. Nanobiosensor includes Au nanowire arrays (AuNWA) and glucose oxidase (Gox) as bioreceptors. The oxidation of glucose molecules produces hydrogen peroxide that is then transferred to AuNWA. A current was measured directly proportional to the glucose concentration (Cui et al. 2014). In another study, Gox was immobilized on platinum NPs and placed on a polyvinyl ferrocenium perchlorate matrix to form an engineered platinum electrode. This system showed signal enhancement by the electrocatalytic effect of platinum NPs (Turkmen et al. 2014). Rhodium NPs and Au NPs are used to make RhNP-modified Pt electrodes, which increase surface area and add active sites to improve activity. In addition, immobilized AuNPs near the enzyme's active site catalyzed the oxidation of peroxide molecules, thereby increasing the glucose sensor sensitivity (Guo et al. 2014). Using Ag, cheaper than Au, scientists constructed a biosensor based on an AgNP-modified TiO_2 nanotube array. The sensor demonstrated a linear range of 50–15.5 mM with a high sensitivity of 1151.98 μA mM^{-1} cm^{-2} (Wang et al. 2014a).

Nano-based biosensors for molecules other than glucose include the myoglobin sensor (Mani et al. 2014), the dopamine sensor (Mir et al. 2015), and the Ag-Au alloy-based vanillin sensor (Zheng et al. 2010). A chemical immunosensor has been established to detect carcinoembryonic antigen (CEA) as a cancer biomarker (Jeong et al. 2013). This sensor performed comparable to traditional ELISA but with significantly improved detection limits. Another study found that Ag-AuNPs composites on gold/mesoporous carbon foam-mediated Ag were also used to detect CEA (Lin et al. 2014). A unique DNA-based nano biosensor was prepared to identify the BCR and ABI fusion genes associated with chronic myelogenous leukemia. The sensor showed good reproducibility but a low detection limit (Wang et al. 2014b). A DNA biosensor with AuNPs linked to a tungsten sulphide (WS2)–graphene (Gr) electrode. This sensor could detect DNA molecules associated with the dengue virus by reduced redox indicator currents directly related to DNA hybridization (Huang et al. 2014). Many more immunosensors, such as enzyme sensors, chemiluminescent, optical, colorimetric and piezoelectric sensors, have been studied for their potential application in medicine (Malekzad et al. 2017).

7.6 Challenges in Nanomedicines Preparation

Desai (2012) reviewed that nanomedicines can be a three-dimensional structure composed of various components with a spatial arrangement suitable for the application. As a result, the superposition of the compound can be undesirably affected due to small changes in process or composition reported earlier (Desai 2012). To support the reproducible nanomedicine manufacturing process, it is important to understand the components via detailed physicochemical characterization and functional testing (Desai 2012).

Before nanomedicines arrive in the clinic, some hurdles begin with detailed study and successful manufacturing of these complex structures. Apart from standard guidelines for acceptable safety, efficacy, and recommendable pharmaceutical properties for most drugs, an ideal nanoparticle system has been used for therapeutic purposes (Desai 2012).

7.7 Conclusions

It has been concluded that some marine microorganisms such as bacteria, yeasts, fungi, and algae can be effective sources of green synthesis of metallic NPs. These metal NPs, such as Au and Ag, have some special properties such as antibacterial, antioxidant, cytotoxic, and anticancer properties, making them suitable for use as nanomedicines. However, there are some hurdles to the manufacture and application of nanomedicine. Therefore, further research is needed to improve the quality and efficiency of nanomedicines.

Acknowledgments All authors are thankful to DST, New Delhi, India, for financial assistance in the form of a major project (DST-SERB file no. EEQ/2018/001202).

References

Abd-Elnaby HM, Abo-Elala GM, Abdel-Raouf UM, Hamed MM (2016) Antibacterial and anticancer activity of extracellular synthesized silver nanoparticles from marine *Streptomyces rochei* MHM13. Egypt J Aquat Res 42(3):301–312

Adibkia K, Omidi Y, Siahi MR, Javadzadeh AR, Barzegar-Jalali M, Barar J, Maleki N, Mohammadi G, Nokhodchi A (2007) Inhibition of endotoxin-induced uveitis by methylprednisolone acetate nanosuspension in rabbits. J Ocul Pharmacol Ther 23(5):421–432

Ahamed M, Alhadlaq HA, Khan MAM, Karuppiah P, Al-Dhabi NA (2014) Synthesis, characterization and antimicrobial activity of copper oxide nanoparticles. J Nanomater 2014:1–4

Ahmad A, Mukherjee P, Senapati S, Mandal D, Khan MI, Kumar R, Sastry M (2003) Extracellular biosynthesis of silver nanoparticles using the fungus *Fusarium oxysporum*. Colloids Surf B Biointerfaces 28(4):313–318

Ajarem JS, Maodaa SN, Allam AA, Taher MM, Khalaf M (2022) Benign synthesis of cobalt oxide nanoparticles containing red algae extract: antioxidant, antimicrobial, anticancer, and anticoagulant activity. J Clust Sci 33(2):717–728

Al-Dhabi NA, Ghilan AM, Arasu MV, Duraipandiyan V (2018) Green biosynthesis of silver nanoparticles produced from marine *Streptomyces* sp. Al-Dhabi-89 and their potential applications against wound infection and drug resistant clinical pathogens. J Photochem Photobiol B 189:176–184

Ali MYS, Anuradha V, Abishek R, Yogananth N, Sheeba H (2017) In vitro anticancer activity of green synthesis ruthenium nanoparticle from *Dictyota dichotoma* marine algae. NanoWorld J 03(4):66–71

Allahverdiyev AM, Abamor ES, Bagirova M, Rafailovich M (2011) Antimicrobial effects of TiO_2 and Ag_2O nanoparticles against drug-resistant bacteria and leishmania parasites. Future Microbiol 6(8):933–940

Almalki MA, Khalifa AYZ (2020) Silver nanoparticles synthesis from Bacillus sp KFU36 and its anticancer effect in breast cancer MCF-7 cells via induction of apoptotic mechanism. J Photochem Photobiol B 204:111786

Andryukov B, Mikhailov V, Besednova N (2019) The biotechnological potential of secondary metabolites from marine bacteria. J Mar Sci Eng 7(6):176

Armentano I, Arciola CR, Fortunati E, Ferrari D, Mattioli S, Amoroso CF, Rizzo J, Kenny JM, Imbriani M (2014) The interaction of bacteria with engineered nanostructured polymeric materials: a review. Sci World J 2014:410423

Aryal S, Park H, Leary JF, Key J (2019) Top-down fabrication-based Nano/microparticles for molecular imaging and drug delivery. Int J Nanomedicine 14:6631–6644

Asmathunisha N, Kathiresan K (2013) A review on biosynthesis of nanoparticles by marine organisms. Colloids Surf B Biointerfaces 103:283–287

Babu B, Palanisamy S, Vinosha M, Anjali R, Kumar P, Pandi B, Tabarsa M, You S, Prabhu NM (2020) Bioengineered gold nanoparticles from marine seaweed *Acanthophora spicifera* for pharmaceutical uses: antioxidant, antibacterial, and anticancer activities. Bioprocess Biosyst Eng 43(12):2231–2242

Baker S, Harini BP, Rakshith D, Satish S (2013) Marine microbes: invisible nanofactories. J Pharm Res 6(3):383–388

Bankar A, Joshi B, Ameeta RK, Zinjarde S (2010a) Banana peel extract mediated synthesis of gold nanoparticles. Colloids Surf B Biointerfaces 80:45–50

Bankar A, Joshi B, Kumar AR, Zinjarde S (2010b) Banana peel extract mediated novel route for the synthesis of silver nanoparticles. Colloids Surf A Physicochem Eng Asp 368(1–3):58–63

Besinis A, De Peralta T, Handy RD (2014) The antibacterial effects of silver, titanium dioxide and silica dioxide nanoparticles compared to the dental disinfectant chlorhexidine on *Streptococcus mutans* using a suite of bioassays. Nanotoxicology 8(1):1–16

Bukka S, Badam R, Vedarajan R, Matsumi N (2019) Photo-generation of ultrasmall Pt nanoparticles on carbon-titanium dioxide nanotube composites: a novel strategy for efficient ORR activity with low Pt content. Int J Hydrog Energy 44(10):4745–4753

Buzea C, Pacheco II, Robbie K (2007) Nanomaterials and nanoparticles: sources and toxicity. Biointerphases 2(4):MR17–MR71

Chang HW, Tsai YC, Cheng CW, Lin CY, Wu PH (2013) Preparation of graphene-supported platinum nanoparticles in aqueous solution by femtosecond laser pulses for methanol oxidation. J Power Sources 239:164–168

Chellapandian C, Ramkumar B, Puja P, Shanmuganathan R, Pugazhendhi A, Kumar P (2019) Gold nanoparticles using red seaweed *Gracilaria verrucose* : green synthesis, characterization and biocompatibility studies. Process Biochem 80:58–63

Claussen JC, Shi J, Rout CS, Artiles MS, Roushar MM, Stensberg MC, Porterfield DM, Fisher TS (2012) Nanosized biosensors for medical applications. In: Biosensors for medical applications. Woodhead Publishing, Cambridge, pp 65–102

Colmer AR, Hinkle ME (1947) The role of microorganisms in acid mine drainage: a preliminary report. Science 106(2751):253–256

Cruz-Luna AR, Cruz-Martínez H, Vásquez-López A, Medina DI (2021) Metal nanoparticles as novel antifungal agents for sustainable agriculture: current advances and future directions. J Fungi (Basel) 7(12):1033

Cui J, Adeloju SB, Wu Y (2014) Integration of a highly ordered gold nanowires array with glucose oxidase for ultra-sensitive glucose detection. Anal Chim Acta 809:134–140

Dakal TC, Kumar A, Majumdar RS, Yadav V (2016) Mechanistic basis of antimicrobial actions of silver nanoparticles. Front Microbiol 7:1831

Dameron CT, Reese RN, Mehra RK, Kortan AR, Carroll PJ, Steigerwald ML, Brus LE, Winge DR (1989) Biosynthesis of cadmium sulphide quantum semiconductor crystallites. Nature 338(6216):596–597

Debbage P, Jaschke W (2008) Molecular imaging with nanoparticles: giant roles for dwarf actors. Histochem Cell Biol 130(5):845–875

Desai N (2012) Challenges in development of nanoparticle-based therapeutics. AAPS J 14(2): 282–295

Dhanaraj S, Thirunavukkarasu S, Allen John H, Pandian S, Salmen SH, Chinnathambi A, Alharbi SA (2020) Novel marine *Nocardiopsis dassonvillei*-DS013 mediated silver nanoparticles characterization and its bactericidal potential against clinical isolates. Saudi J Biol Sci 27(3): 991–995

Dhas TS, Kumar VG, Karthick V, Angel KJ, Govindaraju K (2014) Facile synthesis of silver chloride nanoparticles using marine alga and its antibacterial efficacy. Spectrochim Acta A Mol Biomol Spectrosc 120:416–420

Edison TNJI, Atchudan R, Kamal C, Lee YR (2016) *Caulerpa racemosa*: a marine green alga for eco-friendly synthesis of silver nanoparticles and its catalytic degradation of methylene blue. Bioprocess Biosyst Eng 39(9):1401–1408

Egger S, Lehmann RP, Height MJ, Loessner MJ, Schuppler M (2009) Antimicrobial properties of a novel silver-silica nanocomposite material. Appl Environ Microbiol 75(9):2973–2976

El-Rafie HM, El-Rafie MH, Zahran MK (2013) Green synthesis of silver nanoparticles using polysaccharides extracted from marine macro algae. Carbohydr Polym 96(2):403–410

Forootanfar H, Adeli-Sardou M, Nikkhoo M, Mehrabani M, Amir-Heidari B, Shahverdi AR, Shakibaie M (2014) Antioxidant and cytotoxic effect of biologically synthesized selenium nanoparticles in comparison to selenium dioxide. J Trace Elem Med Biol 28(1):75–79

Gajbhiye M, Kesharwani J, Ingle A, Gade A, Rai M (2009) Fungus-mediated synthesis of silver nanoparticles and their activity against pathogenic fungi in combination with fluconazole. Nanomedicine 5(4):382–386

Goutam SP, Saxena G, Roy D, Yadav AK, Bharagava RN (2020) Green synthesis of nanoparticles and their applications in water and wastewater treatment. In: Saxena G, Bharagava R (eds) Bioremediation of industrial waste for environmental safety. Springer, Singapore

Gruner G (2006) Carbon nanotube transistors for biosensing applications. Anal Bioanal Chem 384(2):322–335

Guo X, Liang B, Jian J, Zhang Y, Ye X (2014) Glucose biosensor based on a platinum electrode modified with rhodium nanoparticles and with glucose oxidase immobilized on gold nanoparticles. Microchim Acta 181(5–6):519–525

Gurunathan S, Han JW, Dayem AA, Eppakayala V, Kim JH (2012) Oxidative stress-mediated antibacterial activity of graphene oxide and reduced graphene oxide in *Pseudomonas aeruginosa*. Int J Nanomedicine 7:5901–5914

Hamed AA, Kabary H, Khedr M, Emam AN (2020) Antibiofilm, antimicrobial and cytotoxic activity of extracellular green-synthesized silver nanoparticles by two marine-derived actinomycete. RSC Adv 10(17):10361–10367

Hamouda RA, Abd El-Mongy M, Eid KF (2019) Comparative study between two red algae for biosynthesis silver nanoparticles capping by SDS: insights of characterization and antibacterial activity. Microb Pathog 129:224–232

Hou W, Dong X, Li Y, Zhang H, Xu L, Tian Y, Jiao A, Chen M (2019) Ultraviolet laser beam-assisted one-step synthesis of clean PtPd nanoarchitectures with excellent electrocatalytic properties for direct methanol fuel cells. Mater Chem Phys 221:409–418

Hu MZ, Zhu T (2015) Semiconductor nanocrystal quantum dot synthesis approaches towards large-scale industrial production for energy applications. Nanoscale Res Lett 10(1):469

Huang X, Jain PK, El-Sayed IH, El-Sayed MA (2007) Gold nanoparticles: interesting optical properties and recent applications in cancer diagnostics and therapy. Nanomedicine (Lond) 2(5):681–693

Huang KJ, Liu YJ, Wang HB, Gan T, Liu YM, Wang LL (2014) Signal amplification for electrochemical DNA biosensor based on two-dimensional graphene analogue tungsten sulfide-graphene composites and gold nanoparticles. Sensors Actuators B Chem 191:828–836

Iavicoli I, Fontana L, Leso V, Bergamaschi A (2013) The effects of nanomaterials as endocrine disruptors. Int J Mol Sci 14(8):16732–16801

Iskandar F (2009) Nanoparticle processing for optical applications-a review. Adv Powder Technol 20(4):283–292

Jadhav L, Phalke V, Panse S, Patil S, Bankar A (2022) Biodiversity of cold-adapted extremophiles from Antarctica and their biotechnological potential. In: Microbial diversity in hotspots. Academic Press, London, pp 231–265

Jan H, Gul R, Andleeb A, Ullah S, Shah M, Khanum M, Ullah I, Hano C, Abbasi BH (2021) A detailed review on biosynthesis of platinum nanoparticles (PtNPs), their potential antimicrobial and biomedical applications. J Saudi Chem Soc 25(8):101297

Jayabalan J, Mani G, Krishnan N, Pernabas J, Devadoss JM, Jang HT (2019) Green biogenic synthesis of zinc oxide nanoparticles using *Pseudomonas putida* culture and its in vitro antibacterial and anti-biofilm activity. Biocatal Agric Biotechnol 21:101327

Jeong B, Akter R, Han OH, Rhee CK, Rahman MA (2013) Increased electrocatalyzed performance through dendrimer-encapsulated gold nanoparticles and carbon nanotube-assisted multiple bienzymatic labels: highly sensitive electrochemical immunosensor for protein detection. Anal Chem 85(3):1784–1791

Jeyarani S, Vinita NM, Puja P, Senthamilselvi S, Devan U, Velangani AJ, Biruntha M, Pugazhendhi A, Kumar P (2020) Biomimetic gold nanoparticles for its cytotoxicity and biocompatibility evidenced by fluorescence-based assays in cancer (MDA-MB-231) and non-cancerous (HEK-293) cells. J Photochem Photobiol B 202:111715

John MS, Nagoth JA, Ramasamy KP, Mancini A, Giuli G, Natalello A, Ballarini P, Miceli C, Pucciarelli S (2020a) Synthesis of bioactive silver nanoparticles by a Pseudomonas strain associated with the Antarctic psychrophilic protozoon euplotes focardii. Mar Drugs 18(1):38

John MS, Nagoth JA, Ramasamy KP, Ballarini P, Mozzicafreddo M, Mancini A, Telatin A, Lio P, Giuli G (2020b) Horizontal gene transfer and silver nanoparticles production in a new Marinomonas strain isolated from the Antarctic psychrophilic ciliate euplotes focardii. Sci Rep 10(1):10218

Jose GP, Santra S, Mandal SK, SenGupta TK (2011) Singlet oxygen mediated DNA degradation by copper nanoparticles: potential towards cytotoxic effect on cancer cells. J Nanobiotechnol 9(1):9

Kango S, Kalia S, Celli A, Njuguna J, Habibi Y, Kumar R (2013) Surface modification of inorganic nanoparticles for development of organic-inorganic nanocomposites-a review. Prog Polym Sci 38(8):1232–1261

Kannan RR, Stirk WA, Staden JV (2013) Synthesis of silver nanoparticles using the seaweed *Codium capitatum* P.C. Silva (Chlorophyceae). S Afr J Sci Bot 86:1–4

Karimi S, Samimi T (2019) Green and simple synthesis route of Ag@AgCl nanomaterial using green marine crude extract and its application for sensitive and selective determination of mercury. Spectrochim Acta A Mol Biomol Spectrosc 222:117216

Karlsson HL, Cronholm P, Hedberg Y, Tornberg M, De Battice L, Svedhem S, Wallinder IO (2013) Cell membrane damage and protein interaction induced by copper containing nanoparticles-importance of the metal release process. Toxicology 313(1):59–69

Kathiraven T, Sundaramanickam A, Shanmugam N, Balasubramanian T (2014) Green synthesis of silver nanoparticles using marine algae *Caulerpa resmosa* and their antibacterial activity against some human pathogens. Appl Nanosci 5:499–504

Kathiresan K, Manivannan S, Nabeel MA, Dhivya B (2009) Studies on silver nanoparticles synthesized by a marine fungus, *Penicillium fellutanum* isolated from coastal mangrove sediment 71. Colloids Surf B Biointerfaces 71(1):133–137

Kathiresan K, Alikunhi NM, Pathmanaban S, Nabikhan A, Kandasamy S (2010) Analysis of antimicrobial silver nanoparticles synthesized by coastal strains of *Escherichia coli* and *Aspergillus niger*. Can J Microbiol 56(12):1050–1059

Khandel P, Shahi S (2016) Microbes mediated synthesis of metal nanoparticles: current status and future prospects. Int J Nanomater Biostruct 6(1):1–24

Kumar D, Karthik L, Kumar G, Roa KB (2011) Biosynthesis of silver nanoparticles from marine yeast and their antimicrobial activity against multidrug resistant pathogens. Pharmacologyonline 3:1100–1111

Lakshmi DS, Trivedi N, Reddy CRK (2017) Synthesis and characterization of seaweed cellulose derived carboxymethyl cellulose. Carbohydr Polym 157:1604–1610

Leung YH, Ng AM, Xu X, Shen Z, Gethings LA, Wong MT, Chan CM, Guo MY, Ng YH (2014) Mechanisms of antibacterial activity of MgO: non-ROS mediated toxicity of MgO nanoparticles towards *Escherichia coli*. Small 10(6):1171–1183

Li G, He D, Qian Y, Guan B, Gao S, Cui Y, Yokoyama K, Wang L (2012) Fungus-mediated green synthesis of silver nanoparticles using *Aspergillus terreus*. Int J Mol Sci 13(1):466–476

Li H, Chen Q, Zhao J, Urmila K (2015) Enhancing the antimicrobial activity of natural extraction using the synthetic ultrasmall metal nanoparticles. Sci Rep 5:11033

Lima E, Guerra R, Lara V, Guzmán A (2013) Gold nanoparticles as efficient antimicrobial agents for Escherichia coli and Salmonella typhi. Chem Cent J 7(1):11

Lin D, Wu J, Ju H, Yan F (2014) Nanogold/mesoporous carbon foam-mediated silver enhancement for graphene-enhanced electrochemical immunosensing of carcinoembryonic antigen. Biosens Bioelectron 52:153–158

Lolina S, Narayanan V (2013) Antimicrobial and anticancer activity of gold nanoparticles synthesized from grapes fruit extract. Chem Sci Trans 2:S105–S110

Luan B, Huynh T, Zhou R (2016) Complete wetting of graphene by biological lipids. Nanoscale 8(10):5750–5754

Maharani V, Sundaramanickam A, Balasubramanian T (2016) In vitro anticancer activity of silver nanoparticle synthesized by *Escherichia coli* VM1 isolated from marine sediments of Ennore southeast coast of India. Enzym Microb Technol 95:146–154

Malekzad H, Zangabad PS, Mirshekari H, Karimi M, Hamblin MR (2017) Noble metal nanoparticles in biosensors: recent studies and applications. Nanotechnol Rev 6(3):301–329

Mani V, Dinesh B, Chen SM, Saraswathi R (2014) Direct electrochemistry of myoglobin at reduced graphene oxide-multiwalled carbon nanotubes-platinum nanoparticles nanocomposite and biosensing towards hydrogen peroxide and nitrite. Biosens Bioelectron 53:420–427

Manivasagan P, Oh J (2015) Production of a novel fucoidanase for the green synthesis of gold nanoparticles by *Streptomyces* sp. and its cytotoxic effect on HeLa cells. Mar Drugs 13(11): 6818–6837

Manivasagan P, Venkatesan J, Senthilkumar K, Sivakumar K, Kim SK (2013) Biosynthesis, antimicrobial and cytotoxic effect of silver nanoparticles using a novel *Nocardiopsis* sp. MBRC-1. Biomed Res Int 2013:287638

Manjunath HM, Joshi CG, Raju NG (2017) Biofabrication of gold nanoparticles using marine endophytic fungus - *Penicillium citrinum*. IET Nanobiotechnol 11(1):40–44

Marchiol L (2012) Synthesis of metal nanoparticles in living plants. Ital J Agronomy 7(3):37

Mashjoor S, Yousefzadi M, Zolgharnein H, Kamrani E, Alishahi M (2019) Phycolinked vs chemogenic magnetite nanoparticles: route selectivity in nanosynthesis, antibacterial and acute zooplanktonic responses. Mater Sci Eng C 102:324–340

Mazari SA, Ali E, Abro R, Khan FSA, Ahmed I, Ahmed M, Nizamuddin S, Siddiqui TH, Hossain N (2021) Nanomaterials: applications, waste-handling, environmental toxicities, and future challenges-a review. J Environ Chem Eng 9(2):105028

Mei W, Wu Q (2018) Applications of metal nanoparticles in medicine/metal nanoparticles as anticancer agents. Met Nanoparticles Synth Appl Pharm Sci 2018:169–190

Mijatovic D, Eijkel J, Van D, Berg A (2005) Technologies for nanofluidic systems: top-down versus bottom-up-a review. Lab Chip 5(5):492–500

Mir TA, Akhtar MH, Gurudatt NG, Kim JI, Choi CS, Shim YB (2015) An amperometric nanobiosensor for the selective detection of K^{+}induced dopamine released from living cells. Biosens Bioelectron 68:421–428

Mistry H, Thakor R, Patil C, Trivedi J, Bariya H (2021) Biogenically proficient synthesis and characterization of silver nanoparticles employing marine procured fungi *Aspergillus brunneoviolaceus* along with their antibacterial and antioxidative potency. Biotechnol Lett 43(1):307–316

Mukherji S, Bharti S, Shukla G, Mukherji S (2018) Synthesis and characterization of size- and shape-controlled silver nanoparticles. Metallic Nanomater (Part B) 1:1–116

Mukhoro OC, Roos WD, Jaffer M, Bolton JJ, Stillman MJ, Beukes DR, Antunes E (2018) Very green photosynthesis of gold nanoparticles by a living aquatic plant: photoreduction of AuIII by the seaweed *Ulva armoricana*. Chemistry 24(7):1657–1666

Murugan K, Samidoss CM, Panneerselvam C, Higuchi A, Roni M, Suresh U, Chandramohan B, Subramaniam J, Madhiyazhagan P (2015) Seaweed-synthesized silver nanoparticles: an eco-friendly tool in the fight against *Plasmodium falciparum* and its vector Anopheles stephensi? Parasitol Res 114(11):4087–4097

Musarrat J, Dwivedi S, Singh BR, Al-Khedhairy AA, Azam A, Naqvi A (2010) Production of antimicrobial silver nanoparticles in water extracts of the fungus *Amylomyces rouxii* strain KSU-09. Bioresour Technol 101(22):8772–8776

Nabila MI, Kannabiran K (2018) Biosynthesis, characterization and antibacterial activity of copper oxide nanoparticles (CuONPs) from actinomycetes. Biocatal Agric Biotechnol 15:56–62

Nagajyothi PC, Lee KD, Sreekanth TVM (2014) Biogenic synthesis of gold nanoparticles (quasi-spherical, triangle, and hexagonal) using *Lonicera japonica* flower extract and its antimicrobial activity. Synth React Inorg Met Org NanoMet Chem 44(7):1011–1018

Nandini B, Hariprasad P, Prakash HS, Shetty HS, Geetha N (2017) Trichogenic-selenium nanoparticles enhance disease suppressive ability of *Trichoderma* against downy mildew disease caused by *Sclerospora graminicola* in pearl millet. Sci Rep 7(1):2612

Neethu S, Midhun SJ, Sunil MA, Soumya S, Radhakrishnan EK, Jyothis M (2018) Efficient visible light induced synthesis of silver nanoparticles by Penicillium polonicum ARA 10 isolated from *Chetomorpha antennina* and its antibacterial efficacy against *Salmonella enterica* serovar *typhimurium*. J Photochem Photobiol B 180:175–185

Oves M, Qari HA, Felemban NM, Khan MZ, Rehan ZA, Ismail IMI (2017) *Marinobacter lipolyticus* from Red Sea for lipase production and modulation of silver nanomaterials for anti-candidal activities. IET Nanobiotechnol 11(4):403–410

Pal S, Tak YK, Song JM (2007) Does the antibacterial activity of silver nanoparticles depend on the shape of the nanoparticle? A study of the gram-negative bacterium Escherichia coli. Appl Environ Microbiol 73(6):1712–1720

Pariona N, Mtz-Enriquez AI, Sánchez-Rangel D, Carrión G, Paraguay-Delgado F, Rosas-Saito G (2019) Green-synthesized copper nanoparticles as a potential antifungal against plant pathogens. RSC Adv 9(33):18835–18843

Park MV, Neigh AM, Vermeulen JP, de la Fonteyne LJ, Verharen HW, Briedé JJ, van Loveren H, de Jong WH (2011) The effect of particle size on the cytotoxicity, inflammation, developmental toxicity and genotoxicity of silver nanoparticles. Biomaterials 32(36):9810–9817

Patil MP, Kim G (2018) Marine microorganisms for synthesis of metallic nanoparticles and their biomedical applications. Colloids Surf B Biointerfaces 172:487–495

Pawar V, Kumar AR, Zinjarde S, Gosavi S (2013) Bioinspired inimitable cadmium telluride quantum dots for bioimaging purposes. J Nanosci Nanotechnol 13(6):3826–3831
Pimprikar PS, Joshi SS, Kumar AR, Zinjarde SS, Kulkarni SK (2009) Influence of biomass and gold salt concentration on nanoparticle synthesis by the tropical marine yeast *Yarrowia lipolytica* NCIM 3589. Colloids Surf B Biointerfaces 74(1):309–316
Princy KF, Gopinath A (2018) Optimization of physicochemical parameters in the biofabrication of gold nanoparticles using marine macroalgae *Padina tetrastromatica* and its catalytic efficacy in the degradation of organic dyes. J Nanostruct Chem 8(3):333–342
Pugazhenthiran N, Anandan S, Kathiravan G, Udaya Prakash NK, Crawford S, Ashokkumar M (2009) Microbial synthesis of silver nanoparticles by *Bacillus* sp. J Nanopart Res 11(7): 1811–1815
Rajamanickam K, Sudha SS, Francis M, Sowmya T, Rengaramanujam J, Sivalingam P, Prabakar K (2013) Microalgae associated *Brevundimonas* sp. MSK 4 as the Nano particle synthesizing unit to produce antimicrobial silver nanoparticles. Spectrochim Acta A Mol Biomol Spectrosc 113: 10–14
Rajendran K, Anwar A, Khan NA, Siddiqui R (2017) Brain-eating amoebae: silver nanoparticle conjugation enhanced efficacy of anti-amoebic drugs against *Naegleria fowleri*. ACS Chem Neurosci 8(12):2626–2630
Rajeshkumar S, Ponnanikajamideen M, Malarkodi C, Malini M, Annadurai G (2014) Microbe-mediated synthesis of antimicrobial semiconductor nanoparticles by marine bacteria. J Nanostruct Chem 4(2):1–7
Ramakrishna M, Rajesh Babu D, Gengan RM, Chandra S, Nageswara Rao G (2016) Green synthesis of gold nanoparticles using marine algae and evaluation of their catalytic activity. J Nanostruct Chem 6(1):1–13
Ramalingam V, Rajaram R, Premkumar C, Santhanam P, Dhinesh P, Vinothkumar S, Kaleshkumar K (2014) Biosynthesis of silver nanoparticles from deep sea bacterium *Pseudomonas aeruginosa* JQ989348 for antimicrobial, antibiofilm, and cytotoxic activity. J Basic Microbiol 54(9):928–936
Reidy B, Haase A, Luch A, Dawson KA, Lynch I (2013) Mechanisms of silver nanoparticle release, transformation and toxicity: a critical review of current knowledge and recommendations for future studies and applications. Materials (Basel) 6(6):2295–2350
Remya RR, Radhika Rajasree SRR, Aranganathan L, Suman TY, Gayathri S (2017) Enhanced cytotoxic activity of AgNPs on retinoblastoma Y79 cell lines synthesised using marine seaweed *Turbinaria ornata*. IET Nanobiotechnol 11(1):18–23
Roni M, Murugan K, Panneerselvam C, Subramaniam J, Nicoletti M, Madhiyazhagan P, Dinesh D, Suresh U, Khater HF (2015) Characterization and biotoxicity of *Hypnea musciformis*-synthesized silver nanoparticles as potential eco-friendly control tool against *Aedes aegypti* and *Plutella xylostella*. Ecotoxicol Environ Saf 121:31–38
Sabzevari A, Adibkia K, Hashemi H, Hedayatfar A, Mohsenzadeh N, Atyabi F, Ghahremani MH, Dinarvand R (2013) Polymeric triamcinolone acetonide nanoparticles as a new alternative in the treatment of uveitis: in vitro and in vivo studies. Eur J Pharm Biopharm 84(1):63–71
Sahayaraj K, Rajesh S, Rathi JAM, Kumar V (2019) Green preparation of seaweed based silver nano-liquid for cotton pathogenic fungi management. IET Nanobiotechnol 13(2):219–225
Sánchez-López E, Gomes D, Esteruelas G, Bonilla L, Lopez-Machado AL, Galindo R, Cano A, Espina M, Ettcheto M (2020) Metal-based nanoparticles as antimicrobial agents: an overview. Nanomaterials (Basel) 10(2):292
Saravanakumar K, Chelliah R, MubarakAli D, Jeevithan E, Oh DH, Kathiresan K, Wang MH (2018) Fungal enzyme-mediated synthesis of chitosan nanoparticles and its biocompatibility, antioxidant and bactericidal properties. Int J Biol Macromol 118(B):1542–1549
Sathiyanarayanan G, Kiran GS, Selvin J (2013) Synthesis of silver nanoparticles by polysaccharide bioflocculant produced from marine *Bacillus subtilis* MSBN17. Colloids Surf B Biointerfaces 102:13–20

Scala PA, Hada A, Astilean S, Vulpoi A, Ginestra G, Marino A, Nostro A, Zammuto V, Gugliandolo C (2019) Marine bacterial exopolymers-mediated green synthesis of noble metal nanoparticles with antimicrobial properties. Polymers 11(7):1157

Senthilkumar P, Surendran L, Sudhagar B, Ranjith Santhosh Kumar DS (2019) Facile green synthesis of gold nanoparticles from marine algae *Gelidiella acerosa* and evaluation of its biological potential. SN Appl Sci 1(4):284

Shankar S, Jaiswal L, Rhim JW (2021) New insight into sulfur nanoparticles: synthesis and applications. Crit Rev Environ Sci Technol 51(20):2329–2356

Sharma D, Kanchi S, Bisetty K (2019) Biogenic synthesis of nanoparticles: a review. Arab J Chem 12(8):3576–3600

Singh M, Kalaivani R, Manikandan S, Sangeetha N, Kumaraguru AK (2013) Facile green synthesis of variable metallic gold nanoparticle using *Padina gymnospora*, a brown marine macroalga. Appl Nanosci 3(2):145–151

Singh P, Pandit S, Mokkapati VRSS, Garg A, Ravikumar V, Mijakovic I (2018) Gold nanoparticles in diagnostics and therapeutics for human cancer. Int J Mol Sci 19(7):1979

Sreelakshmi B, Induja S, Adarsh PP, Rahul HL, Arya SM, Aswana S, Haripriya R, Aswathy BR, Manoj PK (2021) Drought stress amelioration in plants using green synthesised iron oxide nanoparticles. Mater Today Proc 41:723–727

Stincone P, Brandelli A (2020) Marine bacteria as source of antimicrobial compounds. Crit Rev Biotechnol 40(3):306–319

Subbaiya R, Saravanan M, Priya AR, Shankar KR, Selvam M, Ovais M, Balajee R, Barabadi H (2017) Biomimetic synthesis of silver nanoparticles from *Streptomyces atrovirens* and their potential anticancer activity against human breast cancer cells. IET Nanobiotechnol 11(8): 965–972

Talib A, Hui-Fen-Wu (2016) Biomimetic synthesis of lotus leaf extract-assisted silver nanoparticles and shape-directing role of cetyltrimethylammonium bromide. J Mol Liq 220:795–801

Terry LA, White SF, Tigwell LJ (2005) The application of biosensors to fresh produce and the wider food industry. J Agric Food Chem 53(5):1309–1316

Thiruvengadathan R, Korampally V, Ghosh A, Chanda N, Gangopadhyay K, Gangopadhyay S (2013) Nanomaterial processing using self-assembly-bottom up chemical and biological approaches. Rep Prog Phys 76(6):066501

Tiwari PM, Vig K, Dennis VA, Singh SR (2011) Functionalized gold nanoparticles and their biomedical applications. Nanomaterials (Basel) 1(1):31–63

Turkmen E, Bas SZ, Gulce H, Yildiz S (2014) Glucose biosensor based on immobilization of glucose oxidase in electropolymerized poly (o-phenylenediamine) film on platinum nanoparticles-polyvinylferrocenium modified electrode. Electrochim Acta 123:93–102

Vala AK (2015) Exploration on green synthesis of gold nanoparticles by a marine derived fungus *Aspergillus sydowii*. Environ Prog Sustain Energy 34(1):194–197

Venkatesan J, Manivasagan P, Kim SK, Kirthi AV, Marimuthu S, Rahuman AA (2014) Marine algae-mediated synthesis of gold nanoparticles using a novel *Ecklonia cava*. Bioprocess Biosyst Eng 37(8):1591–1597

Vidic J, Stankic S, Haque F, Ciric D, Le Goffic R, Vidy A, Jupille J, Delmas B (2013) Selective antibacterial effects of mixed ZnMgO nanoparticles. J Nanopart Res 15(5):1595

Wang W, Xie Y, Xia C, Du H, Tian F (2014a) Titanium dioxide nanotube arrays modified with a nanocomposite of silver nanoparticles and reduced graphene oxide for electrochemical sensing. Microchim Acta 181(11–12):1325–1331

Wang L, Hua E, Liang M, Ma C, Liu Z, Sheng S, Liu M, Xie G, Feng W (2014b) Graphene sheets, polyaniline and AuNPs based DNA sensor for electrochemical determination of BCR/ABL fusion gene with functional hairpin probe. Biosens Bioelectron 51:201–207

Whitesides GM (2003) The 'right' size in nanobiotechnology. Nat Biotechnol 21(10):1161–1165

Yılmaz Oztürk B, Yenice Gürsu B, Dağ İ (2020) Antibiofilm and antimicrobial activities of green synthesized silver nanoparticles using marine red algae *Gelidium corneum*. Process Biochem 89:208–219

Yun H, Kim JD, Choi HC, Lee CW (2013) Antibacterial activity of CNT-Ag and GO-Ag nanocomposites against Gram-negative and Gram-positive bacteria. Bull Korean Chem Soc 34(11):3261–3264

Zhang X, He X, Wang K, Yang X (2011) Different active biomolecules involved in biosynthesis of gold nanoparticles by three fungus species. J Biomed Nanotechnol 7:245–254

Zheng D, Hu C, Gan T, Dang X, Hu S (2010) Preparation and application of a novel vanillin sensor based on biosynthesis of Au-Ag alloy nanoparticles. Sensors Actuators B Chem 148(1): 247–252

Zhu T, Cloutier SG, Ivanov I, Knappenberger KL, Robel I, Zhang F (2012) Nanocrystals for electronic and optoelectronic applications. J Nanomater 2012:1–2

A Potent Drug L-Asparaginase from Marine Origins: A Comprehensive Review

8

Doniya Elze Mathew, Aastha Soni, Pramod B. Shinde, and Vaibhav A. Mantri

Abstract

L-Asparaginase (E.C.3.5.1.1) is a crucial enzyme with immense applications in the biomedical and food sectors. Due to its ability to convert the amino acid L-asparagine into aspartic acid and ammonia, its potency has been evaluated against acute lymphoblastic leukemia (ALL) and in acrylamide mitigation. L-Asparaginases are widely produced in nature by bacteria, fungi, algae, and plants, but humans lack this. Although microbial L-asparaginases were largely studied for various applications, they later came under assault due to their side effects like hypersensitivity, immunogenic complications, L-asparaginase resistance, glutaminase activity, etc. A vast array of biological and chemical diversity can be obtained from the marine regimes. Marine microbes displaying extreme survival patterns, either novel or with unique and appealing physiological features relative to their terrestrial counterparts, can be encountered. The relevant physiological features such as thermal stability, salt, pressure and pH tolerance, long-term

D. E. Mathew · V. A. Mantri (✉)
Applied Phycology & Biotechnology Division, CSIR-Central Salt and Marine Chemicals Research Institute, Bhavnagar, India

Academy of Scientific and Innovative Research (AcSIR), Ghaziabad, India
e-mail: vaibhav@csmcri.res.in

A. Soni
Applied Phycology & Biotechnology Division, CSIR-Central Salt and Marine Chemicals Research Institute, Bhavnagar, India

P. B. Shinde
Academy of Scientific and Innovative Research (AcSIR), Ghaziabad, India

Natural Products & Green Chemistry Division, CSIR-Central Salt and Marine Chemicals Research Institute, Bhavnagar, India
e-mail: pramodshinde@csmcri.res.in

P. Veera Bramhachari, C. V. Berde (eds.), *Marine Bioactive Molecules for Biomedical and Pharmacotherapeutic Applications*,
https://doi.org/10.1007/978-981-99-6770-4_8

storage properties, and extreme cold adaptability are often seen. Such physico-chemical properties make marine enzymes relevant in biotechnological applications and economic viability. Thus marine L-asparaginase offering unique and enhanced characteristics which overcome the limitations of terrestrial origin can be developed. This review attempts to comprehend the available marine sources, their phylogenetic relationships, upstream and downstream techniques, and various kinetic parameters of L-asparaginase. This review provides collective information on the significance of this enzyme and its application from the viewpoint of marine origin.

Keywords

L-Asparaginase · Marine microbes · Acrylamide · Anticancer · Bioprocess · Fermentation techniques · Chromatography

8.1 Introduction

Infections and diseases are the major bottleneck to any species' survival and establishment on earth, including plants and animals of either marine or terrestrial origin. Developing new pharmaceutical and nutraceutical origins of versatile applications is an essential component of strategies to combat existing and novel infections. L-Asparaginase (LA) (L-asparagine amidohydrolase, E.C.3.5.1.1) is being popularly utilized for its potential to act as an anticancer agent (Benchamin et al. 2019). The catalytic activity of this enzyme over the non-essential amino acid L-asparagine to give aspartic acid and ammonia is the key to its anticancer activity

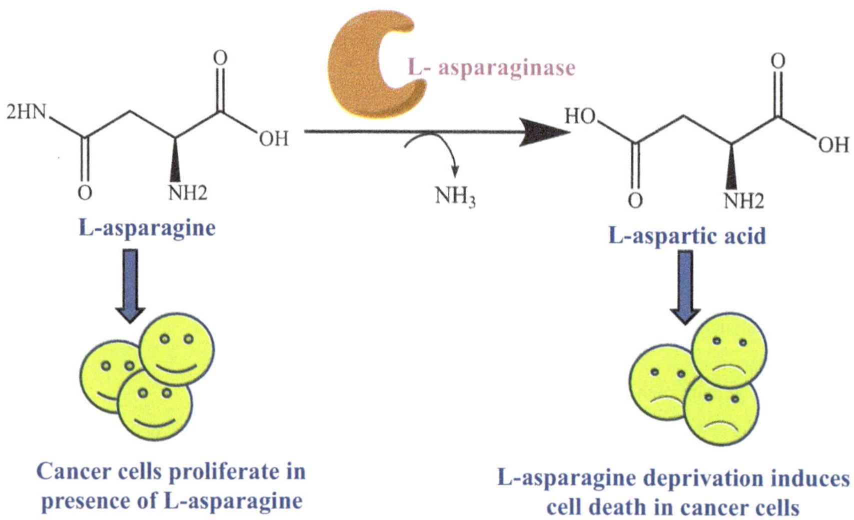

Fig. 8.1 L-Asparaginase catalytic activity on L-ASN

Fig. 8.2 Structure of acrylamide. (Adopted from Blank et al. 2005)

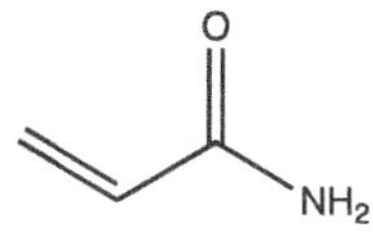

acrylamide

(Fig. 8.1) (Chand et al. 2020; Choi and Coloff 2019; Muneer et al. 2020). The cancerous cells (like acute lymphoblastic leukemia) cannot synthesize L-asparagine. The average human cells have the enzyme L-asparagine synthetase, and they synthesize the asparagine required for growth and metabolism.

On the other hand, leukemic cells depend on the asparagine circulating in the bloodstream to suffice their requirement for protein synthesis. The administration of LA thus reduces the amount of available asparagine in the blood and causes the cancer cells to starve from asparagine depletion and promote their apoptosis and, thus, selectively act on the tumor cells (El-Naggar et al. 2018; Narta et al. 2007; Savitri and Azmi 2003). It was identified that the guinea pig serum can suppress certain types of leukemia and was utilized in the beginning. Later on, the active factor in the serum was identified as LA. Supporting reports subsequently showed that the serum devoid of LA did not display any anticancer activity. Thus, the potent therapeutic for leukemia in guinea pig serum was LA (Boyse et al. 1967).

LA is also known as a potential agent in mitigating acrylamide generation during industrial processes in the food industry (Xu et al. 2016; Dias et al. 2017; Alam et al. 2018; Chand et al. 2020; Muneer et al. 2020). Food heating promotes the formation of toxic substances such as acrylamide (Fig. 8.2) (prop-2-enamide, C_3H_5NO), a potential carcinogen in human cells. According to the International Agency for Research on Cancer (IARC), acrylamide is a carcinogenic potency that may harm human life and health (IARC 1994). It is reported that the thermal processing of foods that contain high-reducing sugars, for instance, French fries, and the presence of L-asparagine leads to acrylamide formation (Dias et al. 2017). Hence, LA's activity can prevent this carcinogenic substance's formation.

Although, LA has several emerging applications in pharmaceuticals, health, and food industry, it has been the major drug for treating ALL. Such a great assortment increases the market demand for this enzyme; hence a better source for enhanced yield and reduced immunogenic drawbacks is required. For instance, the global industrial enzyme demand proves the vast scope of innovations and research in enzymes, which was estimated at $3.3 billion in 2010, excited to $6.4 billion in 2021 and is predicted to be $8.7 billion in 2026 (BCC Research 2021).

Presently two important formulations are utilized commercially for treating ALL, from *Erwinia* sp. and *E.coli* (Chand et al. 2020). *E. coli* was the foremost bacterial contestant producing LA, which presented expected anticancer activity in mice and rats and later exhibited potential in human ALL treatment (Boyse et al. 1967; Peterson and Ceigler 1969). It is stated that the potential of *E. coli* cell extracts on L-asparagine deamination was first reported in 1957 by Tsuji (Cedar and Schwartz 1968). Despite such huge versatility and applications, LA, from its pioneer origins, has lately come under assault because of limited production yield, immunogenic reactions, and hypersensitivity in patients (Peterson and Ceigler 1969), endotoxins, in addition to secondary side effects associated with the glutaminase affinity put forth the necessity of another improved source (Chand et al. 2020). Further studies identified that some other bacteria produce LA in larger quantities than *E. coli*. Among these *Erwinia* sp. was found to release the maximum amount of enzyme from cells. Also, *Erwinia aroideae* produces the enzyme in 8 h (Peterson and Ceigler 1969), a remarkably short time for harvesting any biomolecule for commercial purposes. Recombinant strains of *Aspergillus oryzae* and *Aspergillus niger* are also used in the commercial preparation of LA (Dias et al. 2017). Also, recombinant methods for improvised enzyme production, for example, ScASNase1 encoding *Saccharomyces cerevisiae* L-asparaginase 1, are successfully reported (Costa et al. 2016). Further, the catalytic efficiency and thermostability of LA II from *Bacillus subtilis* (BsAII) were successfully improved by site-directed mutagenesis (Long et al. 2016).

Based on the localization of LA inside the bacterial cell, it occurs in two isoforms: LA I & II. Affinities of both vary over L-asparagine and L-glutamine. However, the specificity, activity level, production yield, and application of LA varies from organism to organism and species to species. LA is distributed widely in nature, so it is found in bacteria, algae, fungi, plants, and many animals (Chand et al. 2020; Vimal and Kumar 2017). It is identified that rodents have this naturally in their blood serum. Humans lack this vital enzyme (Cachumba et al. 2016).

8.2 Marine Sources of L-Asparaginase

Marine natural products play an important role in biomedical research and drug development. Marine microorganisms have the potential to thrive in extreme conditions. Hence, research on marine microbes and their metabolites would help discover novel pharmaceuticals. Organisms flourishing in marine habitats sustain low nutrition, high salinity, and high-pressure conditions that affect their microenvironment and produce biologically unique metabolites/molecules. LA is the most profoundly studied enzyme due to its antileukemic activity in human beings. LA obtained from terrestrial sources for leukemia treatment displayed various side effects. Marine source-derived LA could be the best alternative for the treatment of leukemia as seawater is saline in nature and chemically closer to blood plasma; hence, it could result in fewer side effects (Balakrishnan and Pandey 1996; Nageswara et al. 2014). The characteristics of marine-derived LA from various sources are listed in Table 8.1.

Bacterial Sources Bacterial sources are easy to handle and can abundantly produce LA. They are usually associated with seaweeds, fish, and aquatic plants, floating on the surface in seawater, mangrove soil, and sea sediments. The most abundant marine bacteria producing LA belongs to *Bacillus* sp. and *Pseudomonas* sp. (Qeshmi et al. 2018). The aerial distribution of dormant *Bacillus* sp. spores is extremely

Table 8.1 The characteristics of marine derived L-asparaginase from various sources

No	Class	Organism	Opt. pH	Opt. temp (°C)	Enzyme activity	Molecular weight (kDa)	References
1	Bacteria	*Streptomyces noursei* MTCC 10469	8	5	78.88 IU/mg	10	Dharmaraj (2011)
		Thermococcus kodakaraensis	9.5	85	2350 U/mg	35.5	Chohan and Rashid (2013)
		Streptomyces sp.	7.5	50	49.2 IU/mL	–	Basha et al. (2009)
		Thermococcus gammatolerans	8.5	85	7622 U/mg	36.5	Zuo et al. (2014)
		Bacillus pumilus	–	–	157 IU/mL	136	Sindhwad and Desai (2015)
2	Fungi	*Beauveria bassiana*	9	–	90 U/gds	–	Nageswara et al. (2014)
		Aspergillus niger AKVMKBU	7	30	46.75 IU/mg	90	Vala et al. (2018)
		Aspergillus terreus	6	35	33.86 U/mg	–	Farag et al. (2015)
3	Algae	*Spirulina maxima*	8.5	37	0.51 IU/mL	–	Baky and El Baroty (2016)
		Chlamydomonas sp.	–	55	78.5 IU/mg	–	Paul (1982)

resilient and makes them predominant in marine habitats (Ebrahiminezhad et al. 2011; Qeshmi et al. 2014). Most of these marine bacteria-derived LA are extracellularly produced with high titers of enzymes free from endotoxins. Intracellular LA purification is a tedious task. Hence extracellular LA producing bacteria make downstream processing easy and economically viable (Narayana et al. 2008; El-Naggar et al. 2016). The optimum pH for the production of LA from most marine bacterial species is 7, and the enzyme activity is maximum at the pH range of 6–9. The optimum pH activity of LA from *Bacillus* PG03 is 6 (Rahimzadeh et al. 2016), while from *Mesoflavibacter zeaxanthinifaciens* S86 is 9 (Su-Jin et al. 2015). *Enterobacter hormaechei* produced a maximum amount of LA when cultured at a pH of 9 (Sudha et al. 2017).

Fungal Sources Marine fungi thrive in the sea either as obligates or in facultative mode existing as free-floating entities or lodged on driftwood, sand grains, shells, algae, sponges, corals, mollusks, plants, and fishes (Parte et al. 2017). The marine environment appears to be a prolific source of LA-producing fungi. Fungal L-asparaginases are mostly produced extracellularly hence facilitating the downstream purification process. Fungal-derived LA could be less immunogenic to humans as fungi are evolutionarily closer than bacteria (Vimal and Kumar 2017). LA is secreted by seaweed endophytes of the genera *Alternaria, Chaetomium, Cladosporium, Colletotrichum, Curvularia, Nigrospora, Paecilomyces, Phaeotrichoconis, Phoma,* and *Pithomyces* (Thirunavukkarasu et al. 2011). Sundarmoorthi et al. (2012) selected five fungal strains from the soil of the Arabian Sea that exhibited LA activity. They concluded that around 80% of the fungal strains from marine sources are potential LA producers. Excellent source of LA includes endophytic fungi from tropical seaweeds, as Thirunavukkarasu et al. (2011) reported. The optimum pH for purified LA from *Aspergillus aculeatus* is 9, while the optimum temperature is 30 °C, which is relatively lower than other marine sources (Dange and Peshwe 2011). *Aspergillus* sp. ALAA 2000 yielded two types of LA: AYA-1 and AYA-2, with molecular weights of 25 and 31 kDa, respectively. AYA-1 and AYA-2 are thermostable as they remain active for 1 h at 47 °C and 67 °C and exhibited a pH of 6 and 10, respectively (Mervat et al. 2015).

Algal Sources *Spirulina maxima* (Baky and El Baroty 2016), *Chlamydomonas* sp. (Paul 1982), *Chlorella vulgaris* (Ebrahiminezhad et al. 2014), and *Vaucheria unicata* (Batool et al. 2016) are the algal sources that produce LA. *Chlamydomonas* sp. produce thermotolerant and alkaline LA. LA from *Spirulina* can be therapeutic, but this alga's optimal pH requirements are not yet reported (Paul 1982; Baky and El Baroty 2016). Marine algae can be a good source for producing LA on a large scale under simple culture conditions and in a cost-effective manner.

8.3 Bioprocess Techniques for Marine-Derived L-Asparaginase

Bioprocessing of enzymes is a complex process that involves several complex steps. Efficient bioprocessing of target enzymes requires selecting a suitable source of microorganisms as it directly affects the characteristics of the final product. The selection of robust high-yielding strains for industrial production of LA, its easy accessibility, and economic viability should be considered. Commercially available LA is strongly immunogenic, and has low substrate specificity and high glutaminase activity causing liver dysfunction, pancreatitis, cerebral dysfunction, leucopenia, neurological seizures, and coagulation abnormalities that tend to cause intracranial thrombosis and hemorrhage (Duval et al. 2002). The use of marine sources for extraction of LA is a better alternative as marine organisms (bacteria, fungi, algae, plants) produce enzymes with novel and improved physiological properties such as hyperthermal stability, barophilicity, salt and pH tolerance, adaptability to extreme cold conditions, and novel chemical and stereochemical properties (Mohapatra et al. 1998; Shojaei et al. 2017). The optimal activity of cells is of prime concern and assessed using scale-up and scale-down studies that enable extrapolating and building large-scale industrial fermenters with sufficient confidence and proper functioning with all its behaviors anticipated to achieve high fermentation efficiency.

Screening of marine LA-producing isolates by simple plate method was done using Czapek-Dox or M9 medium containing L-asparagine as the sole nitrogen source combined with phenol red. LA-producing colonies changed from yellow to pink after incubation (18 h—bacteria: 48 h—fungi). The color change was observed as a result of an increase in pH around the colonies when ammonia is released by hydrolysis of L-asparagine (Gulati et al. 1997; Mervat et al. 2015; Nageswara et al. 2014; Qeshmi et al. 2014).

Initially, the experiment is done at a laboratory scale in shaken flasks to check the fidelity of the microorganism. Scale-up processes are then carried out, maintaining the geometry and conditions from pilot to industrial scale for large enzyme production. In bioreactors, the operation mode can generate high productivity by batch, fed-batch, or continuous cultivation (Torres et al. 2016).

8.3.1 Upstream Processing

The upstream bioprocessing of LA requires the optimization of fermentation media for bacterial growth and enzyme production. It involves screening and identifying microorganisms, media preparation, and all the material inputs needed for production. The performance of any bioprocess involves key factors like agitation rate, mass transfer, heat transfer, flow rate, oxygen transfer, pH, temperature, C and N source, bioreactor design, fermentation time, and the type of fermentation (Moorthy et al. 2010). LA derived from different sources differ in their physicochemical and kinetic parameters.

8.3.1.1 Temperature

Biological characteristics of cells and enzymes often impose constraints on bioprocessing; for instance, thermostability properties must be considered while setting the operation temperature of the enzyme reactor. For example, the optimum temperature for the production of LA from *Bacillus licheniformis* is 37 °C, and it shows a significant decrease of 32.19% in the production of LA when the temperature was raised to 47 °C (Chand et al. 2020). Heat-resistant LA can be produced using *Streptomyces gulbargensis and Fusarium equiseti* with temperature stability at 40 and 45 °C, respectively (Chand et al. 2020). Extremely thermostable LA obtained from *Thermococcus gammatolerans* and *Thermococcus kodakaraensis* showed maximum activity at 85 °C (Chohan et al. 2020; Zuo et al. 2014). Mostly, microorganisms require a temperature range of 25–37 °C to produce LA. Enzymes tend to denature above or below their optimal temperature range. Hence proper maintenance of conditions should be provided to maintain the integrity of bioprocessing.

8.3.1.2 pH

pH of the production medium plays an important role in the bioprocessing of enzymes and transport across the cell membrane. The optimum pH for the production of LA is 7 for bacterial sources, 7.5–8.5 for actinomycetes, and 6.2–7.5 for fungal sources. The optimum pH for the bacterium *Pectobacterium carotovorous* is reported to be 7, that for *Streptomyces gulbargensis* was 9, and the same for the fungus *Aspergillus niger* was 6.5. The optimum pH for the production of LA from *Beauveria bassiana* (MSS18/41) *Enterobacter hormaechei* was reported to be 9 (Nageswara et al. 2014; Sudha et al. 2017). The carbon source is also reported to affect the pH of the fermentation medium; for instance, if glucose, sucrose, or mannitol is used as the carbon source, there is production of acid in the fermentation process, decreasing the pH of the medium, which inhibits the synthesis of LA.

8.3.1.3 Fermentation Medium

The most frequently used media for LA production are Luria-Bertani medium, tryptone glucose yeast extract broth, M9 medium, and modified Czapek Dox medium depending upon the type of microorganism used (Baskar and Renganathan 2012; Dharmaraj 2011; Einsfeldt et al. 2016; El-Naggar et al. 2018; Usha et al. 2011; Vimal and Kumar 2017). These media are used when LA is being produced by submerged fermentation. For solid-state fermentation, the substrates used are majorly agricultural wastes that facilitate extracellular enzyme production. Cheap agriculture waste like rice bran, wheat bran, sesame oil cake, soybean meal, gram husk, groundnut oil cake, and tea waste makes the process cost-effective as well as environment-friendly (Vala et al. 2018; Vimal and Kumar 2017). Kumar et al. (2011) obtained 83.16 IU/g LA activity by growing *Serratia marcescens* (NCIM2919) on *Citrus limetta* pulp. El-Bessoumy et al. (2004) obtained LA activity of 1900 IU/mg from *Pseudomonas aeruginosa* 50071 grown on soyabean meal.

8.3.1.4 Carbon Source

The major carbon source microorganisms use to produce LA is dextrose and maltose. While using glucose, fructose, and mannitol as the carbon source, regular monitoring and control of pH is required as the pH of the medium decreases due to acid production during fermentation. This decrease in pH inhibits the production of LA. A microbial culture containing maltose, starch, and trehalose produced a final pH of alkaline nature (7.5–8) (Narayana et al. 2008). Isaac and Abu-Tahon (2016) obtained 438.4 IU/mL activity of LA from *Fusarium solani* AUMC 8615 using maltose as a carbon source. LA activity of 574.24 IU/mg was obtained using glucose as the only carbon source from *Penicillium brevicompactum* NRC 829 by Elshafei et al. (2012).

8.3.1.5 Incubation Time

The optimum cultivation time depends on the type of microorganism used, ranging from 24 to 72 h. *Staphylococcus aureus* NCTC 413 required the lowest optimum incubation period of 10 h, whereas an optimum cultivation period of 96 h was reported in the case of *Pseudomonas aeruginosa* and *Serratia marcescens* NCIM 2919 (Chand et al. 2020). *Spirulina maxima* grown on modified Zarrouk medium reported the highest optimum cultivation period of 432 h for the production of LA (Baky and El Baroty 2016). Incubation time if exceeding the optimum range results in decreased enzyme production, as in the case of *Bacillus licheniformis*, where the yield of the enzyme increased by 30.84% during 36 h of incubation and increased exponentially at 72-h incubation period and after that decreased with an increase in incubation time (Chand et al. 2020). *Pseudomonas aeruginosa* 50071 when incubated for 96 h gave a maximum yield of 1900 IU/mg (El-Bessoumy et al. 2004). *Streptomyces gulbargensis* showed an increased yield of 12.1 IU at 120 h of the incubation period, while a decrease in yield was observed on the increase or decrease of the incubation period (Amena et al. 2008).

8.3.1.6 Agitation Rate

Agitation imparts proper mixing of the culture medium and microorganisms, providing a uniform supply of nutrients and oxygen to every cell. It prevents medium accumulation in the vessel and ensures the fermentation process is not limited to the surface. *Aspergillus terreus* MTCC 1782 showed an increased yield of LA at 24.10 IU/mL with a 160 rpm agitation rate (Baskar and Renganathan 2012). Marine Actinomycetes S3 showed an increased yield of 384.6 IU/mg at an agitation rate of 200 rpm. The optimum agitation speed for LA production in *Streptomyces gulbargensis* was reported to be 200 rpm yielding 12.1 IU of the enzyme. Still, the yield decreased when the agitation rate was increased to 220 rpm (Amena et al. 2008).

8.3.2 Fermentation Techniques

Production of LA by different sources can be done through submerged fermentation and solid-state fermentation techniques (Lopes et al. 2015).

8.3.2.1 Submerged Fermentation

Submerged fermentation is an extensively used method for producing LA and is majorly employed for bacterial enzyme production. This method involves the culture microorganisms in a liquid broth medium in closed containers with high oxygen. In submerged fermentation, manipulating medium components is easier than in solid-state fermentation; no pretreatment of the substrate is required. Hence it could lead to high production yields (Vimal and Kumar 2017). Submerged fermentation can be performed at the laboratory scale (shaken flasks culture and bioreactor up to 10 L) and industrial scale (bioreactor larger than 10 L). Under optimized culture conditions of 35 °C temperature, 6.3 pH, agitation rate of 140 rpm, and incubation period of 58.5 h, *Aspergillus terreus* produced 44.38 U/mL of enzyme (Baskar and Renganathan 2012).

8.3.2.2 Solid-State Fermentation

Solid-state fermentation emerged as an alternative to submerged fermentation, allowing direct use of the crude fermented product as an enzyme source and potentially producing secondary metabolites (Lopes et al. 2015). In this process, the microorganism, i.e., slowly consumes the substrate, and the substrate is utilized for long fermentation periods. The highest LA-specific activity was reported for *A. niger* (344.0 U/g) cultivated in solid-state fermentation on agro-wastes in a column bioreactor using sesame (black) oil cake as a substrate at 32 °C (Uppuluri et al. 2013; Uppuluri and Reddy 2009). Since solid-state fermentation requires complex media, it appears to be still very far from satisfying the high purity requirements of LA for pharmaceutical and medical applications.

8.3.3 Downstream Processing

LA for therapeutic needs to be extracted in a highly purified form, which burdened downstream purification of the enzyme. The downstream processing costs nearly 60–80% of the total investment for biotherapeutic production (Tundisi et al. 2017). The enzyme with high purity generates less toxic and allergic responses. Downstream processing checks the efficiency of protein purification monitored by determining

1. The specific activity, which is a measure of enzyme purity
2. Yield, i.e., the percentage of activity in crude extract
3. Purification fold which quantifies an increase in enzyme purity

The downstream processing involves the steps of (Albertsson 1986) removal of insoluble material (Amena et al. 2008), concentration (Baky and El Baroty 2016), fractionation, and (Balakrishnan and Pandey 1996) purification.

A combination of downstream unit processes is involved in the purification of enzymes that includes non-chromatographic (low-resolution purification steps) and chromatographic separation (high-resolution purification steps) (Cachumba et al. 2016). The non-chromatographic separation process includes centrifugation, fractional precipitation, aqueous biphasic systems, and membrane-based purification, whereas the chromatographic separation includes ion exchange, gel filtration and affinity chromatography (El-Naggar and El-Shweihy 2020; Mahajan et al. 2014; Tundisi et al. 2017).

8.3.3.1 Non-chromatographic Process

Cell Disruption Techniques

Cell disruption is usually performed by three methods (Albertsson 1986): mechanical method including glass bead stirring, sonication, freezing, and drying (Amena et al. 2008); physical method by osmotic shock (Baky and El Baroty 2016); and chemical method using Tween 80, SDS, Toluene, etc. for *E. coli*, *Aspergillus terreus*, and *Leucosporidium muscorum* (Costa-Silva et al. 2018). Kumar et al. (2011) reported sonication to purify LA from *Pectobacterium carotovorum*. Sonication was a better method for LA production in the case of *E. coli* and *A. terreus* (Costa-Silva et al. 2018).

Ultrafiltration and Centrifugation

Commercially preferred technique for the concentration of cell-free filtrate is ultrafiltration, as it is economical and yields little loss of enzyme activity (Chand et al. 2020). A 10 kDa membrane is rinsed with deionized water, cleaned with 500 mL of 0.5 M NaOH, and rinsed with deionized water. Vacuum-filtered crude LA is then concentrated using the membrane, and the LA activity is determined in the retentate and the collected filtrate (Prakash et al. 2021). Extracting extracellular LA from *Streptomyces fradiae* NEAE82, the sample was centrifuged at 11,000 × *g* for 15 min for crude preparation (El-Naggar et al. 2016).

Protein Precipitation

To recover and purify biomolecules, separation by precipitation from an aqueous extract is the most traditional method (Golunski et al. 2011). Precipitation is one of the first steps involved in the downstream processing of enzymes and is usually combined with traditional methods to enhance the purification and yield of enzymes. Ammonium sulfate is the most common precipitant used in the purification of LA (Amena et al. 2010; Basha et al. 2009; Gervais et al. 2013; Manna et al. 1995; Mishra 2006; Moorthy et al. 2010). This method involves temporary disruption of the protein's secondary and tertiary structures, leading to precipitation. The precipitant agent can be recycled in the final process by distillation, heat drying, or evaporation. Partial purification generally starts with adding finely powdered ammonium sulfate to a crude extract containing enzyme up to saturation, causing LA precipitation. Purification of LA obtained from *Streptomyces gulbargensis* increased to 26.9-fold after the precipitate was collected and then centrifuged, and then dialyzed against buffer followed by Sephacryl S-200 column filtration (Amena et al. 2010). Sindhwad and Desai (2015) achieved purification of *Bacillus pumilus* produced LA using 70% ammonium sulfate precipitation yielding 1.5 times purified enzyme with a specific activity of 715–1044 IU/mg respectively for crude and purified extract (Mahajan et al. 2014).

Liquid-Liquid Extraction

It is a process of achieving the concentration of biological products by transferring the desired product from one liquid phase to another. The extraction efficiency depends on the partition coefficient, i.e., the relative distribution of substances between two liquid phases. Aqueous two-phase systems are obtained when mixtures of water-soluble molecules combine with another polymer or certain inorganic salts, surfactants, block copolymers, and ionic liquids above critical concentrations. The isoelectric point, surface hydrophobicity, and molar mass of system components are the physicochemical properties that influence protein partitioning in two-phase systems (Lopes et al. 2015). The protein partitioning also depends on the concentration of polymer or surfactant, pH, and addition of salt (Albertsson 1986). The main drawback of precipitation is the irreversible inactivation of LA in precipitation. Hence, aqueous two-phase systems and membrane processes integration with precipitation are viable processes for purification. An aqueous two-phase system (ATPS) can be designed when two chemically different polymers, viz. dextran and polyethylene glycol (PEG), or a specific salt and one polymer, viz. potassium phosphate and PEG, are mixed at specific concentrations in a solution (Chand et al. 2020).

8.3.3.2 Chromatographic Process

LA requires high levels of purity. Hence it requires a sequence of purification steps. Generally, ionic exchange, gel filtration, size exclusion, and affinity chromatography are used to purify LA (Khushoo et al. 2004). Gel filtration and ion exchange chromatography is the most used purification technique, often preceded by ammonium sulfate precipitation (Lopes et al. 2015). *P. aeruginosa* 50071 obtained by solid-state fermentation, purified by ammonium sulfate fractionation, Sephadex G-100 gel filtration and CM-Sephadex C0 ionic exchange increased specific activity from 17.90 of crude extract to 1900 IU/mg of final preparation (El-Bessoumy et al. 2004). LA purification from different sources requires different purification steps and varied performance in yield and activity.

Ion Exchange Chromatography

LA is generally negatively charged and hence shows higher affinity to anion exchangers (Chand et al. 2020). Diethyl aminoethyl (DEAE)-cellulose chromatography Q-Sepharose and DEAE-Toyopearl column chromatography are the most commonly used matrices for chromatographic LA separation.

Gel Filtration Chromatography
El Bassoumy et al. purified LA from *P. aeruginosa* 50071 using Sephadex G100 gel filtration with a purification fold of 67.7 and recovery of 60.8%. LA was purified from *Cladosporium* sp. using Sepharose 6B gel filtration chromatography in combination with DEAE-cellulose column chromatography and methanol precipitation (Kumar and Manonmani 2013). The major drawback of this technique is that the desired protein gets too diluted.

8.3.4 Purity Analysis of L-Asparaginase

The purity of the extracted LA is analyzed by SDS-PAGE (Sodium Dodecyl Sulfate-Poly Acrylamide Gel Electrophoresis), UPLC (Ultra Performance Liquid Chromatography), DLS-ZP (Dynamic Light Scattering-Zeta Potential), and DSC-TGA (Differential Scanning Calorimetry-Thermogravimetric Analysis). The confirmation of the purity of LA is done by UPLC analysis. The DLS-ZP analysis checks the particle size uniformity and solubility of the protein. The DSC-TGA study phase shifts, heat stability, and moisture content in enzymes (Prakash et al. 2021).

8.3.4.1 SDS-PAGE Analysis

SDS-PAGE is performed to determine LA's purity and molecular weight. The Laemmli method is used for SDS-PAGE separation, which includes a gel system consisting of a stacking gel (5%, pH 6.8), a separating gel (10%, pH 8.8), and SDS (0.1%). After the electrophoresis, protein bands are visualized by staining with Coomassie brilliant blue R-250. De-staining is done using methanol, acetic acid, and water in the ratio 4:1:5. The molecular weight of purified LA is determined using molecular markers of known weight ranging from 9 to 178 kDa. Observation of one distinctive band by SDS separation indicates the presence of homogenous protein.

El-Naggar et al. (2018) obtained a 67 kDa molecular mass band from *Streptomyces brollosae*. Chohan and Rashid (2013) resolved an individual band of 35.5 kDa from *Thermococcus kodakaraensis*.

8.3.4.2 UPLC Analysis

Purified LA is subjected to Ultra Performance Liquid Chromatography which operates at higher pressures of 15,000 psi. It consists of 1.7 μm reverse-phase packing material, allowing smaller particle sizes in a column. UPLC analysis for purified LA is performed with a methanol: water solvent system with 0.1% TFA with a gradient flow of 1 mL/min at 280 nm wavelength. The sample is diluted to a 1:100 ratio in deionized water and performed at 25 °C with a count rate of 210 kbps. Prakash et al. (2021) reported a peak of LA visible at 0.6 min of run from *Bacillus altitudinis*.

8.3.4.3 DLS-ZP Analysis

The DLS analysis determines the size distribution profile of small particles in a suspension. Zeta potential is the indicator of the stability of colloidal dispersions.

The range of transfer of electric charges of the sample to the crude sample is compared for the ZP analysis. For DLS-ZP analysis, the crude and purified sample are dissolved in deionized water in the ratio of 1:100. Prakash et al. (2021) determined the particle size of standard LA and the purified sample from *Bacillus altitudinis* to be 51 and 422 nm, respectively. On comparing with crude LA, a 65% decrease in light intensity was observed, indicating a reduction in average size, ascertaining better dispersion of purified LA. The ZP values of LA bioconjugate compared to standard LA were reported to be −50 to +10 mV by. reported better results with ZP values in the range of +25 to −25 mV (Martins et al. 1996; Prakash et al. 2021).

8.3.4.4 DSC-TGA

DSC analysis is used to check the thermotropic properties of purified LA. It measures how much energy a sample absorbs or releases during heating or cooling. TGA measurements analyze the change in weight with temperature. The pure LA is dehydrated to form dry powder by lyophilization and then subjected to DSC-TGA analysis in a temperature range of 35–600 °C with an increment of 10 °C/min. Prakash et al. (2021) reported initiation in weight loss at 35.86 °C in standard LA and 35.29 °C in the LA purified from *Bacillus altitudinis*. A maximum loss of 83% weight in the standard at 498.12 °C and 65% maximum loss in the purified sample at 796.45 °C was reported, suggesting higher stability of the pure sample.

8.4 Conclusion

Due to the depletion of L-asparagine circulating in the blood, the antileukemic effect associated with the LA enzyme contributes to its high therapeutic value. The chief approved sources are *Escherichia coli* and *Erwinia chrysanthemi,* which are later identified to generate strong immunogenic reactions in around 50% of cancer cases. LA from other microbial sources, including gram-positive bacteria, renders potentially improved LA replacements. Dedicated research on LA from marine bacteria is a promising area for food and biomedical applications owing to their high substrate specificity, novel structure, and low molecular weight. Marine-derived LA displays remarkable stability and activity over pH, temperature, and pressure ranges. Moreover, a high asparagine specificity and low glutaminase affinity highlight the potential of LA from marine sources for treating leukemia with minimal disadvantages.

Acknowledgments The authors thank the Council of Scientific and Industrial Research for funding support. This has CSIR PRIS registration number 112/2022.

References

Albertsson PA (1986) Partition of proteins in liquid polymer–polymer two-phase systems. In: Partition of cell particles and macromolecules, 3rd edn. Wiley Intersciences, New York, pp 709–771

Alam S, Ahmad R, Pranaw K, Mishra P, Khare SK (2018) Asparaginase conjugated magnetic nanoparticles used for reducing acrylamide formation in food model system. Bioresour Technol 269:121–126

Amena S, Vishalakshi N, Prabhakar M, Dayanand A, Lingappa K (2010) Production, purification and characterization of L-asparaginase from Streptomyces gulbargensis. Braz J Microbiol 41:173–178

Amena S, Vishalakshi N, Prabhakar M, Dayanand A, Lingappa K (2008) Production, purification and characterization of L-asparaginase from *Streptomyces gulbargensis*. Braz J Microbiol 41: 173–178

Baky HHAE, El Baroty GS (2016) Optimization of growth conditions for purification and production of L-asparaginase by *Spirulina maxima*. Evid Based Complement Alternat Med 2016: 1785938. https://doi.org/10.1155/2016/1785938

Balakrishnan K, Pandey A (1996) Production of biologically active secondary metabolites in solid state fermentation. J Sci Ind Res 55:365–372

Basha NS, Rekha R, Komala M, Rub S (2009) Production of extracellular anti-leukaemic enzyme L-asparaginase from marine actinomycetes by solid state and submerged fermentation: purification and characterisation. Trop J Pharm Res 8(4):353–360

Baskar G, Renganathan S (2012) Optimization of culture conditions and bench-scale production of L-asparaginase by submerged fermentation of *aspergillus terreus* MTCC 1782. J Microbiol Biotechnol 22(7):923–929

Batool T, Makky EA, Jalal M, Yusoff MM (2016) A comprehensive review on L-asparaginase and its applications. Appl Biochem Biotechnol 178(5):900–923

BCC Research (2021) Global markets for enzymes in industrial applications. BCC Publishing. https://www.bccresearch.com/market-research/biotechnology/global-markets-for-enzymes-in-industrialapplications.html

Benchamin D, Sreejai R, Sujitha S, Jensy Roshan F, Albert C, Rishad KS (2019) Anti-proliferative activity of L-Asparaginase enzyme from fungi on breast cancer. J Pharmacogn Phytochem 8(1): 407–410

Blank I, Robert F, Goldmann T, Pollien P, Varga N, Devaud S, Saucy F, Huynh-Ba T, Stadler RH (2005) Mechanisms of acrylamide formation. In: Chemistry and safety of acrylamide in food. Springer, New York, pp 171–189

Boyse EA, Old LJ, Campbell HA, Mashburn LT (1967) Suppression of murine leukemias by l-asparaginase : incidence of sensitivity among leukemias of various types: comparative inhibitory activities of Guinea pig serum l-asparaginase and *Escherichia coli*L-asparaginase. J Exp Med 125(1):17–31. https://doi.org/10.1084/jem.125.1.17

Cachumba JJM, Antunes FAF, Peres GFD, Brumano LP, Santos JCD, Da Silva SS (2016) Current applications and different approaches for microbial L-asparaginase production, biotechnology and industry microbiology. Braz J Microbiol 47(1):77–85. https://doi.org/10.1016/j.bjm.2016.10.004

Cedar H, Schwartz JH (1968) Production of L asparaginase II by *Escherichia coli*. J Biotechno. 96(6):2043–2048. https://doi.org/10.1128/jb.96.6.2043-2048.1968

Chand S, Mahajan RV, Prasad JP, Sahoo DK, Mihooliya KN, Dhar MS, Sharma G (2020) A comprehensive review on microbial L-asparaginase: bioprocessing, characterization, and industrial applications. Biotechnol Appl Biochem 67(4):619–647. https://doi.org/10.1002/bab.1888

Chohan SM, Rashid N (2013) TK1656, a thermostable L-asparaginase from *Thermococcus kodakaraensis*, exhibiting highest ever reported enzyme activity. J Biosci Bioeng 116(4): 438–443

Chohan SM, Sajed M, Naeem S, Rashid N (2020) Heterologous gene expression and characterization of TK2246, a highly active and thermostable plant type L-asparaginase from *Thermococcus kodakaraensis*. Int J Biol Macromol 147:131–137
Choi B, Coloff JL (2019) The diverse functions of non-essential amino acids in cancer. Cancers 11(5):675. https://doi.org/10.3390/cancers11050675
Costa IM, Schultz L, Pedra DAB, Leite MS, Farsky SH, de Oliveira MA, Pessoa A, Monteiro G (2016) Recombinant L-asparaginase 1 from *Saccharomyces cerevisiae*: an allosteric enzyme with antineoplastic activity. Sci Rep 6:36239. https://doi.org/10.1038/srep36239
Costa-Silva TA, Flores-Santos JC, Freire RK, Vitolo M, Pessoa-Jr A (2018) Microbial cell disruption methods for efficient release of enzyme L-asparaginase. Prep Biochem Biotechnol 48(8):707–717
Dange VU, Peshwe SA (2011) Production, purification and characterization of fungal L-asparaginase. Bionano Front 4(1):162–167
Dharmaraj S (2011) Study of L-asparaginase production by *Streptomyces noursei* MTCC 10469, isolated from marine sponge *Callyspongia diffusa*. Iran J Biotech 9(2):102–108
Dias FFG, Junior SB, Hantao LW, Augusto F, Sato HH (2017) Acrylamide mitigation in French fries using native l-asparaginase from *Aspergillus oryzae* CCT 3940. LWT Food Sci Technol 76 (B):222–229. https://doi.org/10.1016/j.lwt.2016.04.017
Duval M, Suciu S, Ferster A, Rialland X, Nelken B, Lutz P, Benoit Y, Robert A, Manel AM, Vilmer E, Otten J (2002) Comparison of *Escherichia coli*–asparaginase with *Erwinia*-asparaginase in the treatment of childhood lymphoid malignancies: results of a randomized European Organization for Research and Treatment of Cancer–Children's Leukemia Group phase 3 trial. Blood 99(8):2734–2739
Ebrahiminezhad A, Rasoul-Amini A, Ghasemi Y (2011) L-Asparaginase production by moderate halophilic bacteria isolated from Maharloo salt lake. Indian J Microbiol 51:307–311
Ebrahiminezhad A, Rasoul-Amini S, Ghoshoon MB, Ghasemi Y (2014) Chlorella vulgaris, a novel microalgal source for L-asparaginase production. Biocatal Agric Biotechnol 3(2):214–217
Einsfeldt K, Baptista IC, Pereira JCCV, Bruno RE, Mello FV, Costa IC (2016) Recombinant L-Asparaginase from *Zymomonas mobilis*: a potential new antileukemic agent produced in *Escherichia coli*. PLoS One 11(6):e0156692. https://doi.org/10.1371/journal.pone.0156692
El-Bessoumy AA, Sarhan M, Mansour J (2004) Production, isolation, and purification of L-asparaginase from *Pseudomonas aeruginosa* 50071 using solid-state fermentation. J Biochem Mol Biol 37(4):387–393. https://doi.org/10.5483/BMBRep.2004.37.4.387
El-Naggar NEA, El-Shweihy NM (2020) Bioprocess development for L-asparaginase production by *Streptomyces rochei*, purifcation and in-vitro efficacy against various human carcinoma cell lines. Sci Rep 10:7942. https://doi.org/10.1038/s41598-020-64052-x
El-Naggar NE, Deraz SF, Soliman HM, El-Deeb NM, El-Ewasy SM (2016) Purification, characterization, cytotoxicity and anticancer activities of L-asparaginase, anti-colon cancer protein, from the newly isolated alkaliphilic *Streptomyces fradiae* NEAE-82. Sci Rep 6:32926
El-Naggar NE, Deraz SF, El-Ewasy SM, Suddek GM (2018) Purification, characterization and immunogenicity assessment of glutaminase free L-asparaginase from *Streptomyces brollosae* NEAE-115. BMC Pharmacol Toxicol 19(1):51
Elshafei AM, Hassan MM, Abouzeid ME, Mahmoud DA, Elghonemy DH (2012) Purification, characterization and antitumor activity of L-asparaginase from *Penicillium brevicompactum* NRC 829. Br Microbiol Res J 2(3):158–174
Farag AM, Hassan SW, Beltagy EA, El-Shenawy MA (2015) Optimization of production of anti-tumor L-asparaginase by free and immobilized marine *Aspergillus terreus*. Egypt J Aquat Res 41(4):295–302
Gervais D, Allison N, Jennings A, Jones S, Mark T (2013) Validation of a 30-year old process for the manufacture of L-asparaginase from *Erwinia chrysanthemi*. Bioprocess Biosyst Eng 36: 453–460

Golunski S, Astolfi V, Carniel N, de Oliveira D, Luccio DM, Mazutti MA, Treichel H (2011) Ethanol precipitation and ultrafiltration of inulinases from *Kluyveromyces marxianus*. Sep Purif Technol 78:261–265

Gulati R, Saxena R, Gupta R (1997) A rapid plate assay for screening L-asparaginase producing microorganisms. Lett Appl Microbiol 24(1):23–26

IARC (1994) Acrylamide. In: IARC monographs on the evaluation of the carcinogenic risk of chemicals to humans, vol 60. International Agency for Research on Cancer, Lyon, pp 389–433. http://www.inchem.org/documents/iarc/ vol60/m60-11.html

Isaac GS, Abu-Tahon MA (2016) Production of extracellular antileukemic enzyme L-asparaginase from *Fusarium solani* AUMC 8615 grown under solid-state fermentation conditions: purification and characterization of the free and immobilized enzyme. Egypt J Bot 56(3):799–816. https://doi.org/10.21608/EJBO.2016.3776

Khushoo A, Pal Y, Singh BN, Mukherjee KJ (2004) Extracellular expression and single step purification of recombinant *Escherichia coli* L-asparaginase II. Protein Expr Purif 38:29–36

Kumar NS, Manonmani HK (2013) Purification, characterization and kinetic properties of extracellular L-asparaginase produced by Cladosporium sp. World J Microbiol Biotechnol 29(4): 577–587

Kumar S, Veeranki VD, Pakshirajan K (2011) Assessment of physical process conditions for enhanced production of novel glutaminase free L-asparaginase from *Pectobacterium carotovorum* MTCC 1428. Appl Biochem Biotechnol 163(3):327–337. https://doi.org/10.1007/s12010-010-9041-x

Long S, Zhang X, Rao Z, Chen K, Xu M, Yang T, Yang S (2016) Amino acid residues adjacent to the catalytic cavity of tetramer L-asparaginase II contribute significantly to its catalytic efficiency and thermostability. Enzyme Microb Technol 82:15–22

Lopes AM, Oliveira-Nascimento LD, Ribeiro A, Tairum CA Jr, Breyer CA, de Oliveira MA, Monteiro G, de Souza-Motta CM, Magalhaes PO, Avendano JGF, Cavaco-Paulo AM, Mazzola PG, Rangel-Yagui CO, Sette LD, Converti A, Pessoa A (2015) Therapeutic L-asparaginase: upstream, downstream and beyond. Crit Rev Biotechnol 37(1):82–99. https://doi.org/10.3109/07388551.2015.112070

Mahajan RV, Kumar V, Rajendran V, Saran S, Ghosh PC, Saxena RK (2014) Purification and characterization of a novel and robust L-asparaginase having low glutaminase activity from *Bacillus licheniformis*: in vitro evaluation of anti-cancerous properties. PLoS One 9(6):e99037. https://doi.org/10.1371/journal.pone.0099037

Manna S, Sinha A, Sadhukhan R, Chakrabart SL (1995) Purification, characterization and antitumor activity of L-asparaginase isolated from *Pseudomonas stutzeri* MB-405. Curr Microbiol 30:291–298

Martins MBAF, Goncalves APV, Cruz MEM (1996) Biochemical characterization of an L-asparaginase bioconjugate. Bioconjug Chem 7(4):430–435. https://doi.org/10.1021/bc9600269

Mervat MA, Nageh A, Taher T, Fareed H (2015) Production, purification and characterization of L-asparaginase from marine endophytic *Aspergillus* sp: ALAA-2000 under submerged and solid state fermentation. J Microb Biochem Technol 7(3):165–172

Mishra A (2006) Production of L-asparaginase, an anticancer agent, from *Aspergillus niger* using agricultural waste in solid state fermentation. Appl Biochem Biotechnol 135:33–42

Mohapatra B, Banerjee U, Bapuji M (1998) Characterization of a fungal amylase from *Mucor* sp: associated with the marine sponge *Spirastrella* sp. J Biotechnol 60(1):113–117

Moorthy V, Ramalingam A, Sumantha A, Shankaranaya RT (2010) Production, purification and characterisation of extracellular L-asparaginase from a soil isolate of *Bacillus* sp. Afr J Microbiol Res 4:1862–1867. https://doi.org/10.5897/AJMR.9000601

Muneer F, Siddique MH, Azeem F, Rasul I, Muzammil S, Zubair M, Afzal M, Nadeem H (2020) Microbial L-asparaginase: purification, characterization and applications. Arch Microbiol 202: 967–981

Nageswara S, Guntuku G, Tadimalla P (2014) Production of L-asparaginase by solid state fermentation using marine fungus. Biomed Res 1(1):1–9

Narayana KJP, Kumar KG, Vijayalakshmi M (2008) L-asparaginase production by *Streptomyces albidoflavus*. Indian J Microbiol 48:331–336

Narta UK, Kanwar SS, Azmi W (2007) Pharmacological and clinical evaluation of l-asparaginase in the treatment of leukemia. Crit Rev Oncol Hematol 61(3):208–221. https://doi.org/10.1016/j.critrevonc.2006.07.009

Parte S, Sirisha V, D'Souza J (2017) Biotechnological applications of marine enzymes from algae, bacteria, fungi, and sponges. Adv Food Nutr Res 80:75–106

Paul JH (1982) Isolation and characterization of a ChlamydomonasL-asparaginase. Biochem J 203(1):109–115

Peterson RE, Ciegler A (1969) L-asparaginase production by Erwinia aroideae. Appl Microbiol 18 (1):64–67

Prakash P, Chandrayan S, Tiwari P, Singh HR, Jha SK (2021) Development of a downstream process for purification and purity analysis of glutaminase free L-asparaginase using UPLC, DLS-ZP and DSC-TG. J Taibah Univ Sci 15(1):458–467. https://doi.org/10.1080/16583655.2021.1984694

Qeshmi FI, Javadpour S, Malekzadeh K, Tamadoni JS, Rahimzadeh M (2014) Persian gulf is a bioresource of potent L-asparaginase producing bacteria: isolation & molecular differentiating. Int J Environ Res 8(3):813–818

Qeshmi FI, Homaei A, Fernandes P, Javadpour S (2018) Marine microbial L-asparaginase: biochemistry, molecular approaches and applications in tumor therapy and in food industry. Microbiol Res 208:98–112

Rahimzadeh M, Poodat M, Javadpour S, Qeshmi FI, Shamsipour F (2016) Purification characterization and comparison between two new L-asparaginases from *Bacillus* PG03 and *Bacillus* PG04. Open Biochem J 10:35–45

Savitri AN, Azmi W (2003) Microbial L-asparaginase: a potent anti-tumour enzyme. Indian J Biotechnol 02(2):184–194

Shojaei F, Homaei A, Taherizadeh MR, Kamrani E (2017) Characterization of biosynthesized chitosan nanoparticles from *Penaeus vannamei* for immobilization of *P. vannamei* protease: an eco-friendly nanobiocatalyst. Int J Food Prop 20(2):1413–1423

Sindhwad P, Desai K (2015) Media optimization, isolation and purification of L-asparaginase from marine isolate. Asian Pac J Health Sci 2(3):72–82

Sudha K, Rani J, Sivansankari TR (2017) Extracellular L-asparaginase production by halotolerant strain of *Enterobacter hormaechei* isolated from marine fishes. Int J Adv Res 5(1):2619–2625

Su-Jin L, Youngdeuk L, Gun-Hoo P, Navaneethaiyer U, Soo-Jin H, De ZM, Won-Kyo J, Dae-Won L, Hanjun K, Do-Hyung K, Chulhong O (2015) A newly identified glutaminase-free L-asparaginase (L-ASPG86) from the marine bacterium *Mesoflavibacter zeaxanthinifaciens*. J Microbiol Biotechnol 26(6):1115–1123. https://doi.org/10.4014/jmb.1510.10092

Sundarmoorthi C, Rajakumari R, Dharamsi A, Vengadeshprabhu K (2012) Production and immobilization of L-asparaginase from marine source. Int J Pharm Pharm Sci 4(4):229–232

Thirunavukkarasu N, Suryanarayanan TS, Murali TS, Ravishankar JP, Gummadi SN (2011) L-asparaginase from marine derived fungal endophytes of seaweeds. Mycosphere 2011:147–155

Torres FAE, Zaccarim BR, Novaes LC, Jozala AF, dos Santos CA, Teixeira MFS, Santos-Ebinuma VC (2016) Natural colorants from filamentous fungi. Appl Microbiol Biotechnol 100:2511–2521. https://doi.org/10.1007/s00253-015-7274-x

Tundisi LL, Coelho DF, Zanchetta B, Moriel P, Pessoa A Jr, Tambourgi EB, Mazzola PG (2017) L-asparaginase purification. Sep Purif Rev 46(1):35–43

Uppuluri KB, Reddy DSR (2009) Optimization of L-asparaginase production by isolated *Aspergillus niger* using sesame cake in a column bioreactor. J Pure Appl Microbiol 3:83–90

Uppuluri KB, Dasari RKVR, Sajja V, Jacob AS, Reddy DSR (2013) Optimization of L-asparaginase production by isolated *Aspergillus niger* C4 from sesame (black) oil cake under SSF using Box–Behnken design in column bioreactor. Int J Chem React Eng 11:103–109

Usha R, Mala KK, Venil CK, Palaniswamy M (2011) Screening of actinomycetes from mangrove ecosystem for L-asparaginase activity and optimization by response surface methodology. Pol J Microbiol 60(3):213–221

Vala AK, Sachaniya B, Dudhagara D, Panseriya HZ, Gosai H, Rawal R, Dave BP (2018) Characterization of L-asparaginase from marine-derived *Aspergillus niger* AKV-MKBU, its antiproliferative activity and bench scale production using industrial waste. Int J Biol Macromol 108:41–46. https://doi.org/10.1016/j.ijbiomac.2017.11.114

Vimal A, Kumar A (2017) Biotechnological production and practical application of L-asparaginase enzyme. Biotechnol Genet Eng Rev 33(1):40–61. https://doi.org/10.1080/02648725.2017.1357294

Xu F, Oruna-Concha MJ, Elmore JS (2016) The use of asparaginase to reduce acrylamide levels in cooked food. Food Chem 210:163–171

Zuo S, Xue D, Zhang T, Jiang B, Mu W (2014) Biochemical characterization of an extremely thermostable L-asparaginase from *Thermococcus gammatolerans* EJ3. J Mol Catal B Enzym 109:122–129. https://doi.org/10.1016/j.molcatb.2014.08.021

Marine Environment: A Treasure Trove of Natural Polymers for Tissue Engineering

9

Prajakta P. Bhende, Aesha Sharma, Anasuya Ganguly, and Judith M. Bragança

Abstract

The human body is subjected to wear and tear. The cells proliferate and replace dead cells and are sloughed off or damaged due to trauma. The body can "repair" the damage wholly or partly. The repair or healing of the tissues is done either by regenerating cells of the same type or by replacing the damaged organ or part. The cells of certain tissues have a better proliferative and self-regenerative ability than others. However, there is a need for replacement in case of severe damage or trauma to the permanent tissues. The ever-increasing transplant waiting list versus the unavailability of organs for transplantation has facilitated exploring alternative methods like tissue engineering and regenerative medicine. The tissue regeneration technology is based on three factors: scaffolds, cells, and growth. A scaffold is a template on which the appropriate cells and growth factors are added. To be used as a scaffold, the material should be porous and nontoxic, possess appropriate mechanical strength, and be biocompatible, i.e., to help the cells adhere, grow, migrate, and differentiate. In other words, it should mimic the extracellular matrix in vitro. The polymers used as scaffolds can be broadly classified into two types based on the nature of the material, i.e., proteinaceous or polysaccharide-based. Frequently used polymers for tissue engineering are collagen and gelatin. Collagen is the most abundant animal protein and is isolated from animal tissues. It is a matter of concern as there is a possible risk of transmission of viral and prion diseases. The use of recombinant technology is not feasible economically. Therefore, researchers are exploring the marine ecosystem for the supply of natural polymers. Collagen and gelatin are now being

P. P. Bhende · A. Sharma · A. Ganguly · J. M. Bragança (✉)
Department of Biological Sciences, Birla Institute of Technology and Science Pilani, K K Birla Goa Campus, Sancoale, Goa, India
e-mail: judith@goa.bits-pilani.ac.in

P. Veera Bramhachari, C. V. Berde (eds.), *Marine Bioactive Molecules for Biomedical and Pharmacotherapeutic Applications*,
https://doi.org/10.1007/978-981-99-6770-4_9

extracted from fish waste, i.e., bones, skin, scales, and fins. Other potential sources of collagen are jellyfish (*Aurelia* spp.) and sea stars (*Acanthaster* spp.). The invasive species can be targeted in such cases, lessening their adverse environmental impact. Polysaccharide-based polymers include agarose, mostly obtained from red algae (*Gracilaria* spp.), and alginate from macroalgae (*Saccharina* spp.). These collagen and gelatin exhibit excellent gelling properties and are used in fabricating hydrogel scaffolds. Chitin and chitosan are present in the exoskeleton of marine arthropods and mollusks. Chitosan has excellent biocompatible properties and antibacterial activity and thus finds applications in wound-healing dressings and artificial skin grafts. The marine ecosystem can be called "the womb" for life began and evolved from the oceans, and it continues to sustain life both in water and on land.

Keywords

Tissue engineering · Regenerative medicine · Scaffolds · Polymers · Collagen · Gelatin · Agarose · Alginate · Chitosan

9.1 Introduction

The emerging field of tissue engineering and regenerative medicine aims to develop substitutes to restore or regenerate damaged tissues or organs. It is an interdisciplinary field that combines engineering and cell science principles to fabricate bioactive scaffolds, cell therapy, smart drug delivery systems, and wound-healing mats. Tissue engineering comprises three main components (Fig. 9.1) (1): the scaffold material, which serves as a template; (2) the cells, which will grow and differentiate to form a tissue; and (3) the growth factors, which are required for the healthy growth of cells (Kumar et al. 2018). Nontherapeutic applications include the use of tissues as biosensors to detect chemical or biological threats, as well as the development of organs-on-a-chip for toxicity screening of experimental medications (Oprea and Voicu 2020).

A scaffold is a framework on which the cells adhere, grow, and differentiate. An ideal scaffold should be able to mimic the extracellular matrix (ECM) of the tissue. Apart from this, the scaffold should have a mechanically stable architecture to support the developing tissue. It should possess good cytocompatibility to allow the cells to grow in in vitro and in vivo conditions. It should be able to interact with the cells and growth factors. The scaffold should be porous to provide a larger surface area for the cells to grow and differentiate. It should also be biodegradable with nontoxic degradation products (Fig. 9.2) (Zarrintaj et al. 2018; Chan and Leong 2008).

Scaffolds can be fabricated from either natural materials or synthetic materials. Natural materials have good biocompatibility but often lack mechanical strength. The synthetic ones have tuneable properties but do not degrade with time. A

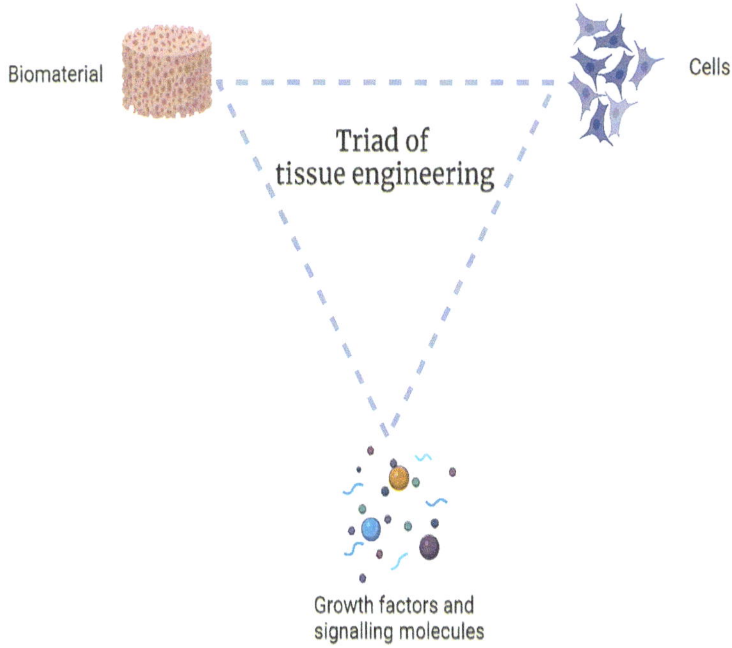

Fig. 9.1 Triad of tissue engineering. (Created with Biorender.com)

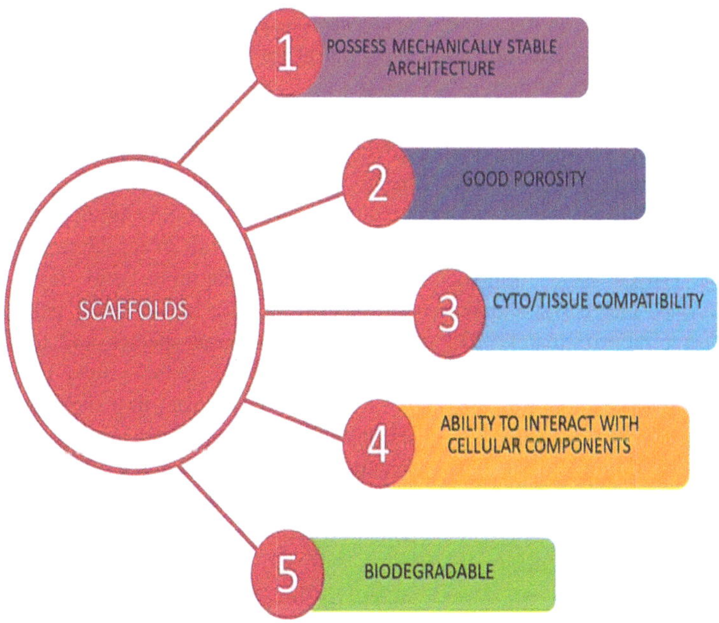

Fig. 9.2 Characteristics of an ideal scaffold

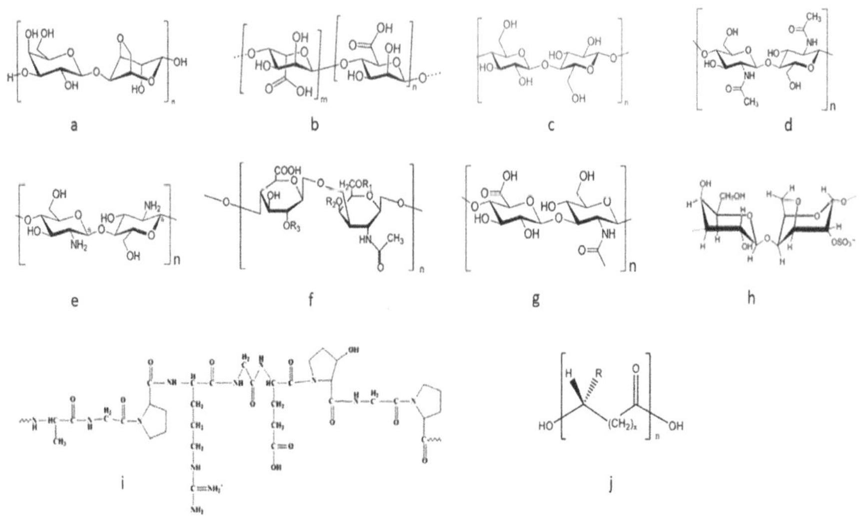

Fig. 9.3 Structural formula of (**a**) agarose, (**b**) alginate, (**c**) cellulose, (**d**) chitin, (**e**) chitosan, (**f**) chondroitin sulfate, (**g**) hyaluronic acid, (**h**) α-carrageenan, (**i**) gelatin, (**j**) polyhydroxyalkanoate. (Source: Wikipedia)

composite scaffold of natural and synthetic materials is a good alternative, with each material complementing the other. A scaffold can be a 2D or a 3D structure, with the latter having an advantage. A 3D system provides an increased surface area of interaction and closely duplicates the natural ECM microenvironment. Natural scaffolds, like fibrin, amniotic membrane, etc., are a viable option because they already contain a diverse range of signaling molecules that play an important role in cellular morphogenesis and function development. However, the composition varies among the species and is difficult to reproduce (Shin et al. 2016).

The marine environment has tremendous potential and is an untapped source of biomaterials. The sources of marine biomaterials include macroalgae, planktons, sponges, crustaceans, mollusks, gastropods, cartilaginous and bony fishes, echinoderms, etc. Polymers can be found in marine waste from fisheries and aquaculture. Approximately 132 million tons of marine species and aquaculture (mostly fish, mollusks, and crustaceans) constitute the global catch. Out of this, more than 35% of the total weight is a by-product or waste, like skeletons, heads, viscera, etc., created in seafood production or rejects such as species, sizes, or quality having no commercial value. Typically, the production of such wastes is concentrated around coastlines, posing the problem of environmental degradation caused by inadequate residue management. There is a need to harness this waste to extract polymers (Vázquez et al. 2013).

This chapter focuses on these materials' sources, extraction, and application in tissue engineering. The polymers (Fig. 9.3) are broadly divided into two groups:

(1) polysaccharide-based and (2) protein-based polymers. This chapter also discusses polyesters and silicates as potential biomaterials.

9.2 Marine-Derived Polysaccharide-Based Polymers

9.2.1 Agarose

Agarose polysaccharide is a transparent, linear, neutrally charged, thermoreversible polysaccharide. It consists of a disaccharide β-D-galactose and 3,6-anhydro-α-L-galactopyranose units in repetition. It accounts for the major portion of agar. The main source of agar is macroalgae of class Rhodophyceae (commonly identified as red seaweeds), mainly belonging to the genera *Gelidium* and *Gracilaria*. *Gelidium* produces high-quality agar but is very difficult to culture. Commercially, agar is produced from *Gracilaria* (Fig. 9.4) (Trivedi and Kumar 2014), *Gelidium*, *Ceramium*, *Akantkopeltis*, and *Pterocladia* spp. (Zucca et al. 2016). Agarose is widely used in molecular biology and biomedical research because it is inexpensive and nontoxic and has good water absorption abilities, excellent gelling characteristics, high porosity (average pore size of 100–300 nm) to facilitate the transport of molecules through the gel, and controllable stiffness for tuning the mechanical properties of the scaffold. Porosity and water absorbance limit are

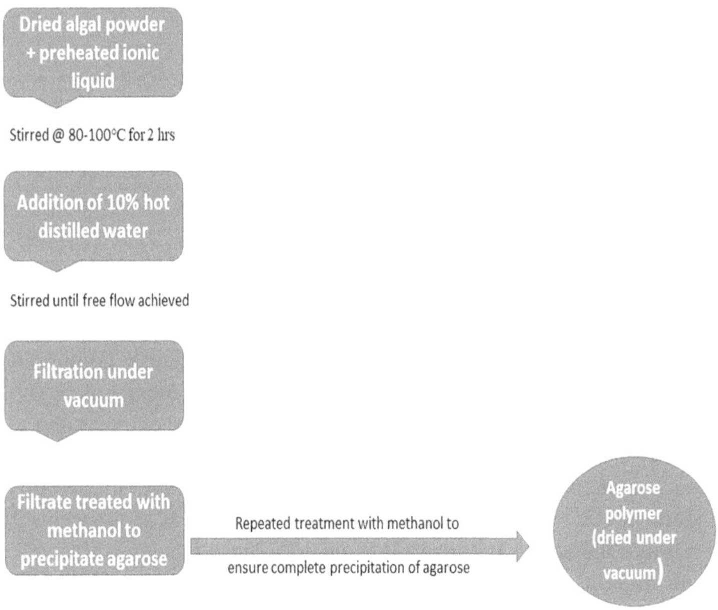

Fig. 9.4 Extraction of agarose from dried red seaweeds (*Gracilaria dura*) (Trivedi and Kumar 2014)

necessary for cell adhesion, growth, and multiplication as they promote the diffusion of oxygen and growth supplements in the scaffold. Agarose also exhibits pH and electro-responsive properties and thermal gelling properties. The presence of ionizable groups, such as polyelectrolytes, maintains the pH. In in vitro 3D culture systems, agarose hydrogels have been shown to enhance cell growth and retain cellular phenotypes (Salati et al. 2020; Shin et al. 2016).

Agarose has been extensively used as a scaffold in tissue engineering and regenerative medicine. In an experimental study by Vivcharenko et al. (2020), a biocomposite of chitosan/agarose incorporated with vitamin C was synthesized for skin cell growth and regeneration. Vitamin C is a strong antioxidant and causes fibroblast proliferation, migration, and collagen deposition. The scaffold was tested on normal human epidermal keratinocytes transformed with HPV-16 (HEK001), normal human skin fibroblasts (BJ), and human acute monocytic leukemia cells (THP-1). This study has a potential application as a superabsorbent external wound bandage.

Agarose composites are also used in enamel regeneration. Agarose is blended with other polymers like collagen and chitosan to enhance its mechanical properties. Its application is restricted at the cellular level to repair enamel loss. It provides a microenvironment for the regeneration of dentine that gets coated by ionic ($CaCl_2$) and nonionic agarose hydrogel (Zarrintaj et al. 2018).

Kumar et al. (2018) synthesized a biocomposite of agarose chitosan-coated silver nanoparticles for engineering soft tissue like the pancreas. A long-term study on the biocompatibility of the scaffold was evaluated on HeLa cells. Since the scaffold has a chitosan-coated nano-silver component, it also showed good antibacterial properties.

Apart from the various advantages, there are certain limitations too. When incubated with cells for a long time, the highly porous agarose structure cannot maintain its mechanical strength. Hence, it is challenging to use it in animal models. Therefore, instead of using agarose directly, it is combined with other polymers to overcome this drawback (Singh et al. 2016).

9.2.2 Alginate

Alginate is a linear, water-soluble polysaccharide that is mainly composed of α-L-guluronic acid (G) and β-D-mannuronic acid (M) copolymers linked by a 1→4 glycosidic linkage. The monomers are arranged in blocks of the same polymers (homopolymeric—GG or MM) or different polymers (heteropolymeric—GM or MG). The structure of the polymer is dependent on these monomeric units, e.g., a GG arrangement will give a buckled ribbonlike structure; MM sequences will lead to a flat ribbonlike appearance; and a GM and MG arrangement will give rise to a helical conformation (Peteiro 2018). In the presence of metal cations like Na^+, Ca^{2+} alginate undergoes cross-linking reaction, stabilizing the emulsions and forming hydrogels. At high pH (>10), alginates are stable but at low pH exhibit extremely low solubility.

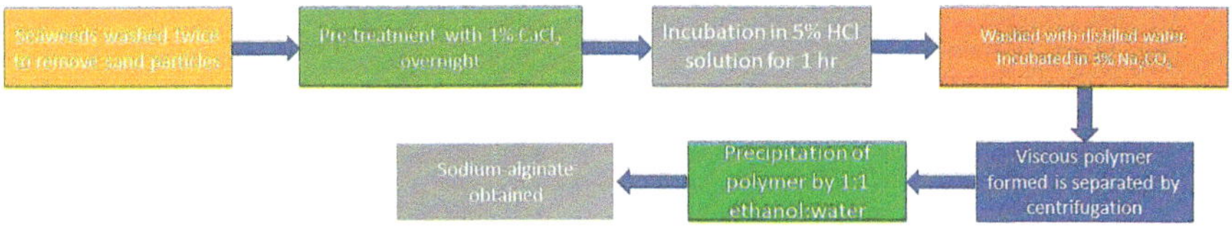

Fig. 9.5 Extraction of alginates from seaweeds (Hasnain et al. 2020)

Alginate is found in the marine brown algae belonging to the class Phaeophyceae. It is the main structural polymer in seaweeds and macroalgae, located between the cell walls and intracellular matrix, imparting mechanical stability. It constitutes almost 45–50% of its dry weight. The marine algae involved in large-scale extraction of alginates are *Laminaria* spp., *Sargassum*, *Ascophyllum nodosum*, *Ecklonia*, *Saccharina japonica*, and *Macrocystis pyrifera* (Fig. 9.5) (Venkatesan and Kim 2015; Hasnain et al. 2020).

Alginates are biocompatible and non-immunogenic hydrogels with various applications in tissue engineering and regenerative medicine. They have been used as composite scaffolds, wound dressings, and encapsulating materials for drugs and nanoparticles. Diaz-Rodriguez et al. (2018) fabricated a composite scaffold of alginate and $CaCO_3$ derived from mussels and oyster shells for bone tissue engineering. The cytocompatibility of this scaffold was assessed on the human osteoblast cell line, Saos-2. All the alginate scaffold samples showed good cell viability. However, alginate doped with $CaCO_3$ from the shells of oysters and mussels showed better biocompatibility than the ones with just alginate (Diaz-Rodriguez et al. 2018). Alginates have tremendous potential in 3D printing as bioinks or cell inks. Govindharaj et al. (2019) devised a 3D printable scaffold from alginates and collagen extracted from the discarded skins of eels. Evaluation of biocompatibility was performed on human umbilical cord-derived cells. The hydrogels showed more cell viability and attachment than the hydrogels sans collagen. Venkatesan et al. (2014) developed a composite of alginate and chitosan with fucoidan by liquid displacement method. The scaffold formed had a high porosity (>90%). Degradation of the alginate-chitosan-fucoidan scaffold was also found to be higher than the alginate-chitosan scaffold after 72 h. The in vitro cytotoxicity was evaluated on the human osteoblast-like cells, MG-63. The results showed a cell viability of 70–80% even after 48 h.

9.2.3 Cellulose and Ulvans

The most abundant biopolymer on earth, cellulose, has terrestrial, marine, and microbial (bacterial cellulose) origin. It provides the biomechanical ability; this is why 90% of the plant cell wall is made up of cellulose and helps in the vertical development of plants when with lignin and hemicellulose. As a biomaterial, cellulose provides structural support for forming wound dressings, bone tissue, tendons, blood vessels, urinary tracts, and larynx tissues. Cell adhesion and

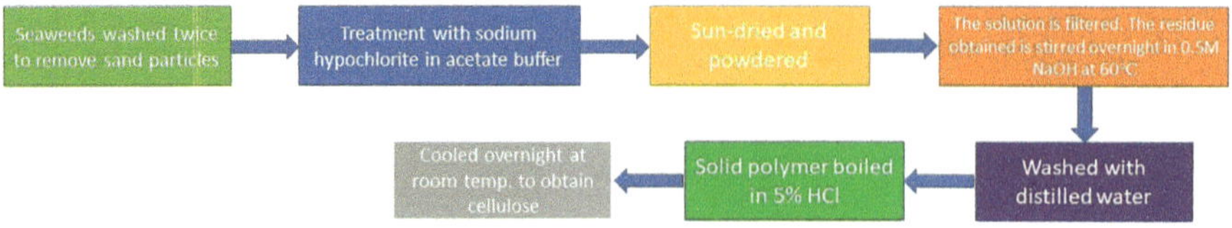

Fig. 9.6 Extraction of cellulose from seaweeds (Bar-Shai et al. 2021)

proliferation can be boosted only when cellulose hydrophilicity and fluid absorption create a moist environment (Bar-Shai et al. 2021). The marine sources of cellulose are mainly seaweeds or macroalgae from the genus *Ulva* (*Ulva lactuca*, *Ulva rigida*, *Ulva pertusa*) (Fig. 9.6).

The structural component of the macroalgal cells is made up of a water-soluble polymer, ulvan, and water-insoluble cellulose. The cellulose from macroalgae has higher crystallinity and is mechanically, chemically, and biologically stable. Cellulose is a high molecular weight biopolymer that is made of repeating units of β-D-glucopyranosyl connected by a 1–4 glycosidic linkage. At the same time, ulvan comprises glucuronic acid, iduronic acid, and sulfated L-rhamnose, glucose, and xylose (Cesário et al. 2018). Exploring the applications of biopolymers obtained from seaweeds is a fairly recent field of study. Bar-Shai et al. (2021) extracted cellulose from *Ulva* sp. and *Cladophora* sp. The whole-tissue cellulose scaffold was used to culture mouse embryonic fibroblast NIH-3T3, which expresses GFP-actin. This fluorescent tag helped determine cell viability, cell attachment, and morphology with the help of confocal microscopy.

Madub et al. (2021) reported cellulose extraction from the genus *Ulva* in their study. Electrospun fibrous mesh was fabricated from the algal cellulose and also composites with polydioxanone (PDX), poly-L-lactide (PLLA), or poly-DL-lactide (PDLLA). The in vitro biocompatibility was evaluated on L929 fibroblasts by MTT assay. The acetylated derivative of the extracted cellulose showed lower viability than the composites. The in vivo biocompatibility of 100% ulvan-cellulose scaffolds was assessed on Wistar rats over 4 weeks. At the end of 4 weeks, the tissue surrounding the scaffolds was harvested and analyzed. A dense fibrous capsule formation was observed with minimal vascularization. This material finds application in diabetic wound healing.

Bettaieb et al. (2015) extracted cellulose from the *Posidonia oceanica* biomass, a type of seagrass endemic to the Mediterranean region. The extracted cellulose was converted to cellulose nanocrystals (CNC). TEM analyzed the synthesized nanocrystals and found that the shape was rodlike, elongated, and well dispersed. The CNC from *P. oceanica* leaves an average diameter of 6.8 and 7 nm and length of 520 and 338 nm, while that synthesized from *P. oceanica* balls is much smaller, around an average diameter of 8.1 and 8 nm and length of 290 and 276 nm. CNCs can be used as a prospective biomaterial.

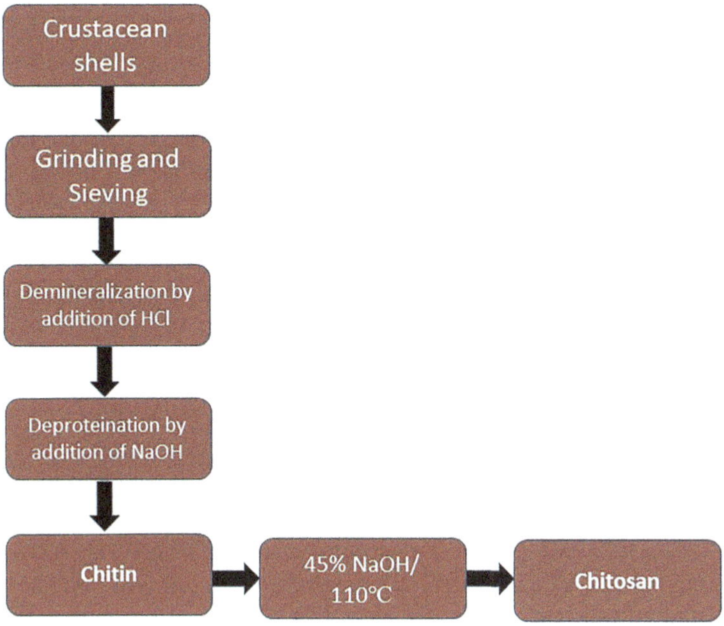

Fig. 9.7 Extraction of chitin and chitosan from crustacean shells (Vázquez et al. 2013)

9.2.4 Chitin and Chitosan

The second most abundant naturally occurring polysaccharide, chitin, is a polymer of D-glucosamine and *N*-acetyl-D-glucosamine units linked together by a β-1→4 glycosidic linkage. It is a structural polymer that makes up the exoskeleton of phylum Arthropoda (insects, crustaceans). It is also found in algae, zooplankton, marine and freshwater sponges, mollusks, and fungi. Commercially, chitin is extracted from the exoskeletons of shrimps, crabs, lobsters, and bivalves (Fig. 9.7). Chitin exists in nature in three different forms: α-chitin, found in crustaceans; β-chitin, from the cephalopods like squids; and γ-chitin, which is isolated from fungi. Chitin is a compact and water-insoluble polymer that cleaves the *N*-acetyl group at the C2 position on enzymatic or acidic deacetylation, giving rise to a more soluble polymer derivative, chitosan (Silva et al. 2012). The extraction process of chitin and chitosan from crustaceans, as described by Sagheer et al. (2009), is as follows:

(a) The shells were separated from the tissue, washed, dried, ground, and sieved through a 250 μm sieve.
(b) Demineralization using dilute HCl to remove the inorganic $CaCO_3$ debris.
(c) Deproteination treatment with NaOH to remove the bound proteins.
(d) Boiling treatment in hot ethanol and acetone to remove any impurities.

(e) The chitin obtained after the boiling treatment is further heated with 45% NaOH to yield chitosan.

Chitosan is biodegradable and biocompatible and also has an antibacterial capability. Therefore, it is an attractive polymer in tissue engineering used to regenerate organs like the liver, kidney, pancreas, etc. (Kumar et al. 2018). Chitosan also can enhance cell adhesion. It also forms hydrogen bonds with other polymers due to polar groups in its structure (–OH and $-NH_2$) (Saeedi Garakani et al. 2020).

In a study by Mutsenko et al. (2017), chitin was extracted from marine demosponge *Aplysina aerophoba*, and a 3D scaffold was prepared for culturing human mesenchymal stromal cells (hMSCs). After 7 days of incubation, the authors reported a twofold increase in cell metabolic activity and proliferation. The cells differentiated into the chondrogenic, adipogenic, and osteogenic lineages, respectively, when provided with the differentiation media.

Machałowski et al. (2022) used a 3D chitinous matrix from a demosponge, *Aplysina fistularis*. A biocomposite was fabricated using an ex vivo deposition of $CaCO_3$ using the hemolymph from a cultivated mollusk, *Cornu aspersum*. The physical characterization showed a fourfold increase in mechanical resistance and surface roughness, making it a promising candidate for bone tissue engineering. In another study by the same author, the scaffold was tested on murine fibroblasts (Balb/3T3), human dermal fibroblasts (NHDF), human keratinocyte (HaCaT), and human neuronal (SH-SY5T) cells. The results showed increased wound healing and also antibacterial properties (Machałowski et al. 2021).

Extraction of chitin and chitosan from the shells of blue crabs (*Callinectes sapidus*) by Demir et al. (2016) is another example of how biopolymers can be obtained from discarded marine waste. The chitosan obtained was then used for the synthesis of cryogel scaffolds.

9.2.5 Chondroitin Sulfate

Chondroitin sulfate is a sulfated mucopolysaccharide belonging to the glycosaminoglycan group. It consists of repeating units of *N*-acetylgalactosamine and glucuronic acid alternatingly. The sulfate groups are located on either the fourth or sixth position of the *N*-acetylgalactosamine unit. In chondroitin sulfate, there is a diversity in the chain length and the number and position of sulfate groups present. Chondroitin sulfate is classified into five subgroups as follows:

(a) CS-A (chondroitin-4-sulfate)—tracheal cartilage
(b) CS-B (chondroitin-2,4 sulfate/dermatan sulfate)
(c) CS-C (chondroitin-6-sulfate)—shark cartilage
(d) CS-D (chondroitin-2,6-sulfate)—shark cartilage
(e) CS-E (chondroitin-4,6-sulfate)

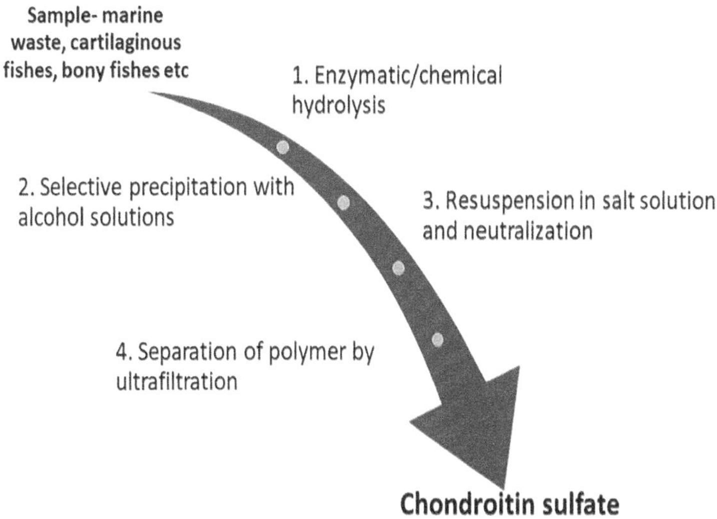

Fig. 9.8 Extraction of chondroitin sulfate from marine waste (Vázquez et al. 2013)

The marine sources of chondroitin sulfate are sea cucumbers, octopuses, arms, skin, head, and cornea of squids; the fish waste, especially fins, a head portion, and skeleton; cartilages and fins of small-spotted catshark (*Scyliorhinus canicula*), blue shark (*Prionace glauca*), and shortfin mako shark (*Isurus oxyrinchus*); the skin, scales, and skeleton of carp, tuna, and salmon; and the cartilages of rays and skates (Abdallah et al. 2020). Extraction of chondroitin sulfate from marine organisms is shown in Fig. 9.8. The general procedure that is followed includes the following steps:

1. Hydrolysis of the cartilage by chemical agents
2. Proteoglycan core disruption
3. Removal of proteinaceous debris and recovery of chondroitin sulfate
4. Purification by ultrafiltration, dialysis, and ion exchange chromatography

Chondroitin sulfate has excellent biocompatibility. It is one of the components of the extracellular matrix (ECM) of the connective tissues like cartilage, blood vessels, tendons, neuronal bodies, the brain, etc. (Bishnoi et al. 2016). It improves the elasticity of the articular cartilage and facilitates cell adhesion, growth, and differentiation of chondrocytes. López-Senra et al. (2020), in an attempt to provide a regenerative treatment to osteoarthritis, evaluated the effect of chondroitin sulfate isolated from the blue shark (*Prionace glauca*), small-spotted catshark (*Scyliorhinus canicula*), and thornback skate (*Raja clavata*) on the proliferation of osteochondral cell lines (MG-63 and T/C-28a2) and the chondrogenic differentiation of mesenchymal stromal cells (MSCs) and compared it to the chondroitin sulfate extracted from bovine sources. It was noted that after 72 h of incubation, the cytotoxicity was negligible for chondroitin sulfate concentrations below 0.5 mg/ml, and therefore, it was assigned as the cytotoxicity threshold value. T/C-28a2 proliferation was significantly improved by intermediate ratios of chondroitin sulfate from *S. canicula* and *R. clavata*. Since chondroitin sulfate is a polyanionic polymer, a composite scaffold with metallic ions that influence physiological activities (like calcium, magnesium, etc.) is beneficial in bone tissue engineering. Chondroitin sulfate is also known to produce a temporary matrix facilitating cell growth and interaction between cells and ECM, thereby influencing wound healing. Chondroitin sulfate is nontoxic. But care should be taken when used on patients belonging to vulnerable categories like:

1. Pregnancy/breastfeeding
2. Hypersensitive to seafood
3. Patients on blood thinners/anticoagulants (because chondroitin sulfate can act antagonistically to the blood thinners and lead to the formation of clots) (Sharma et al. 2022; Bishnoi et al. 2016)

9.2.6 Hyaluronic Acid

Hyaluronic acid, like chondroitin sulfate, is also a glycosaminoglycan linked by β-(1→3) and β-(1→4) glycosidic bonds but is not sulfated. It is abundantly present in the hyaline cartilage, synovial fluid, vitreous humor in the eyes, umbilical cord, etc. The current commercialization of hyaluronic acid was carried out using the abovementioned mammalian substrates of pig or cattle origin and also from rooster combs. Marine alternatives such as the eyes of discarded fishes/wastes (Fig. 9.9) and cartilaginous skeletons of fishes belonging to class Chondrichthyes (sharks, rays, and skates) are being explored. Hyaluronic acid was extracted from the eyes of swordfishes (*Xiphias gladius*) and shark species (*Prionace* sp.) by Murado et al. (2010) (Fig. 9.9). The vitreous humor was drained from the frozen eyeballs and homogenized. The debris was separated from the supernatant by centrifugation and purified by ultrafiltration.

The obtained retentates were salted using NaCl, and the polymer and some proteinaceous debris were precipitated using ethanol under vigorous agitation. The residue was then redissolved in a hydroalcoholic solution and centrifuged. The hyaluronic acid polymer was obtained in the supernatant, while the proteinaceous pellet was discarded. Hyaluronic acid is commonly used as a hydrogel for osteochondral tissue engineering. In 3D bioprinting, hyaluronic acid is used as an additive in bio-ink to improve cell attachment and viability (Zhang et al. 2019). Correia et al. (2011) fabricated a blend of hyaluronic acid with chitosan to study the viability and proliferation of chondrocytes. It was noted that an increase in the concentration of hyaluronic acid provided a better adhesion, proliferation, and formation of a matrix than pure chitosan scaffolds. Wang et al. (2006) constructed a bilayered hyaluronic acid scaffold with chondroitin sulfate and gelatin to mimic the skin microenvironment. The bottom-most layer was seeded with dermal fibroblasts, a feeder layer over which the keratinocyte layer was inoculated. After 21 days, the

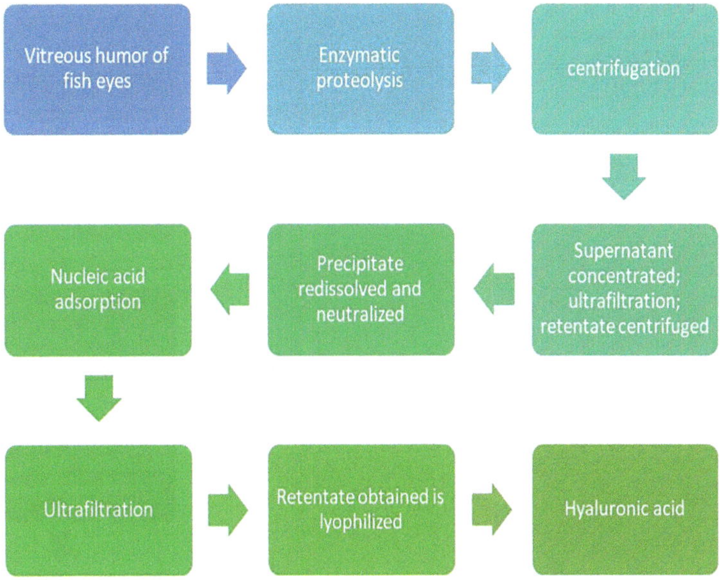

Fig. 9.9 Extraction and purification of hyaluronic acid from fish eyeballs (Vázquez et al. 2013)

keratinocyte layer formed the epidermis, while the fibroblasts gave rise to the dermis-like structure.

9.2.7 Carrageenan

Carrageenan is a naturally occurring sulfated polysaccharide that is a prospective contender in tissue engineering and regenerative medicine due to its similarity to native glycosaminoglycans. Carrageenan has many uses, from pharmaceutical medication formulations to tissue engineering scaffolds (Yegappan et al. 2018).

The marine red algae family is a source of underutilized carrageenan with functional capabilities and a renewable polymer source (Fig. 9.10) (Gereniu et al. 2018). Carrageenan (CRG) is different from naturally occurring glycosaminoglycans; it is an anionic polygalacton that has long linear chains of -1,3 D-galactose and -1, 4 3, 6-anhydro-galactose (3, 6-AG) with ester sulfates (15–40%). Kappa, Iota, Lambda, Mu, Nu, and Theta CRG are the six basic forms of seaweed-derived polysaccharides. As discussed by Sharma et al. (2013), a 3D super-macroporous scaffold was prepared in which different ratios of carrageenan/gelatin were used to make a solution and cross-linked using glutaraldehyde and 1-ethyl-3-(3-dimethyl aminopropyl) carbodiimide (EDC), to form cryogel matrices. A great improvement was reported in the properties of cell adhesion, proliferation, and secretion of ECM when tested on the Cos-7 fibroblast cell line.

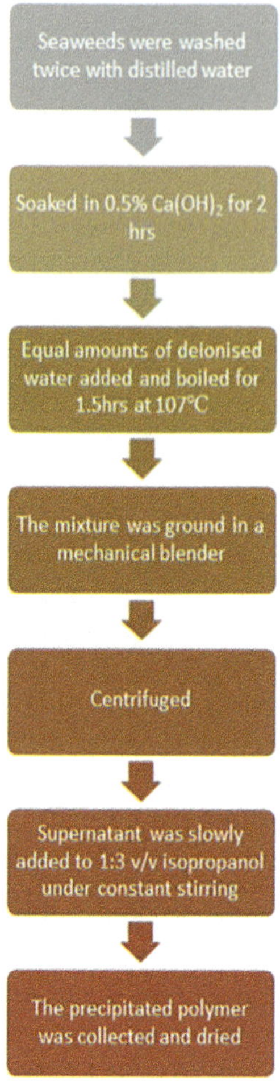

Fig. 9.10 Extraction of carrageenan from seaweeds by Collin Rudolf Nobbs Gereniu et al. (2018)

Polysaccharides derived from marine ulvans have great potential to be used as a scaffold for osteoinductivity. Ulvan blended with alpha carrageenan and polysaccharide was used for cell adhesion and viability of mesenchymal stem cells by Kikionis et al. (2021). Good mechanical strength and viscoelastic behavior were observed in the matrices.

9.3 Marine-Derived Protein-Based Polymers

9.3.1 Collagen

Collagen is the dominant component of the extracellular matrix and also makes up the cartilage, tendons, ligaments, etc. To date, 29 types of collagens have been identified and named Type I, Type II, Type III, and so on. Type I collagen is abundant in the body, almost 80–85%, and occurs mainly in tendons, ligaments, periosteum, cornea, and dermis. Type II collagen makes up 90% of the hyaline cartilages; Type III collagen is distributed in the parenchyma of the lungs, liver, spleen, and uterus and in the adventitia of blood vessels and also occurs in the dermis in association with the Type I (Silva et al. 2014). Collagen is a fibrous protein with a molecular weight of around 300 kDa. The collagen structure is organized into triple-helical domains with repeating glycine-X-Y residues, with X and Y often proline and 4-hydroxyproline, respectively. Collagen satisfies all the criteria required to be a scaffold material. It is highly biocompatible, is weakly immunogenic, decreases the chances of immune rejection, and is biodegradable (Felician et al. 2018). Apart from this, collagen has a good surface and water absorption properties. The extraction of collagen in the past was isolated from bovine and porcine sources. However, this had some major drawbacks:

(a) Possible chances of viral and prion contamination, e.g., foot and mouth disease virus and bovine spongiform encephalopathy.
(b) Substandard purification methods led to contamination with other immunogenic substances leading to inflammatory responses.
(c) Religious and social constraints arising from using collagen from cattle and pig sources (Coppola et al. 2020).

Alternative sources of collagen are marine organisms such as invertebrates, chordates, and vertebrates, including marine mammals. Marine collagen is highly abundant in the by-product or waste arising from aquaculture and fishery industries, such as the fish scales, skin, fins, and skeleton, and also in the accidental catches like the inedible echinoderms and jellyfish (Fig. 9.11) (Hema et al. 2013). Marine collagens belong to the following types: Type I accounts for 85% of the collagen found in the body of marine organisms, e.g., jellyfish, sea urchins, squids, and bony fishes like cod, salmon, sardines, etc. Type II collagen is found in jellyfish and cartilages of sharks; sponges contain collagen type IV. Type V-like collagen is found in marine crustaceans and mollusks. Sponges (phylum Porifera) have a jellylike connective tissue composed of sulfated glucans, collagen, etc. (Langasco et al. 2017; Zierer and Mourão 2000).

Langasco et al. (2017) investigated the potential of this connective tissue for wound-healing applications. The drug was loaded onto the powdered sponge, and this was later injected into a copolymer giving rise to a drug-loaded polymer film. Liu et al. (2010) isolated collagen from giant red sea cucumbers (*Parastichopus californicus*). Collagen was extracted from the sea cucumbers' skin using chemicals

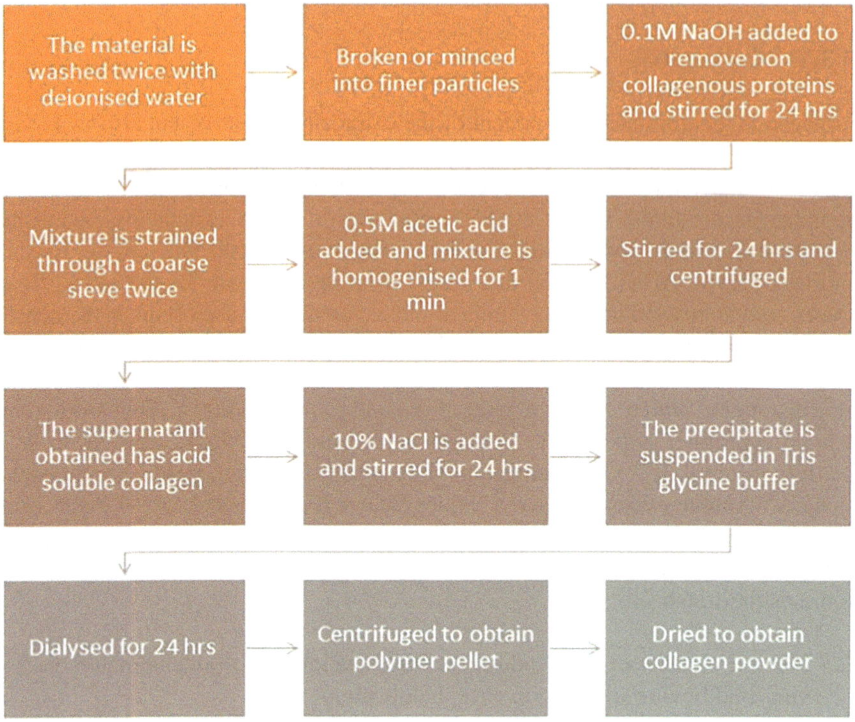

Fig. 9.11 Extraction of collagen from fish skin by Hema et al. (2013)

like EDTA, 0.1 M Tris-HCl, and digestion by pepsin enzyme. A higher yield of 20–25% per dry weight was obtained by pepsin solubilization and served as a promising candidate to replace the use of mammalian collagen. Ferrario et al. (2017) studied the potential of echinoderm (sea stars, sea cucumbers, and sea urchins) collagen as a potential "blue alternative" to mammalian ones. The in vitro studies were performed on primary human dermal fibroblasts. The cells showed an elongated morphology compared to the control. The echinoderm-derived collagen membranes also exhibited a higher resistance and mechanical stability. Thus, this biomaterial finds applications especially in engineering mechanically demanding tissues such as cartilage, bone, tendon, etc. Matsumoto et al. (2015) evaluated the effects of fibril formation of tilapia fish scale collagen membrane on osteogenic differentiation of human mesenchymal stem cells (hMSCs). It was found that the tilapia collagen upregulated the early osteogenic markers, thereby accelerating the differentiation into osteogenic lineage. Thus, marine collagen-based membranes are attractive biomaterials for osteological tissue engineering.

9.3.2 Gelatin

Gelatin does not occur in nature on its own. The physical, chemical, and thermal denaturation of collagen produces it. Collagen-gelatin together makes up the main frame of the extracellular matrix (ECM). The drawback of collagen, i.e., its poor water solubility, is one of the main advantages of gelatin as it has a reversible sol-gel transition. It provides a 3D bioactive microenvironment for the cells and helps in growth, cell viability, and proliferation. Gelatin is also used to encapsulate therapeutic agents, proteins, nanoparticles, etc., due to its ability to interact with charged particles, thereby creating polyionic complexes (Echave et al. 2019). Like collagen, extraction of gelatin from cattle and pigs may lead to zoonotic transmission of diseases. Therefore, marine-origin gelatin is a safer alternative and eco-friendly, as most of it is derived from the by-product of the fishery industry. Denaturation of mammalian collagen to form gelatin requires a higher temperature than that of marine-derived collagen (Carvalho et al. 2021). Marine organisms used in gelatin extraction are squids (*Loligo*), jellyfish (*Cassiopea*), tilapia fish (*Oreochromis*), giant grouper (*Epinephelus lanceolatus*), cod (*Gadus morhua*), and African catfish (*Clarias gariepinus*). Alves et al. (2022) tested the gelatin material from shark skin (*Prionace glauca*) on C57BL mice and checked the immunogenicity on bone marrow-derived macrophage cells in vitro and also evaluated the in vivo biocompatibility by injecting the material intraperitoneally in the mice. No acute inflammation was found even after 24 h of injection, and therefore, the material did not provoke an aggravated immunological response. Extraction of gelatin from squids (*Loligo*) can lead to contamination with the inherent pigments. This problem was studied by Nagarajan et al. (2013), and it was resolved by bleaching the squid skin with 2% H_2O_2. The treated films had a lower water vapor permeability but showed increased extensibility.

In a Gaspar-Pintiliescu et al. (2019) study, gelatin from a marine snail *Rapana venosa* was investigated as a potential biomaterial. The soft tissue of the snail was subjected to chemical hydrolysis using NaOH and acetic acid. The in vitro biocompatibility was evaluated on human keratinocytes (HaCaT cells). The cell viability was found to be >80% after 48 h.

Gelatin, like collagen, can be obtained copiously from fish refuse/waste and jellyfish (Fig. 9.12) (Cho et al. 2014), lowering the extraction cost. The weak mechanical strength of gelatin can sometimes be a drawback if the tissue demands more elasticity. In such cases, adding composite materials or cross-linking agents can improve the elastic nature of the polymer.

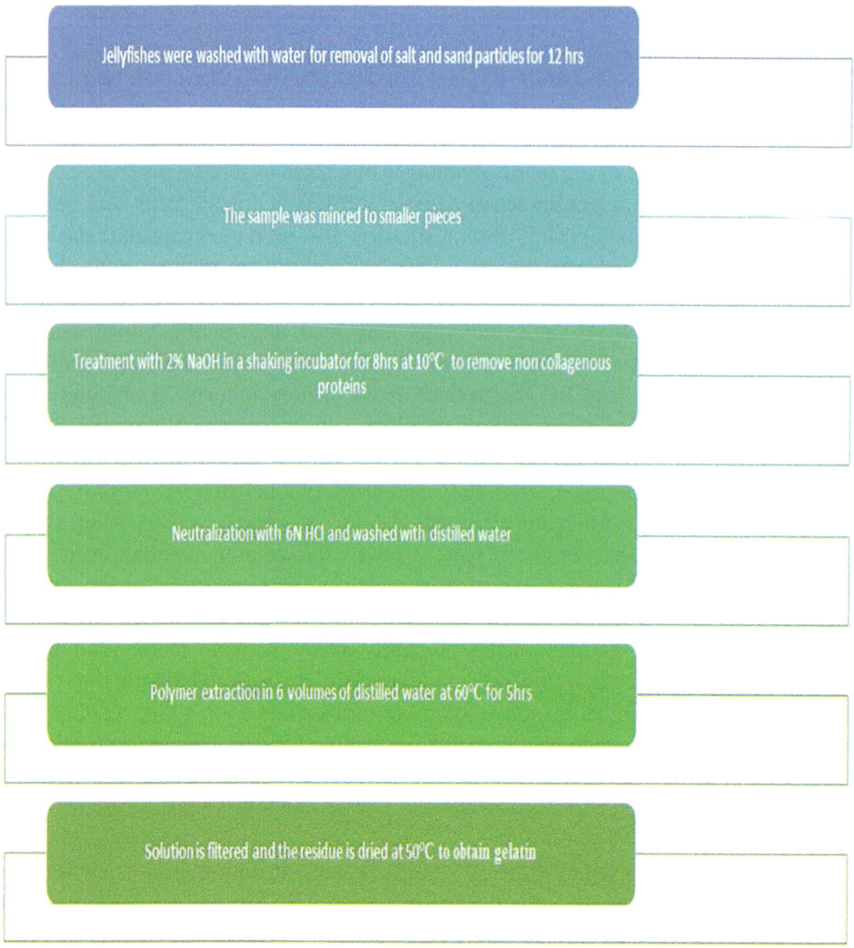

Fig. 9.12 Extraction of gelatin from jellyfish by Suengmok Cho, Ju-Ryun Ahn et al. (2014)

9.4 Other Marine-Derived Biopolymers

9.4.1 Polyhydroxyalkanoates

Polyhydroxyalkanoates (PHAs) are a class of polyester compounds synthesized intracellularly by various organisms in response to stress conditions like the limitation of nutrients such as nitrogen, phosphorus, oxygen, and sulfur but excess of carbon. These carbonosomes are synthesized in the late log phase of growth by microorganisms, stored as lipid inclusion bodies, and depolymerized on demand. PHAs are biodegradable compounds and are broken down into CO_2 and water

(Getachew and Woldesenbet 2016; Mohapatra et al. 2017). These microorganisms occur in diverse habitats. Still, in this case, the emphasis will be laid on the marine and associated habitats like coastal waters, sediments, mangroves, estuarine, and salterns. The structure of PHA (Fig. 9.2j) consists of hydroxy acid monomers connected by an ester bond. The R represents the alkyl side chain; the x denotes the number of repeating $–CH_2$ units, usually 1–4, while n stands for the number of times that the entire monomeric unit is repeated—100 and above.

Based on the number of carbon atoms and the length of the chain, the PHAs are classified into:

1. Short-chain length (SCL)—3–5 carbon atoms
2. Medium-chain length (MCL)—6–14 carbon atoms

The carbon source that is provided to the microorganisms determines whether the polymer will be a homopolymer, i.e., poly3-hydroxybutyric acid (P3HB) and poly4-hydroxybutyric acid (P4HB), or a copolymer, e.g., poly 3-hydroxybutyrate-co-3-hydroxy valerate (P(3HB-co-3HV)), poly3-hydroxybutyrate-co-4-hydroxybutyrate (P(3HB-co-4HB), poly(3hydroxyhexanoate-co-3hydroxy octanoate) (P(3HHx-co-3HO), etc. (Albuquerque and Malafaia 2018; Miu et al. 2022; Możejko-Ciesielska and Kiewisz 2016).

PHAs are synthesized by both Gram-positive and Gram-negative bacteria. The latter includes organisms like *Cupriavidus necator*, *Pseudomonas* spp., *Aeromonas* spp., and *Vibrio* spp. The extensively studied halophilic bacteria belong to the *Halomonas* genus. A study showed that *Halomonas* spp. could utilize various carbon sources like glucose, sucrose, xylose, sodium acetate, etc. and synthesize PHAs up to 81% of CDW (Angra et al. 2022; Ganapathy et al. 2018). The halophilic archaea which produce PHAs include *Haloferax* spp., *Halogeometricum* spp., *Halococcus* spp., *Haloarcula* spp., *Haloterrigena* spp., etc. PHA content varies from 16 to 80% of CDW. *Halogeometricum borinquense* is reported to synthesize a copolymer poly3-hydroxybutyrate-co-3-hydroxy valerate (Salgaonkar and Bragança 2017).

The properties of PHAs can be modified according to the application. PHAs are nontoxic, biocompatible, noncarcinogenic, and biodegradable. The PHB is brittle and has a higher crystallinity. This can be altered by introducing certain polymers to enhance the overall properties of the materials. Peschel et al. (2008) blended P3HB and P4HB with hyaluronic acid and chitosan. The porous films were developed as scaffolds for the replacement of damaged skin. The scaffolds were tested on the adherent HaCaT cell line. It was reported that chitosan- and hyaluronic acid-blended PHAs promoted the growth and adherence of keratinocytes. Shishatskaya et al. (2016) designed a nanofibrous mesh of P(3HB/4HB) as a wound dressing mesh in Wistar rats. These meshes promoted wound healing and skin regeneration 3.5 times faster than control, achieving complete wound healing by day 14.

PHA polymers are reported to show good biocompatibility after prolonged incubation in blood and hence find applications as vascular patches. A study by Antonova et al. (2016) showed the accelerated formation of functional monolayer endothelial cells and improved graft compression strength when VEGF and RGD peptides were incorporated into PHBV/PCL composite. Mohanna et al. (2005) used PHB conduits incorporated with glial growth factor or alginate matrix to bridge a 2–4 cm of the peroneal nerve in rabbits. The PHB-GGF conduit was reported to cause an increase in nerve regeneration for up to 63 days. The quantity of Schwann cells and axonal regeneration was also found to increase more than the control conduits. PHA polymers have been used for cartilage and bone tissue engineering by blending with hydroxyapatite (Guo et al. 2022).

PHA is an interesting biomaterial whose property can be altered according to the requirements. The cost of production and extraction of PHAs can be reduced by using cheaper substrates like spent oil and agrowastes and/or by genetically modifying the organisms, thereby increasing the overall yield.

9.4.2 Biosilica or Bioglass

Biosilica or bioglass is an inorganic material composed of silicon dioxide, disodium oxide, calcium oxide, and phosphorus pentoxide. These are amorphous ceramics that possess the ability to interact with living tissue depending on its composition. This is due to a hydroxy carbonate apatite layer forming on its surface, resembling a bone's surface. It is specifically suited for hard tissue engineering like bone grafts, dental implants, etc. Biosilicates can be either osteoproductive, where the hydroxy carbonate apatite layer is formed quickly, and binding occurs between both soft tissue and bone, or it can be osteoconductive, where the hydroxy carbonate apatite layer formation is delayed, and binding between the tissues is not facilitated (Al-Harbi et al. 2021).

Gabbai-Armelin et al. (2019) isolated biosilica from marine sponges—*Tedania ignis* and *Dragmacidon reticulatum*. The organic matter was removed using a mixture of nitric acid/sulfuric acid in a ratio of 1:4. The in vitro cytotoxicity was evaluated on MC3T3-E1 cells. No significant cytotoxicity was seen when compared to control. The viability of cells on the biosilica scaffold showed an increase on day 6 compared to the control.

Tamburaci and Tihminlioglu (2018) investigated the biocompatible properties of biosilica/chitosan composite. The source of the biosilica in this study was diatomite. The cytotoxicity studies were performed on cell lines of MG 63 (human osteosarcoma) and Saos-2 (human osteosarcoma). Scanning electron microscopy revealed good cell attachment and proliferation. The in vitro cytotoxic analyses reported no significant cytotoxicity when the diatomite concentration was between 5 and 20%. It was noted that the diatomite frustules altered the surface topography and showed improved cell attachment. Biosilica also finds applications in prosthodontics as it can be used for enamel remineralization, dental wound healing, dental pulp caps, etc. (Al-Harbi et al. 2021).

9.5 Conclusion

The chapter highlights how the marine ecosystem is a source of natural polymers used for tissue engineering. However, this has to be approached in a very sustainable manner. The world is no stranger to the anthropological damages caused due to overexploitation of resources. An ideal approach would be a collaborative effort with the fisheries industries to turn its "waste into wealth." Aquaculture techniques include growing and breeding favorable organisms like sponges, jellyfish, fishes, etc. in artificial ponds. Instead of using nearly threatened organisms like sharks, the focus should be on utilizing fish waste and jellyfishes. Jellyfishes are commonly caught in the fishermen's net and are discarded on the beach. This will also tackle the problem of jellyfish blooms. Strict quality control is required when harnessing marine-derived polymers as biomaterials. The presence of immunogenic substances, endotoxins, and pyrogenic substances can render it useless for the said purpose. If the above criteria are fulfilled, the field of tissue engineering and regenerative medicine will get a necessary boost from marine-derived polymers.

Acknowledgments Prajakta P Bhende would like to acknowledge BITS Pilani KK Birla Goa Campus for the Institute Fellowship. Anasuya Ganguly and Judith M. Braganca thank Department of Biotechnology, Ministry of Science and Technology, Government of India, for funding under the DBT Builder—BITS Pilani, K K Birla Goa Campus Interdisciplinary Life Science Programme for Advance Research and Education. Level III (BT/INF/22/SP2543/2021).

The authors would also like to acknowledge Biorender.com for the graphical abstract and Fig. 9.1.

Declaration of Conflict of Interest The authors declare no conflict of interest.

References

Abdallah MM, Fernández N, Matias AA, do Rosário Bronze M (2020) Hyaluronic acid and chondroitin sulfate from marine and terrestrial sources: extraction and purification methods. Carbohydr Polym 243:116441

Albuquerque PBS, Malafaia CB (2018) Perspectives on the production, structural characteristics and potential applications of bioplastics derived from polyhydroxyalkanoates. Int J Biol Macromol 107(Part A):615–625. https://doi.org/10.1016/j.ijbiomac.2017.09.026

Al-Harbi N et al (2021) Silica-based bioactive glasses and their applications in hard tissue regeneration: a review. Pharmaceuticals 14(2):1–20. /pmc/articles/PMC7909272/ (24 June 2022)

Alves AL et al (2022) Study of the immunologic response of marine-derived collagen and gelatin extracts for tissue engineering applications. Acta Biomater 141:123–131

Angra V, Sehgal R, Gupta R (2022) Trends in PHA production by microbially diverse and functionally distinct communities. Microb Ecol 85:572. https://doi.org/10.1007/s00248-022-01995-w

Antonova LV et al (2016) Conjugation with RGD peptides and incorporation of vascular endothelial growth factor are equally efficient for biofunctionalization of tissue-engineered vascular grafts. Int J Mol Sci 17(11):1920. https://www.mdpi.com/1422-0067/17/11/1920/htm. Accessed 25 Aug 2022

Bar-Shai N et al (2021) Seaweed cellulose scaffolds derived from green macroalgae for tissue engineering. Sci Rep 11(1):1–17. https://www.nature.com/articles/s41598-021-90903-2. Accesses 18 July 2022

Bettaieb F et al (2015) Preparation and characterization of new cellulose nanocrystals from marine biomass Posidonia oceanica. Ind Crop Prod 72:175–182

Bishnoi M, Jain A, Hurkat P, Jain SK (2016) Chondroitin sulphate: a focus on osteoarthritis. Glycoconjugate J 33(5):693–705. https://link.springer.com/article/10.1007/s10719-016-9665-3. Accessed 7 July 2022

Carvalho DN, Reis RL, Silva TH (2021) Marine origin materials on biomaterials and advanced therapies to cartilage tissue engineering and regenerative medicine. Biomater Sci 9(20): 6718–6736. https://pubs.rsc.org/en/content/articlehtml/2021/bm/d1bm00809a. Accessed 23 June 2022

Cesário M, Teresa MM, da Fonseca R, Marques MM, de Almeida MCMD (2018) Marine algal carbohydrates as carbon sources for the production of biochemicals and biomaterials. Biotechnol Adv 36(3):798–817

Chan BP, Leong KW (2008) Scaffolding in tissue engineering: general approaches and tissue-specific considerations. Eur Spine J 17(suppl 4):467. /pmc/articles/PMC2587658/ (19 July 2022)

Cho S, Ahn JR, Koo JS, Kim SB (2014) Physicochemical properties of Gelatin from Jellyfish *Rhopilema hispidum*. Published online 2014. https://doi.org/10.5657/FAS.2014.0299

Coppola D et al (2020) Marine collagen from alternative and sustainable sources: extraction, processing and applications. Mar Drugs 18(4):214

Correia CR et al (2011) Chitosan scaffolds containing hyaluronic acid for cartilage tissue engineering. Tissue Eng Part C Methods 17(7):717–730. https://pubmed.ncbi.nlm.nih.gov/21517692/. Accessed 9 July 2022

Demir D, Öfkeli F, Ceylan S, Bölgen N (2016) Extraction and characterization of chitin and chitosan from blue crab and synthesis of chitosan cryogel scaffolds. J Turk Chem Soc Sect A Chem 3(3):131

Diaz-Rodriguez P et al (2018) Mineralized alginate hydrogels using marine carbonates for bone tissue engineering applications. Carbohydr Polym 195:235–242

Echave MC et al (2019) Recent advances in gelatin-based therapeutics. Expert Opin Biol Ther 19(8):773–779

Felician FF, Xia C, Qi W, Hanmei X (2018) Collagen from marine biological sources and medical applications. Chem Biodivers 15(5):e1700557

Ferrario C et al (2017) Marine-derived collagen biomaterials from echinoderm connective tissues. Mar Environ Res 128:46–57

Gabbai-Armelin PR et al (2019) Characterization and cytotoxicity evaluation of a marine sponge biosilica. Mar Biotechnol 21(1):65–75. https://link.springer.com/article/10.1007/s10126-018-9858-9. Accessed 24 June 2022

Ganapathy K, Ramasamy R, Dhinakarasamy I (2018) Polyhydroxybutyrate production from marine source and its application. Int J Biol Macromol 111:102–108. https://doi.org/10.1016/j.ijbiomac.2017.12.155

Gaspar-Pintiliescu A et al (2019) Physicochemical and biological properties of gelatin extracted from marine snail Rapana venosa. Mar Drugs 17(10):589. https://www.mdpi.com/1660-3397/17/10/589/htm. Accessed 23 June 2022

Gereniu CRN, Saravana PS, Chun BS (2018) Recovery of carrageenan from Solomon Islands red seaweed using ionic liquid-assisted subcritical water extraction. Sep Purif Technol 196:309–317

Getachew A, Woldesenbet F (2016) Production of biodegradable plastic by polyhydroxybutyrate (PHB) accumulating bacteria using low cost agricultural waste material. BMC Res Notes 9(1): 1–9

Govindharaj M, Roopavath UK, Rath SN (2019) Valorization of discarded marine eel fish skin for collagen extraction as a 3D printable blue biomaterial for tissue engineering. J Clean Prod 230: 412–419

Guo W et al (2022) Polyhydroxyalkanoates in tissue repair and regeneration. Eng Regen 3(1): 24–40. https://doi.org/10.1016/j.engreg.2022.01.003

Hasnain MS, Jameel E, Mohanta B, Dhara AK, Alkahtani S, Nayak AK (2020) Chapter 1 - Alginates: sources, structure, and properties. In: Nayak AK, Hasnain MS (eds) Alginates in drug delivery. Academic Press, pp 1–17, ISBN 9780128176405. https://doi.org/10.1016/B978-0-12-817640-5.00001-7

Hema GS et al (2013) A simple method for isolation of fish skin collagen-biochemical characterization of skin collagen extracted from Albacore Tuna (Thunnus Alalunga), Dog Shark (Scoliodon Sorrakowah), and Rohu (Labeo Rohita). Ann Biol Res 4(1):271–278. www.scholarsresearchlibrary.com. Accessed 22 July 2022

Kikionis S et al (2021) The marine polysaccharide ulvan confers potent Osteoinductive capacity to PCL-based scaffolds for bone tissue engineering applications. Int J Mol Sci 22(6):1–20

Kumar N et al (2018) Biocompatible agarose-chitosan coated silver nanoparticle composite for soft tissue engineering applications. Artif Cells Nanomed Biotechnol 46(3):637–649

Langasco R et al (2017) Natural collagenic skeleton of marine sponges in pharmaceutics: innovative biomaterial for topical drug delivery. Mater Sci Eng C 70:710–720. https://doi.org/10.1016/j.msec.2016.09.041

Liu Z, Oliveira ACM, Yi Cheng S (2010) Purification and characterization of pepsin-solubilized collagen from skin and connective tissue of Giant red sea cucumber (Parastichopus Californicus). J Agric Food Chem 58(2):1270–1274

López-Senra E et al (2020) Impact of prevalence ratios of chondroitin sulfate (CS)- 4 and -6 isomers derived from marine sources in cell proliferation and chondrogenic differentiation processes. Mar Drugs 18(2):94. https://www.mdpi.com/1660-3397/18/2/94/htm. Accessed 8 July 2022

Machałowski T et al (2021) Naturally formed chitinous skeleton isolated from the marine demosponge Aplysina fistularis as a 3d scaffold for tissue engineering. Materials 14(11):2992

Machałowski T et al (2022) Naturally prefabricated 3D chitinous skeletal scaffold of marine demosponge origin, biomineralized ex vivo as a functional biomaterial. Carbohydr Polym 275:118750

Madub K et al (2021) Green seaweeds ulvan-cellulose scaffolds enhance in vitro cell growth and in vivo angiogenesis for skin tissue engineering. Carbohydr Polym 251:117025

Matsumoto R et al (2015) Rapid oriented fibril formation of fish scale collagen facilitates early osteoblastic differentiation of human mesenchymal stem cells. J Biomed Mater Res Part A 103(8):2531–2539

Miu DM, Eremia MC, Moscovici M (2022) Polyhydroxyalkanoates (PHAs) as biomaterials in tissue engineering: production, isolation, characterization. Materials 15(4):1410

Mohanna PN, Terenghi G, Wiberg M (2005) Composite PHB-GGF conduit for long nerve gap repair: a long-term evaluation. Scand J Plast Reconstr Surg Hand Surg 39(3):129–137

Mohapatra S et al (2017) Bacillus and biopolymer: prospects and challenges. Biochem Biophy Rep 12:206–213

Możejko-Ciesielska J, Kiewisz R (2016) Bacterial polyhydroxyalkanoates: still fabulous? Microbiol Res 192(2016):271–282

Murado MA et al (2010) Preparation of highly purified chondroitin sulphate from skate (Raja Clavata) cartilage by-products.process optimization including a new procedure of alkaline hydroalcoholic hydrolysis. Biochem Eng J 49(1):126–132

Mutsenko VV et al (2017) 3D chitinous scaffolds derived from cultivated marine demosponge Aplysina aerophoba for tissue engineering approaches based on human mesenchymal stromal cells. Int J Biol Macromol 104:1966–1974

Nagarajan M et al (2013) Film forming ability of gelatins from splendid squid (Loligo formosana) skin bleached with hydrogen peroxide. Food Chem 138(2–3):1101–1108

Oprea M, Voicu SI (2020) Cellulose composites with graphene for tissue engineering applications. Materials 13(23):1–24

Peschel G et al (2008) Growth of keratinocytes on porous films of poly(3-hydroxybutyrate) and poly(4-hydroxybutyrate) blended with hyaluronic acid and chitosan. J Biomed Mater Res Part A 85(4):1072–1081
Peteiro C (2018) Alginate production from marine macroalgae, with emphasis on kelp farming. Springer Ser Biomater Sci Eng 11:27–66
Saeedi Garakani S et al (2020) Fabrication of chitosan/agarose scaffolds containing extracellular matrix for tissue engineering applications. Int J Biol Macromol 143:533–545
Sagheer FA, Al MA, Al-Sughayer SM, Elsabee MZ (2009) Extraction and characterization of chitin and chitosan from marine sources in Arabian Gulf. Carbohydr Polym 77(2):410–419
Salati MA et al (2020) Agarose-based biomaterials: opportunities and challenges in cartilage tissue engineering. Polymers 12(5):1150
Salgaonkar BB, Bragança JM (2017) Utilization of sugarcane bagasse by Halogeometricum borinquense strain E3 for biosynthesis of poly(3-hydroxybutyrate-co-3-hydroxyvalerate). Bioengineering 4(2):50
Sharma A et al (2013) Three-dimensional supermacroporous carrageenan-gelatin cryogel matrix for tissue engineering applications. Biomed Res Int 2013:478279
Sharma R et al (2022) Chondroitin sulfate: emerging biomaterial for biopharmaceutical purpose and tissue engineering. Carbohydr Polym 286:119305
Shin S et al (2016) Alginate-marine collagen-agarose composite hydrogels as matrices for biomimetic 3D cell spheroid formation. RSC Adv 6(52):46952–46965
Shishatskaya EI, Nikolaeva ED, Vinogradova ON, Volova TG (2016) Experimental wound dressings of degradable PHA for skin defect repair. J Mater Sci Mater Med 27(11):165. https://doi.org/10.1007/s10856-016-5776-4
Silva TH, Duarte ARC, Moreira-Silva J, Mano JF, Reis RL (2012) Biomaterials from marine-origin biopolymers. In: Mano JF (ed) Biomimetic approaches for biomaterials development. https://doi.org/10.1002/9783527652273.ch1
Silva TH et al (2014) Marine origin collagens and its potential applications. Mar Drugs 12(12): 5881–5901
Singh YP, Bhardwaj N, Mandal BB (2016) Potential of agarose/silk fibroin blended hydrogel for in vitro cartilage tissue engineering. ACS Appl Mater Interfaces 8(33):21236–21249
Tamburaci S, Tihminlioglu F (2018) Biosilica incorporated 3D porous scaffolds for bone tissue engineering applications. Mater Sci Eng C 91:274–291
Trivedi TJ, Kumar A (2014) Efficient extraction of agarose from red algae using ionic liquids. Green Sustain Chem 4(4):190–201
Vázquez JA et al (2013) Chondroitin sulfate, hyaluronic acid and chitin/chitosan production using marine waste sources: characteristics, applications and eco-friendly processes: a review. Mar Drugs 11(3):747–774
Venkatesan J, Kim SK (2015) Marine biomaterials. In: Kim SK (ed) Springer handbook of marine biotechnology. Springer handbooks. Springer, Berlin. https://doi.org/10.1007/978-3-642-53971-8_53
Venkatesan J, Bhatnagar I, Kim SK (2014) Chitosan-alginate biocomposite containing fucoidan for bone tissue engineering. Mar Drugs 12(1):300–316. https://www.mdpi.com/1660-3397/12/1/300/htm. Accessed 13 July 2022
Vivcharenko V, Wojcik M, Przekora A (2020) Cellular response to vitamin C-enriched chitosan/agarose film with potential application as artificial skin substitute for chronic wound treatment. Cell 9(5):1185
Wang TW et al (2006) Biomimetic bilayered gelatin-chondroitin 6 sulfate-hyaluronic acid biopolymer as a scaffold for skin equivalent tissue engineering. Artif Organs 30(3):141–149. https://onlinelibrary.wiley.com/doi/full/10.1111/j.1525-1594.2006.00200.x. Accessed 9 July 2022
Yegappan R, Selvaprithiviraj V, Amirthalingam S, Jayakumar R (2018) Carrageenan based hydrogels for drug delivery, tissue engineering and wound healing. Carbohydr Polym 198: 385–400

Zarrintaj P et al (2018) Agarose-based biomaterials for tissue engineering. Carbohydr Polym 187: 66–84
Zhang Y et al (2019) Biomaterials based on marine resources for 3D bioprinting applications. Mar Drugs 17(10):555. https://www.mdpi.com/1660-3397/17/10/555/htm. Accessed 9 July 2022
Zierer MS, Mourão PAS (2000) A wide diversity of sulfated polysaccharides are synthesized by different species of marine sponges. Carbohydr Res 328(2):209–216
Zucca P, Fernandez-Lafuente R, Sanjust E (2016) Agarose and its derivatives as supports for enzyme immobilization. Molecules 21(11):1577. /pmc/articles/PMC6273708/ (10 July 2022)

Exploration of Bioactive Functional Molecules from Marine Algae: Challenges and Applications in Nutraceuticals

10

Thummala Chandrasekhar, Khateef Riazunnisa, Degati Vijaya Lakshmi, Vankara Anu Prasanna, and Pallaval Veera Bramhachari

Abstract

Marine organisms provide a rich source of chemically diverse compounds which have the potential to develop as novel therapeutic agents. Specifically, marine algae, a big group of organisms, were targeted to figure their primary and secondary metabolites such as polysaccharides, peptides, fatty acids, polyunsaturated fatty acids, minerals, vitamins, and pigments including phycobilins, carotenoids, carotene, xanthophylls, chlorophylls, phenolics, etc., which have a broad spectrum of natural bioactivities. These bioactive compounds of marine algae possess anti-obesity, anti-inflammatory, antioxidant, antibacterial, antifungal, antiviral, anticancer, and other functional potentials. Hence, we emphasized using these marine algal metabolites as nutraceuticals and human food. In conclusion, this chapter deals with the potentialities of marine algal metabolites for future nutraceutical applications.

T. Chandrasekhar (✉)
Department of Environmental Science, Yogi Vemana University, Kadapa, Andhra Pradesh, India
e-mail: tcs@yogivemanauniversity.ac.in

K. Riazunnisa
Department of Biotechnology and Bioinformatics, Yogi Vemana University, Kadapa, Andhra Pradesh, India

D. Vijaya Lakshmi
Department of Microbiology, Yogi Vemana University, Kadapa, Andhra Pradesh, India

V. Anu Prasanna
Department of Zoology, Yogi Vemana University, Kadapa, Andhra Pradesh, India

P. Veera Bramhachari
Department of Biotechnology, Krishna University, Machilipatnam, Andhra Pradesh, India

P. Veera Bramhachari, C. V. Berde (eds.), *Marine Bioactive Molecules for Biomedical and Pharmacotherapeutic Applications*,
https://doi.org/10.1007/978-981-99-6770-4_10

Keywords

Marine algae · Nutraceuticals · Bioactive compounds · Diseases · Health benefits

10.1 Introduction

Bioactive molecules with diverse chemical structures are useful as food, fodder, cosmetics, nutraceuticals, pharmaceuticals, and medicines. These molecules are involved in physical functions as well as several biochemical pathways. Moreover, the number of emerging diseases in present society urges us to focus on new bioactive compounds (Suleria et al. 2015). Industrial demand is also one of the reasons for searching for novel bioactive compounds from different sources to cure various diseases. Searching for bioactive compounds from various plant and microbial sources is a continuous process (Šimat et al. 2020). Among all the sources, both freshwater and marine algae are one of the best nutraceutical resources. Specifically, these algal species possess polysaccharides, peptides, certain lipids/fatty acids, pigments, vitamins, and particular interesting molecules such as polyphenolic constituents, alkaloids, terpenoids, tannins, tocopherols, tocotrienols, etc. (Menaa et al. 2021). Also, certain marine algal species are used as functional foods and dietary supplements.

Algae are considered primitive plants or photosynthetic organisms, which can be further divided into microalage (unicellular) and macroalgae (multicellular) forms. Algal species are habituated in terrestrial and aquatic systems (Paramesh et al. 2018). Further, these aquatic algae are classified into freshwater and marine species. Marine and freshwater algal species are available in bulk numbers among aquatic species. Depending on the pigment accumulation, algae are further divided into green, blue-green, brown and red groups (Croft et al. 2006). Algae are a crucial staple for food, fodder, medicine, fertilizer, and energy industries. Some known algae contain significant amounts of polysaccharides, peptides, lipids, vitamins, pigments, minerals, and other secondary metabolites (Hentati et al. 2020). Both green and blue-green (cyanobacteria) algae are popular species for generating various products due to their potential photosynthetic pathway. In algal species, the nutrient content may differ depending on season, location, temperature, etc. But certain countries neglect algae, which become often orphan species (Varaprasad et al. 2021).

Unicellular microalgae mainly possess simple structures with efficient photosynthetic capacity. Hence, chlorophyll accumulation in algal species is an important key issue for further growth and development. Moreover, it is very important to focus on biomass, which decides the yield (Raga Sudha et al. 2021). The biochemical methods involved in biomass conversion into bioproducts are well established using different algal species in the recent past. The biggest advantage of algae is its ability to grow in non-potable water sources (Paramesh and Chandrasekhar 2020). Hence, they are being explored as a potential high-yield feedstock for bioactive compound production. Some earlier studies focused on secondary metabolite production using different algal species.

The marine ecosystem is still an unexploited reservoir of biologically active compounds, which have considerable potential to supply food ingredients for developing new functional foods (Ande et al. 2017). Particularly, marine algae have been used as food or dietary supplement in human consumption. Algal biomass was incorporated into various food items such as gums, bars, candy, snacks, pickles, and beverages (Juneja et al. 2013). The content of algal biomass is a crucial factor for generating bioactive compounds. The qualitative and quantitative accumulation of compounds varies from species to species.

Moreover, the continuous population explosion, increased soil pollution, and reduction of arable land urge us to focus on alternate food supplement sources such as marine algae (Saadaoui et al. 2020). The bioactive compounds of marine algae possess a wide range of health benefits, including antibacterial, antiviral, antioxidant, and anticancer properties (Pal et al. 2014). Also, these molecules are involved in treating obesity problems, cardiac diseases, arthritis, high blood pressure, avoiding cell damage, etc. (Gullón et al. 2020). Therefore, certain marine algal species have been considered functional foods recently. Several algal species, including *Spirulina*, *Chlorella*, *Gracilaria*, *Dunaliella*, *Tetraselmis*, etc., are used for nutraceutical purposes (Plaza et al. 2010; Menaa et al. 2021). Moreover, it is easy to handle the algal culture, which is considered eco-friendly and cost-effective. By observing several health benefits, this chapter intends to summarize the list of marine algal species and their nutraceutical potentials. Figure 10.1 illustrates the bioactive compounds isolated from algal samples.

10.2 Marine Algae-Derived Carbohydrates and Polysaccharides

Marine algae are rich in functional bioactive compounds such as carbohydrates, including specific polysaccharides. Table 10.1 includes the various marine algae used to isolate different bioactive compounds. The accumulation of carbohydrates depends on the species, age, season, location, and salinity levels (Hentati et al. 2020). Polysaccharides are sugar polymers used as functional foods and cosmetics and are involved in drug development (De Jesus et al. 2015). These bioactive polysaccharides possess both sulfated and non-sulfated groups (Venugopal 2019). Sulfated compounds including agar-agar, carrageenans, galactans, and the non-sulfated group contain alginates and laminarins. Majorly, brown algae possess laminarin, alginate, and fucoidan, while ulvan and carrageenan are found in green and red algae, respectively (Pierre et al. 2019). These high-molar polymers have varied structures, including homo-polysaccharides (monosaccharides) and hetero-polysaccharides (two or more monosaccharides). These bioactive molecules have anticoagulant, antitumor, antidiabetic, antioxidant, antiallergic, antibacterial, antiviral, immunomodulatory, and neuroprotective properties (Šimat et al. 2020). Specifically, these marine polysaccharides have antioxidant properties against reactive oxygen species (ROS).

A water-soluble polysaccharide called ulvan (xylorhamnoglycuronans) is found in various green marine algal species. Majorly, it contains sulfate, xylose, rhamnose,

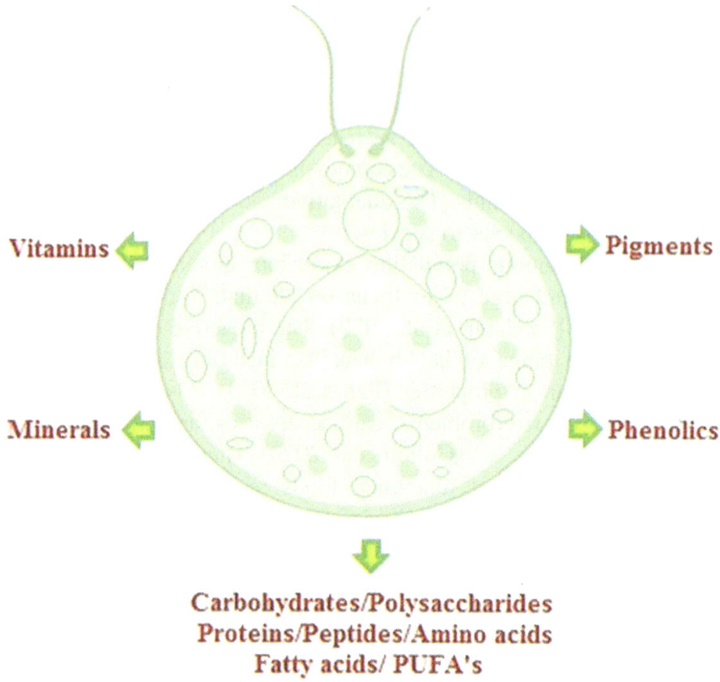

Fig. 10.1 Bioactive compounds present in marine algal species

Table 10.1 Marine algal species used for isolation of bioactive compounds

Bioactive Compounds	Green algae	Red algae	Brown algae	Blue Green Algae
Carbohydrates/ Polysaccharides, Proteins/Peptides/ Amino acids, Fatty acids/PUFAs, Pigments, Minerals, Phenolic compounds and Vitamins	*Ulva* sps *Caulerpa* sps *Chlorella* sps *Dunaliella* sps *Chlorococcum* *Tetraselmis* sps	*Eucheuma* sps *Porphyra* sps *Gracilaria* sps *Gelidium* sps *Gelidiella* sps	*Laminaria* sps *Sargassum* sps *Ascophyllum* sps *Undaria* sps *Colpomenia* sps *Padina* sps	*Spirulina* sps *Microcystis* sps *Leptolynbgya* sps

glucuronic acid, etc. (Lahaye and Robic 2007). Ulvan polysaccharides are generally found in *Ulva* or *Enteromorpha* species which have antiviral, anticancer, antioxidant, and antihyperlipidemic properties (Kidgell et al. 2019). Specifically, it involves several molecular signaling methods and reduces cholesterol levels. The polysaccharides of *Chlorella* are involved in immune stimulation and free radical reduction (Matos et al. 2017). Carrageenan is a sulfated polysaccharide that contains α-1,3-galactose and β-1,4,3,6-anhydrogalactose with ester sulfates. These are the structural components of red algal cell membranes (Prajapati et al. 2014). Generally, *Eucheuma* species were used for carrageenan production. Due to its gelling properties, it has been used in drug delivery, wound healing, and regeneration of bone and cartilage tissues. It also has anticancer, antiviral, antithrombotic, and immunomodulatory functions (Campo et al. 2009). Fucoidan is another important bioactive polysaccharide isolated from brown algae (Zhao et al. 2016). This compound contains sulfate groups attached to fucose units. It also possesses galactose, mannose, xylose, and glucose. It is highly used as a food item and is generally called a nutraceutical. It is one of the best remedies for various cancers, including breast, colon, and bladder cancers (Fitton et al. 2019).

Laminarin is another important brown algal polysaccharide with low molecular weight which is found in various species of *Laminaria*, *Sargassum*, *Ascophyllum*, *Fucus*, *Cystoseira*, etc. (Stiger-Pouvreau et al. 2016). Laminarin comprises (1,3)-β-D-glucan and some β-(1,6)-intrachain links. These bioactive compounds are involved in anticancer activities (Ji and Ji 2014). Laminarin of some of the species is involved in wound healing and drug delivery processes. Alginate is another brown algal cell wall polysaccharide that contains (1-4)-linked β-D-mannuronic acid (M) and α-L-guluronic acid (G) monomers for different structures (Hentati et al. 2020). These molecules are used in drug delivery and wound dressings. Alginates are used in food protection, dental impression material preparation, and preventing gastric reflux (Šimat et al. 2020). Alginates are also used as gelling, coating, thickening, emulsifying, and stabilizing agents in food products.

10.3 Marine Algae Is a Source of Essential Fatty Acids

Marine algae are considered a source of essential fatty acids, including polyunsaturated fatty acids (PUFAs). Green algae consist of alpha-linolenic acid (ALA), and blue-green algae possess gamma-linolenic acid (GLA). Also, other polyunsaturated fatty acids such as eicosapentaenoic acid (EPA), docosahexaenoic acid (DHA), octadecatetraenoic acid (SDA), and arachidonic acid (AA) are available in marine algal species (Borowitzka 2013). These PUFAs are useful in treating skin diseases, rheumatism, and inflammation of the mucosa of the gastrointestinal tract (Vidyashankar et al. 2015). Certain marine algal species, such as *Chlorella*, *Ulva*, *Gracilaria*, *Laminaria*, *Dunaliella*, *Tetraselmis*, *Nannochloropsis*, etc., contain various fatty acid compositions depending on the age, location, season, and salinity levels (da Vaz et al. 2016).

ALA is one of the essential polyunsaturated omega-3 fatty acids and is a precursor for producing EPA and DHA (Ahlgren et al. 1992). Alpha-linolenic acid is useful in fighting against inflammation, blood pressure, and cardiovascular diseases. Both AA and DHA contribute major polyunsaturated fatty acids of brain cell phospholipids, and the earlier one is involved in infant development. DHA is essential to treat eye, brain, and heart-related diseases. EPA is useful in treating heart and inflammatory diseases, arthritis, migraine headaches, asthma, and psoriasis (Galano et al. 2018). AA is essential for membrane phospholipids and also useful for muscle development. These essential fatty acids are useful for heart and brain diseases and neurological functions like neurogenesis and neurotransmission. Certain sterols, such as fucosterol, cholesterol, clionasterol etc., are the main nutritional constituents in marine algal weeds (Luo et al. 2015). Sterols have antioxidant, anticancer, anti-obesity, and antiviral properties, along with being effective against cardiovascular diseases.

10.4 Marine Algae-Derived Essential Proteins and Pigments

Marine algae also possess high protein content, including both peptides and amino acids. Red and green marine algae contain more proteins when compared to brown algal species (Cerná 2011). Red algae *Porphyra tenera* and *Palmaria palmata* contain 35% and 47% proteins, respectively, whereas green alga *Ulva pertusa* contains 26% protein. Along with proteins, certain amino acids, including glycine, proline, arginine, alanine, aspartic, and glutamic acids, also existed in several marine algal species (Admassu et al. 2015). A blue-green alga (cyanobacterium), *Spirulina*, is considered a single-cell protein that contains high protein along with other contents such as thiamine, riboflavin, β-carotene, minerals, and linolenic acid (Hentati et al. 2020).

Marine algal cells also produce both primary and secondary light-harvesting pigments, i.e., chlorophylls and carotenoids and phycobilins, along with other pigments like lutein, astaxanthin, and zeaxanthin (Kathiresan et al. 2007). Chlorophyll is useful for fastening cell growth and increasing hemoglobin. It is also used in the treatment of liver ulcers and repairing cells. Carotenoids include carotenes, lycopene, fucoxanthin, astaxanthin, zeaxanthin, neoxanthin, lutein, and violaxanthin (Šimat et al. 2020). Among carotenes, there was more β-carotene production in *Dunaliella salina*. Astaxanthin is one of the essential natural antioxidants of the xanthophyll group of the carotenoid family, having the higher antioxidant potential on earth (Aryee et al. 2018). Astaxanthin is essential for humans to protect their skin from UV light and is involved in anti-tumor therapy. Lutein is another carotenoid pigment available in marine algae and is used in treating cardiovascular diseases, retinal degeneration, and cancer. *Sargassum siliquosum* and *Sargassum polycystum* are rich sources of fucoxanthin which have shown inhibitory activities on ACE, α-amylase, and α-glucosidase and reduce CVD-related risk. *Spirulina* is considered one of the rich sources of phycobiliproteins such as phycocyanin, phycoerythrin, and allophycocyanin (Tavanandi et al. 2018). These have a wide range of applications as nutraceuticals and are useful for cancer treatment. Apart from blue-green algae, phycobiliproteins are also found in red algae.

10.5 Marine Algae-Derived Phenolic Compounds and Minerals

Phenolic compounds are important substances with aromatic rings having one or more hydroxyl groups (Liu et al. 2011). Though terrestrial plants have more phenolic compounds, marine algae also contain phenolic acids, flavonoids, tannins, phlorotannins, and catechins. The accumulation of phenolic compounds depends on the species, salinity levels, nutrient availability, and temperature, including UV radiation (Gómez-Guzmán et al. 2018). Brown marine algal species are the main sources of phlorotannins and complex polymers composed of oligomers of phloroglucinol. Phlorotannins are used for food preservation and work as antiaging products. Also, these compounds are used in the pharmaceutical and nutraceutical industries. Both red and green marine algae are rich in flavonoids, phenolic acids, and bromophenols (Hentati et al. 2020). Generally, phenolics are antiallergic and antidiabetic compounds used as functional foods for treating neurodegenerative disease and CVD. Also, polyphenols isolated from marine algae are useful because of their antimicrobial, antiviral, anti-obesity, antiproliferative, anti-inflammatory, antioxidant, and anticancer properties (Gullón et al. 2020).

Marine algae are also significant sources of essential minerals such as Na, P, Ca, Mg, and K (macroelements) and Fe, Zn, Mn, and Cu (microelements). In addition, these algal species are the best sources of heavy metals such as Pb, As, Sr, Cr, etc. (Lorenzo et al. 2017). Specifically, Ca, I, and Na levels are high when compared to land plants. The green marine alga *Ulva clathrata*, which originated in Mexico, possesses a total mineral content of 49.6% of dry matter. Mineral content is higher in brown and red marine algae (Hentati et al. 2020). Minerals and polysaccharides play a crucial role in human tissue formation and act as cofactors of many metalloenzymes. Therefore, marine algal species are important functional foods and supplements (Balina et al. 2016). Iodine is an anti-goiter, antioxidant, and anticancer agent and an important nutrient in metabolic regulation.

10.6 Marine Algae Are Sources of Vitamins

Marine algal species are also sources of different vitamins used as functional foods and nutraceuticals (Wells et al. 2017). They possess fat-soluble vitamins such as A, D, E, and carotenoids (provitamin forms of vitamin A) and water-soluble vitamins B (B1, B2, B3, B6, B12), C, niacin, folic acid, pantothenic acid, and riboflavin (Croft et al. 2006). A green marine alga, namely, *Ulva*, is a rich source of vitamin A. *Caulerpa* is one of the best sources of vitamins A and E. Vitamin B12 is more abundant in green marine algal species compared to red and brown algal species (Grossman 2016). Vitamins C and B3 are accumulated in all the algal species. However, it is important to isolate using new strategies for potential health properties (Šimat et al. 2020).

Along with the bioactive molecules, antioxidants and cosmetic and fuel compounds are abundant in marine algal species. Many diversified marine algae have enormous potential in various fields for different applications due to the presence of a wide variety of compounds (Tandon et al. 2017). To increase products, much research has been focused on using biotechnological tools to discover and produce new compounds. Overall, marine algal species are useful for nutraceutical purpose.

10.7 Conclusion

Marine algae possess carbohydrates, novel proteins and peptides, polyunsaturated fatty acids, pigments, vitamins, minerals, polyphenols, etc. A number of research works were carried out to know the action of marine algal molecules on various organisms, including human health. Working on isolating and characterizing novel biomolecules will give human society an advantage, specifically in treating various diseases. It is also a new opportunity to understand the action and mechanism of these novel molecules in various metabolic activities. In this chapter, we highlighted marine algal species' bioactive compounds that possess various functional and nutraceutical properties. Particularly, we discussed the effect of these compounds on various disease-treating methods.

Acknowledgments The authors are thankful to Prof. Mathivanan, University of Madras, Chennai, India, for algal culture to our laboratory and valuable suggestions to initiate algal research.

References

Admassu H, Zhao W, Yang R, Gasmalla M, Alsir E (2015) Development of functional foods: seaweeds (algae) untouched potential and alternative resource-a review. Int J Sci Technol Res 4: 108–115

Ahlgren G, Gustafsson IB, Boberg M (1992) Fatty acid content and chemical composition of freshwater microalgae 1. J Phycol 28:37–50

Ande MP, Syamala K, Srinivasa Rao P, Murali Mohan K, Lingam SS (2017) Marine nutraceuticals. Aquaculture Times 3:6–9

Aryee AN, Agyei D, Akanbi TO (2018) Recovery and utilization of seaweed pigments in food processing. Curr Opin Food Sci 19:113–119

Balina K, Romagnoli F, Blumberga D (2016) Chemical composition and potential use of *Fucus vesiculosus* from Gulf of Riga. Energy Procedia 95:43–49

Borowitzka MA (2013) High-value products from microalgae—their development and commercialisation. J Appl Phycol 25:743–756

Campo VL, Kawano DF, da Silva DB, Jr Carvalho I (2009) Carrageenans: biological properties, chemical modifications and structural analysis-a review. Carbohydr Polym 77:167–180

Cerná M (2011) Seaweed proteins and amino acids as nutraceuticals. Adv Food Nutr Res 64:297–312

Croft MT, Warren MJ, Smith AG (2006) Algae need their vitamins. Eukaryot Cell 5:1175–1183

De Jesus Raposo M, de Morais A, de Morais R (2015) Marine polysaccharides from algae with potential biomedical applications. Mar Drugs 13:2967–3028

Fitton JH, Stringer DN, Park AY, Karpiniec SS (2019) Therapies from fucoidan: new developments. Mar Drugs 17:571

Galano JM, Roy J, Durand T, Lee JC-Y, Le Guennec JY, Oger C, Demion M (2018) Biological activities of non-enzymatic oxygenated metabolites of polyunsaturated fatty acids

(NEO-PUFAs) derived from EPA and DHA: new anti-arrhythmic compounds? Mol Asp Med 64:161–168
Gómez-Guzmán M, Rodríguez-Nogales A, Algieri F, Gálvez J (2018) Potential role of seaweed polyphenols in cardiovascular-associated disorders. Mar Drugs 16:250
Grossman A (2016) Nutrient acquisition: the generation of bioactive vitamin B12 by microalgae. Curr Biol 26:319–321
Gullón B, Gagaoua M, Barba FJ, Gullón P, Zhang W, Lorenzo JM (2020) Seaweeds as promising resource of bioactive compounds: overview of novel extraction strategies and design of tailored meat products. Trends Food Sci Technol 100:1–18
Hentati F, Tounsi L, Djomdi D, Pierre G, Delattre C, Ursu AV, Fendri I, Abdelkafi S, Michaud P (2020) Bioactive polysaccharides from seaweeds. Molecules 25:3152
Ji CF, Ji YB (2014) Laminarin-induced apoptosis in human colon cancer LoVo cells. Oncol Lett 7: 1728–1732
Juneja A, Ceballos RM, Murthy GS (2013) Effects of environmental factors and nutrient availability on the biochemical composition of algae for biofuels production: a review. Energies 6:4607–4638
Kathiresan S, Sarada R, Bhattacharya S, Ravishankar GA (2007) Culture media optimisation for growth and phycoerythrin production from *Porphyridium purpureum*. Biotechnol Bioeng 96: 456–463
Kidgell JT, Magnusson M, de Nys R, Glasson CRK (2019) Ulvan: a systematic review of extraction, composition and function. Algal Res 39:101422
Lahaye M, Robic A (2007) Structure and functional properties of ulvan, a polysaccharide from green seaweeds. Biomacromolecules 8:1765–1774
Liu M, Hansen PE, Lin X (2011) Bromophenols in marine algae and their bioactivities. Mar Drugs 9:1273–1292
Lorenzo J, Agregán R, Munekata P, Franco D, Carballo J, Sahin S, Lacomba R, Barba F (2017) Proximate composition and nutritional value of three macroalgae: *Ascophyllum nodosum*, *Fucus vesiculosus* and *Bifurcaria* bifurcata. Mar Drugs 15:360
Luo X, Su P, Zhang W (2015) Advances in microalgae-derived phytosterols for functional food and pharmaceutical applications. Mar Drugs 13:4231–4254
Matos J, Cardoso C, Bandarra NM, Afonso C (2017) Microalgae as healthy ingredients for functional food: a review. Food Funct 8:2672–2685
Menaa F, Wijesinghe U, Thiripuranathar G, Althobaiti NA, Albalawi AE, Khan BA, Menaa B (2021) Marine algae-derived bioactive compounds: a new wave of nanodrugs? Mar Drugs 19: 484
Pal A, Kamthania MC, Kumar A (2014) Bioactive compounds and properties of seaweeds-a review. Open Access Libr J 1:1–17
Paramesh K, Chandrasekhar T (2020) Improvement of photobiological hydrogen production in *Chlorococcum minutum* using various oxygen scavengers. Int J Hydrog Energy 13:7641–7646
Paramesh K, Lakshmana Reddy N, Shankar MV et al (2018) Enhancement of biological hydrogen production using green alga *Chlorococcum minutum*. Int J Hydrog Energy 43:3957–3966
Pierre G, Delattre C, Dubessay P, Jubeau S, Vialleix C, Cadoret JP, Probert I, Michaud P (2019) What is in store for EPS microalgae in the next decade? Molecules 24:4296
Plaza M, Santoyo S, Jaime L, García-Blairsy Reina G, Herrero M, Señoráns FJ, Ibáñez E (2010) Screening for bioactive compounds from algae. J Pharm Biomed Anal 2010(51):450–455
Prajapati VD, Maheriya PM, Jani GK, Solanki HK (2014) Carrageenan: a natural seaweed polysaccharide and its applications. Carbohydr Polym 105:97–112
Raga Sudha N, Varaprasad D, Bramhachari PV et al (2021) Effects of various factors on biomass, bioethanol and biohydrogen production in green alga *Chlamydomonas reinhardtii*. J Appl Biol Biotechnol 9:152–156
Saadaoui I, Rasheed R, Abdulrahman N, Bounnit T, Cherif M, Al Jabri H, Mraiche F (2020) Algae-derived bioactive compounds with anti-lung cancer potential. Mar Drugs 18:197

Šimat V, Elabed N, Kulawik P, Ceylan Z, Jamroz E, Yazgan H, Čagalj M, Regenstein JM, Özogul F (2020) Recent advances in marine-based nutraceuticals and their health benefits. Mar Drugs 18: 627

Stiger-Pouvreau V, Bourgougnon N, Deslandes E (2016) Carbohydrates from seaweeds. In: Seaweed in health and disease prevention. Academic Press, London, pp 223–247

Suleria HA, Osborne S, Masci P, Gobe G (2015) Marine-based nutraceuticals: an innovative trend in the food and supplement industries. Mar Drugs 13:6336–6351

Tandon P, Jin Q, Huang L (2017) A promising approach to enhance microalgae productivity by exogenous supply of vitamins. Microb Cell Factor 16:219

Tavanandi HA, Mittal R, Chandrasekhar J, Raghavarao K (2018) Simple and efficient method for extraction of C-Phycocyanin from dry biomass of *Arthrospira platensis*. Algal Res 31:239–251

Varaprasad D, Narasimham D, Paramesh K et al (2021) Improvement of production of ethanol using green alga *Chlorococcum minutum*. Environ Technol 42:1383–1391

da Vaz BS, Moreira JB, De Morais MG, Costa JAV (2016) Microalgae as a new source of bioactive compounds in food supplements. Curr Opin Food Sci 7:73–77

Venugopal V (2019) Sulfated and non-sulfated polysaccharides from seaweeds and their uses: an overview. ECronicon Nutr 2:126–141

Vidyashankar S, Sireesha E, Chauhan VS, Sarada R (2015) Evaluation of microalgae as vegetarian source of dietary polyunsaturated fatty acids under autotrophic growth conditions. J Food Sci Technol 52:7070–7080

Wells ML, Potin P, Craigie JS, Raven JA, Merchant SS, Helliwell KE, Smith AG, Camire ME, Brawley SH (2017) Algae as nutritional and functional food sources: revisiting our understanding. J Appl Phycol 29:949–982

Zhao X, Guo F, Hu J, Zhang L, Xue C, Zhang Z, Li B (2016) Antithrombotic activity of oral administered low molecular weight fucoidan from *Laminaria Japonica*. Thromb Res 144:46–52

Properties of Violacein: A Promising Natural Pharmaceutical Secondary Metabolite from Marine Environment with Emphasis on Its Anticancer Activity

11

Yogini Kanade, Rajashree Patwardhan, and Pragati Abhyankar

Abstract

Violacein, an indole-derived purple color natural pigment, has been reported from bacteria *Janthinobacterium lividum*, *Chromobacterium violaceum*, *Pseudoalteromonas luteoviolacea*, and *Collimonas* isolated from seawater, marine sponges, and coral reefs. It is formed by the condensation of two molecules of tryptophan. Factors like carbon source, pH, temperature, rate of agitation, and incubation time are known to affect violacein production. It exhibits extensive biological properties such as antibacterial, antiviral, antifungal, and antinematodal properties. It is an antioxidant in nature. It has immunomodulatory and anticancer activities, making it a promising natural pharmaceutical agent. Understanding violacein's function and mechanism of action has relevance to those displaying its therapeutic benefits. Violacein displayed anticancer activity in different types of cancers such as breast, colon, prostate, melanoma, lung, cervical, ovarian, central nervous, and renal cancers. Violacein can be established as a reliable natural drug efficient in causing apoptosis in cancerous cells and slowing down the proliferation rate of tumor cells by regulating the expressions of the cell cycle and tumor suppressor proteins. This review describes various pharmaceutical properties of violacein, highlighting its anticancer activities.

Keywords

Violacein · Marine metabolite · Anticancer · Bacterial pigment

---

Y. Kanade · R. Patwardhan (✉) · P. Abhyankar
Department of Microbiology, Haribhai V. Desai College of Arts, Science and Commerce, Pune, India

P. Veera Bramhachari, C. V. Berde (eds.), *Marine Bioactive Molecules for Biomedical and Pharmacotherapeutic Applications*,
https://doi.org/10.1007/978-981-99-6770-4_11

11.1 Introduction

Marine bacteria are exposed to changing environmental conditions, reactive oxygen species, UV radiation, and heavy metals that produce stress-relieving metabolites like pigments, peptides, enzymes, etc. (Pawar et al. 2018). Symbiotic marine microorganisms are a common source to engrossed for bioactive pigment-producing microorganisms, as the symbionts coevolve with their specific host and are thus rich in secondary metabolites with medically and commercially striking bioactivities (Brotosudarmo 2021). The potential of marine natural compounds as a pharmaceutical candidate can be seen from the fact that different phases of clinical trials of 13 natural products obtained from marine microorganisms are under investigation, and many other products are engrossed for preclinical investigations (Soliev et al. 2011). Furthermore, evidence suggests that marine bacteria can be explored to develop novel natural pharmaceutical secondary metabolites (Srinivasan et al. 2021).

Fifty to 80% of all life on Earth is harbored in the ocean. Marine entities' massive genetic and ecological diversity makes the oceans a unique reservoir of bioactive compounds that can be used in pharmaceutical industries. More than 20,000 natural compounds have been identified over the last 50 years from the marine environment. Marine plants, animals, their associated microbes, and the microbes isolated from marine sediments and marine water serve as a complex marine ecosystem source. The natural compounds obtained from the marine environment have an advantage over the ones found in the terrestrial environment in having several unique structural characteristics. The branch of marine drug discovery has expanded over the last 25 years with more than 35,000 research articles on natural chemical compounds derived from marine origins. Adequate screening and scrutiny of marine bacteria might provide us with novel antimicrobials that can help to fight drug-resistant pathogens over the next 100 years. Hence, the marine environment is expected to provide the pharmaceutical sector with the next generation of antimicrobial agents (Srinivasan et al. 2021).

Soil microorganisms are a rich source of secondary metabolites with antibacterial, antifungal, and anticancer properties. The biosynthetic capabilities of the soil microbes have been harnessed over the past few decades to produce novel molecules that can be used as potential pharmaceutical agents. Molecular genetic analyses of such microbes display that the genes responsible for the secondary metabolite biosynthetic pathways are clustered and range in size from 5 to greater than 100 kb. The revelation of these natural product biosynthetic pathways encourages more groundbreaking efforts to produce these secondary metabolites in heterologous hosts like *E. coli* that are more suitable for physiological and genetic manipulations. However, the nonfunctioning of the enzymes involved in the biosynthetic pathway of these secondary metabolites impose problems due to the absence of essential cofactors, precursors, or accessory proteins in the recombinant host, thereby making the recombinant expression of these metabolites difficult in the heterologous host (August et al. 2000).

Violacein is a purple pigment produced as a secondary metabolite by various bacteria. It is a bisindole characterized as a $C_{20}H_{13}N_3O_3$ (Cortes-Osorio et al. 2017). This versatile secondary metabolite is reported to show antibacterial (Duran et al. 2016; Rokade and Pethe 2017), antiviral (May et al. 1991; Duran et al. 2007; Duran et al. 2021), antifungal (Harris et al. 2009; Sasidharan et al. 2015), antiparasitic (Lopes et al. 2009; Bilsland et al. 2018), antioxidant (Konzen et al. 2006, Vishnu and Palaniswamy 2018), immunostimulatory (Venegas et al. 2019), and anticancer properties (Hashimi et al. 2015; Vishnu and Palaniswamy 2017). This chapter will explore several aspects of these nifty pigments, emphasizing their anticancer properties.

11.2 Characteristics of Violacein

Violacein (3-[1,2-dihydro-5-(5-hydroxy-1H-indol-3-yl)-2-oxo-3H-pyrrol-3-ylidene]-1,3-dihydro-2H-indol-2-one) (Fig. 11.1) is a secondary metabolite. It is a purple-colored bisindole (Fuller et al. 2016) with a molecular mass of 343.3. It is insoluble in water, slightly soluble in ethanol, moderately soluble in dioxane and acetone, and soluble in DMSO, methanol, and ethyl acetate. Its melting point is >290 °C. The UV-Vis spectrum exhibits maximum absorbances at 258, 372, and 575 nm in ethanol (Duran et al. 2007). Deoxyviolacein, a deoxy analogue of violacein, is a minor product. 5-Hydroxyindole, oxindole, and 2-pyrrolidone nuclei are the three structural units of this purple pigment. All the carbon skeleton is biosynthesized by condensing two molecules of L-tryptophan with decarboxylation process. Molecular oxygen acts as a source for the oxygen atoms in the skeleton of violacein. All the carbon, hydrogen, and nitrogen atoms of violacein come from the tryptophan molecule (Antonio and Creczynski-Pasa 2004). Rearrangement of the indole ring occurs during the synthesis of violacein. Monooxygenase and oxygenase enzymes catalyze the hydroxylation at the sixth position and incorporation of all oxygen atoms of violacein, respectively. Analogues of violacein, proviolacein, and prodeoxyviolacein accumulated with the addition of the copper-chelating agent *N*′, *N*′-diethyldithiocarbamate (Momen and Hoshino 2000).

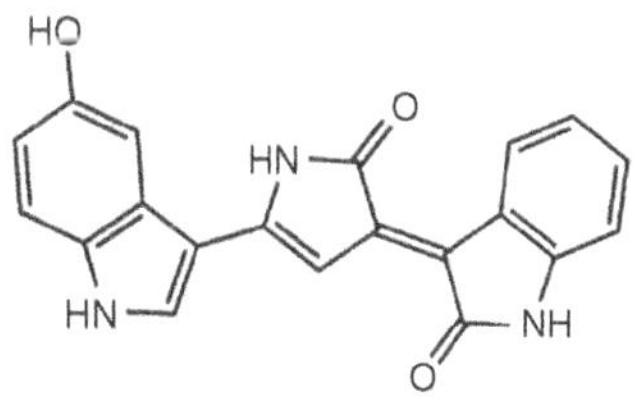

Violacein **(104)**
$C_{20}H_{13}N_3O_3$

Fig. 11.1 Chemical structure of violacein (Chatragadda et al. 2019a, b)

11.3 Violacein Production

11.3.1 Genes Involved in Violacein Production

Violacein biosynthesis is allied with the secondary metabolic pathway. Violacein production is regulated by a 7.3-kb-long DNA fragment covering, vio operon. Vio operon genes are highly conserved in *J. lividum* (Friedrich et al. 2020). Vio operon consists of five enzyme-coding genes, namely, vioA, vioB, vioC, vioD, and vioE, which are transcribed in one direction (Fig. 11.2). VioB, VioC, and VioD are more important than VioA and VioE in the violacein production. The analysis of the sequence of violacein gene cluster in plasmid pVIO4 displays an insert of 10,094 bp encoding five open reading frames (ORFs). These are orf1(1629 bp), vioA (1257 bp), vioB (2997 bp), vioC (1290 bp), vioD (1122 bp), and orf2 (partial). BLAST analysis revealed that the product of orf1 was weak like phospholipase C precursor protein. No significant homology to any known proteins was shown by proteins VioA (418 amino acids), VioC (429 amino acids), and VioD (373 amino acids), which are encoded by vioA, vioC, and vioD genes, respectively. The investigation by profile scan displayed a conserved NAD(P)H binding domain in VioA (amino acid 7–39), VioC (amino acid 3–33), and VioD (amino acid 4–32) (Fig. 11.3). The result of the analysis of these proteins with the sensitive homology search tool PSI-BLAST proposed that proteins VioA, VioC, and VioD are related to the PheA (phenol hydroxylase)/TfdB (2,4 dichlorophenol hydroxylase) family of FAD monooxygenases. This family of monoamine oxidases catalyzes the oxidative deamination of hydrophobic and aromatic L-amino acids (August et al. 2000; Antonio and Creczynski-Pasa 2004; Duran et al. 2016; Kothari et al. 2017). HPLC analysis of the structure of the intermediates produced by viomutants suggests the role of viogene products in violacein biosynthesis as described in Table 11.1.

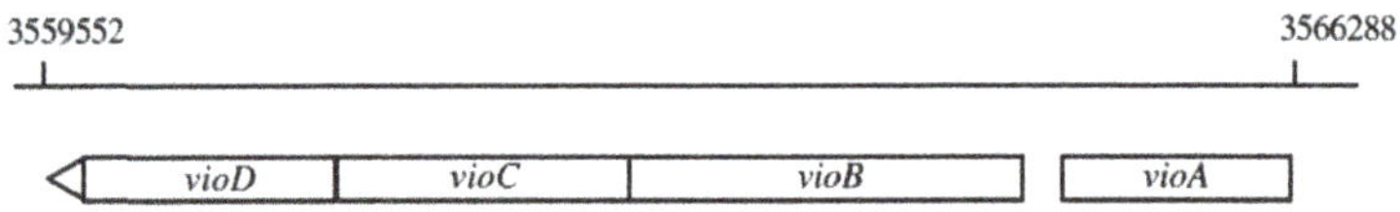

Fig. 11.2 Schematic drawing of structural genes of the violacein biosynthesis operon (Antonio and Creczynski-Pasa 2004)

Fig. 11.3 Alignment of the VioA, VioB, and VioC proteins revealing a conserved GXGXXG nucleotide-binding motif (August et al. 2000)

Table 11.1 *vio* operon genes and their function in violacein biosynthesis (August et al. 2000; Antonio and Creczynski-Pasa 2004; Duran et al. 2016; Kothari et al. 2017)

Gene	Gene product	Enzyme codes for	Role in the violacein biosynthesis	Molecular weight
*vio*A	VioA	Tryptophan 2-monooxygenase	Generation of an indole pyruvate intermediate	48
*vio*B	VioB	Considered to be a polyketide synthase	Catalyzes the 1,2-indole shift of tryptophan as well as the condensation reaction to generate the violacein pyrrole ring	111
*vio*C	VioC	FAD-dependent monooxygenase	Acts on prodeoxyviolacein and hydroxylates the indole ring to generate the violacein oxindole subunit	48
*vio*D	VioD	Flavin-dependent monooxygenase	Additional oxygen atom incorporation on violacein oxindole subunit	42
*vio*E	VioE	Conversion of flavanone to isoflavone	Conversion of short-lived dimer indole-3-pyruvic acid imine (IPA) into protodeoxyviolaceinic acid (PVA)	22

11.3.2 Violacein Biosynthetic Pathway

Tryptophan 2-monooxygenase, a flavin-dependent enzyme that is the product of gene *vioA*, catalyzes the oxidation of tryptophan to indole-3-pyruvic acid imine (IPA). With this step, the five-step biosynthesis of violacein begins (Fig. 11.4). Simultaneously, the reduction of cofactor FAD to $FADH_2$ takes place during the first step of violacein synthesis. In the second step, the heme-containing oxidase VioB, a product of the gene *vioB*, which is proposed to be a polyketide synthase catalyzes the oxidative coupling of the two imines forming a short-lived imine dimer. This short-lived dimer has two fates, either it instantly transformed into chromopyrrolic acid involved in indolocarbazole biosynthesis or VioE product of *vioE* gene acts on IPA imine dimer and catalyzes a 1,2 shift of the indole ring to produce proto-deoxy-violaceinic acid. Structural and biochemical researches for VioE display that this rearrangement of the indole does not require any cofactor or metal ion. The successive catalysis by two flavin-dependent oxygenases VioD and VioC completes the violacein biosynthesis. Proviolacein is formed by hydroxylation of one indole ring at the fifth position by VioD. Oxidation of NADPH to NADP is coupled with this reaction. VioC subsequently hydroxylates the second indole ring at the two positions with the oxidation of NADPH to $NADP^+$. Spontaneous oxidative decarboxylation following the action of VioD and VioC then produces violacein with the release of water and carbon dioxide. Under physiological conditions, deoxyviolacein, which lacks the hydroxyl group at the fifth position, is synthesized from the violacein biosynthetic pathway without the contribution of VioD. In summary, violacein biosynthesis is a six-step process involving five enzymatic reactions and a nonenzymatic oxidative decarboxylation step (August et al. 2000; Antonio and Creczynski-Pasa 2004; Duran et al. 2016; Fuller et al. 2016; Kothari et al. 2017).

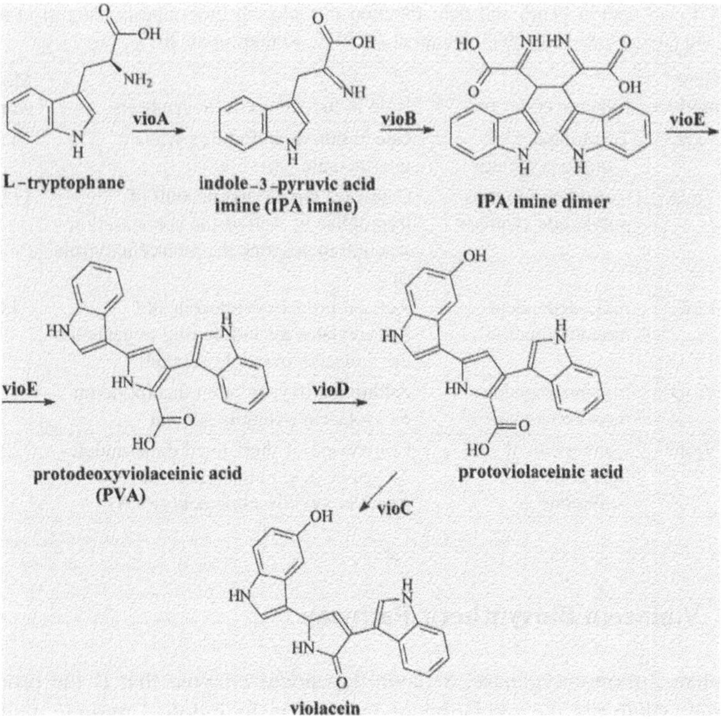

Fig. 11.4 Biosynthetic pathway of violacein (Duran et al. 2016)

11.3.3 Regulation of Violacein Production

Tryptophan acts as a precursor molecule for the biosynthesis of violacein (Fig. 11.5). Erythrose 4-phosphate (E4P) and phosphoenolpyruvate intermediates of hexose monophosphate pathway (HMP) and glycolysis, respectively, are required for the biosynthesis of aromatic amino acids, the former acting as a limiting substrate for the biosynthesis of tryptophan. Both these precursors lead to the production of chorismate through several steps. In turn, chorismate can be directed to the tryptophan pathway, with anthranilate as the first molecule of the pathway. *C. violaceum* lacks genes encoding for 6-phosphogluconate dehydrogenase of HMP. Thus, in *C. violaceum*, it is proposed that the linkage between fructose 6-phosphate and glyceraldehyde 3-phosphate, intermediates of the glycolysis, through the activity of transketolases and trans-aldolases provides E4P for the tryptophan biosynthesis. With these characteristics, *C. violaceum* should be able to convert large amounts of glucose to aromatic amino acids (Antonio and Creczynski-Pasa 2004).

Bacteria's response to the environment with the production of specific molecules is governed by a cooperative and coordinated metabolism known as quorum sensing. In quorum sensing, small molecules are released into the environment, which acts as a ligand and binds to its receptor, thereby directly or indirectly amending cellular

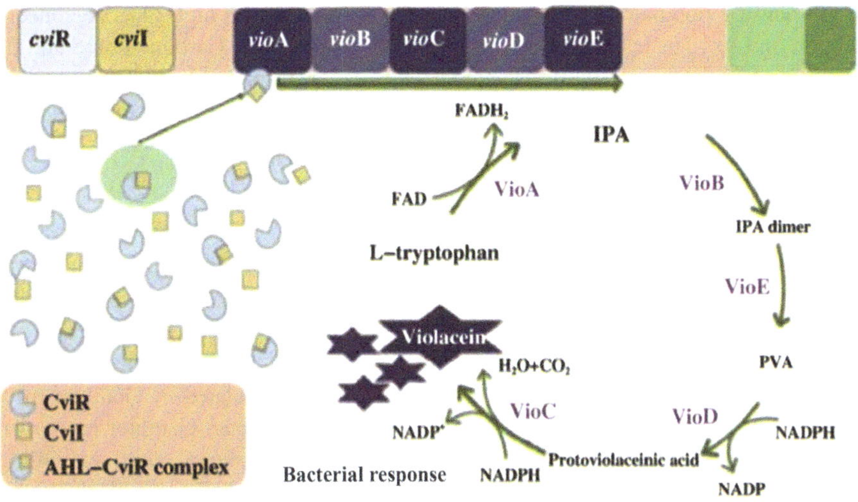

Fig. 11.5 Schematic of violacein biosynthesis (Kothari et al. 2017)

processes and biosynthetic pathways, thereby further affecting the quorum sensing system of *Chromobacterium violaceum* ATCC 31532 made up of CviI synthase, acyl homoserine lactone (AHL), a diffusible molecule, CviR-type receptor, and some target genes like *cviI* responsible for AHL synthesis and violacein operon (*vio*ABCDE). In *C. violaceum*, CviI synthase catalyzes the conversion of *S*-adenosyl methionine and fatty acids into AHLs. Once a certain concentration of AHL is reached, it triggers the quorum sensing signal and binds to CviR, a constitutive protein that inspires the transcription of the vioABCDE operon (Duran et al. 2016).

N-Acyl-homoserine-lactone (AHL) is a diffusible signal molecule known to control violacein biosynthesis. In *C. violaceum*, an AHL identified as *N*-hexanoyl-homoserine lactone (HHL) has been found. An evolutionarily conserved transcriptional regulator including two adjacent genes, *cvi*I, encoding an auto inducer (HHL) synthase, and p*cvi*R, encoding a receptor (regulator protein) as a part of quorum sensing system of *C. violaceum* was reported to regulate the production of violacein. When the growth of the *C. violaceum* reaches the stationary phase, large amount of constitutive CviI (product of gene *cviI*) accumulates HHL. A stable protein-ligand complex forms on the binding of HHL to *cvi*R operator leading to its expression. Increased amount of CviR (product of gene *cvi*R) allows it to bind to the promoter site of violacein operon, thereby increasing pigment production. In strain CV026 of *C. violaceum*, violacein production is inducible by AHL with *N*-acyl side chains of C4–C8 length. On contrary, *N*-acyl side chains from C10 to C14 inhibit violacein production (McClean et al. 1997). A similar result was reported for the *Pseudoalteromonas ulvae* TC14, where C6-, C12-, 3-oxo-C8, and 3-oxo-C12-HSLs (homoserine lactones) upregulated violacein production, while 3-oxo-C6-HSL downregulated pigment production (Antonio and Creczynski-Pasa 2004; Morohoshi et al. 2010; Aye et al. 2015; Kothari et al. 2017; Kanade et al. 2022).

In densely populated culture of *C. violaceum* grown with high glucose concentration and oxygen levels, violacein production is retarded. Carbon source and its concentration in the medium also impact the violacein production. These experimental observations suggest a control mechanism on violacein production apart from HHL similar to the one mediated by cyclic AMP through catabolic activator protein (Antonio and Creczynski-Pasa 2004).

11.4 Bacteria Producing Violacein

Hamilton and Austin were one of the first to report violacein production from the marine bacterium *Chromobacterium marinum* isolated from open ocean waters. Sixteen violet-pigmented heterotrophic bacilli were later isolated by Gauthier from Mediterranean coastal waters as *Alteromonas luteo-violaceus*. Gauthier et al. also isolated six bacterial species from the French Mediterranean coast, exhibiting similarities with *Alteromonas* species and producing light, dark, and black shades of violacein. Several *Alteromonas* species producing purple pigment were isolated from Kinko Bay in Kagoshima Prefecture, Japan. Violacein-like pigment-producing 13 Gram-negative bacilli were isolated from the Pacific Ocean at a depth of 320 m off the coast of Cape Muroto, Kochi Prefecture, Japan (Soliev et al. 2011).

Pseudoalteromonas luteoviolacea was isolated from the marine sponge *Acanthella cavernosa* (Yang et al. 2007a, b). *Pseudoalteromonas luteoviolacea*, *Pseudoalteromonas tunicata*, and *Pseudoalteromonas* sp.520P1are the bacteria reported to produce violacein isolated from marine environment (Soliev et al. 2011). Violacein-producing *Pseudoalteromonas* sp. 520P1, a Gram-negative bacterium, was isolated from the Pacific coast of Japan (Dang et al. 2014). *Pseudoalteromonas ulvae* TC14 was isolated from the Mediterranean Sea (Aye et al. 2015). Members of the genus *Pseudoalteromonas* are widely distributed in different habitats and are well adapted to marine environments like coastal, open and deep seawaters, marine invertebrates, fish, sediments, and algae. Two novel species of *Pseudoalteromonas amylolytica* sp. nov. producing violacein was isolated from the surface water of the Arabian Sea (E67° N24°) (Wu et al. 2017). Two strains of *Pseudoalteromonas luteoviolacea* designated as P42 and M64, producing deep purple pigments, were isolated from coral reef regions of Minicoy Lagoon in the Lakshadweep Sea and Palk Bay, off Olaikuda village in Tamil Nadu, India (Jayasree et al. 2021) (Fig. 11.6).

Collimonas species producing violacein were isolated from the sea surface microlayer at the coast of Trondelag, Norway. One extremophilic species of *Collimonas* was isolated from the submarine ikaite columns in Greenland (Hakvag et al. 2009). Violacein and deoxyviolacein-producing novel bacterium *Massilia* sp. BS-1 was isolated from a soil sample collected from Akita Prefecture, Japan (Agematu et al. 2011). Violacein-producing *Duganella violaceinigra* str. NI28 was isolated from the temperate forest soil sample from Ulsan, South Korea (Choi et al. 2015a, b). Violacein-producing *Iodobacter* sp. designated as BJB302 was isolated from Hudson Valley freshwater sources (Doing et al. 2018). A psychrotolerant

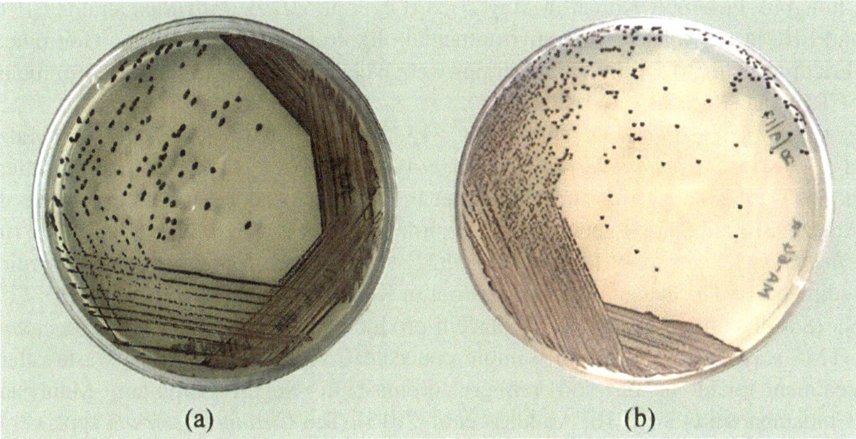

Fig. 11.6 Colonies of *P. luteoviolacea* (**a**) P42 and (**b**) M64 strains grown on marine agar medium (Jayasree et.al. 2021)

violacein-producing *Iodobacter* sp. 7MAnt was isolated from King George Island near Eduardo Frei Montalva, Chilean Air Base (62°12′0″S58°57′51″W), Antarctic Peninsula (Atalah et al. 2020). A psychrophilic *Rugamonas violacea* sp. nov. type strain P5900^T producing violacein was isolated from streams and lakes located in a deglaciated northern part of James Ross Island, Antarctica (Sedlacek et al. 2021). A group of seven psychrophilic bacterial strains representatives of the genus *Massilia* producing violacein pigment was isolated from freshwater samples collected in deglaciated parts of James Ross Island and Eagle Island, Antarctica (Sedlacek et al. 2022).

Bluish-purple pigment-producing *Janthinobacterium lividum* was isolated from the wet silk thread (Shirata et al. 2000). *Janthinobacterium lividum* CM37 was isolated from mesotrophic lake (Matz et al. 2004). *Janthinobacterium lividum* is a violacein-producing organism commonly isolated from forest soil, Antarctic glaciers, lake, and liver water. *Janthinobacterium lividum* strain was reported to be isolated from the skin of the salamander *Hemidactylium scutatum* from the George Washington National Forest, Rockingham Co., VA, USA (Harris et al. 2009). A psychrotrophic strain of *Janthinobacterium lividum* was isolated from the soil core obtained from a site on an island in the Tanana River in the Bonanza Creek Long-Term Ecological Research site approximately 30 km southwest of Fairbanks, Alaska, USA (64°40.24′N, 148°14.46′W) (Schloss et al. 2010). *Janthinobacterium* strain SMN 33.6 was isolated from a soil sample in Fildes Peninsula (62°12′S, 58° 57′W) located on King George Island, Antarctica (Asencio et al. 2014). A novel psychrophilic *Janthinobacterium lividum* MMPP4 was isolated from Manimahesh Lake of Chamba district, Himachal Pradesh, India (Suman et al. 2015). A new variant of *Janthinobacterium lividum*, *Janthinobacterium lividum*-ROICE173, was isolated from the Antarctic snow (Baricz et al. 2018). Violacein-producing strain was isolated from the rainbow trout fish *Oncorhynchus mykiss* farm located in

Chungbuk Province, Korea, in May 2018 (Oh et al. 2019). *J. lividum* strains EIF1 and EIF2 were isolated from environmental oligotrophic water ponds in Gottingen (Friedrich et al. 2020). *J. lividum* strains were isolated from the tropical environment of Trinidad (Ramdass et al. 2021).

Chromobacterium violaceum ATCC 31532 was isolated from fresh water (Matz et al. 2004). *Chromobacterium violaceum*-type strain ATCC 12472^T and 24 other strains were isolated from the fresh water from sites closed to Manaus city, state of Amazon, Brazil (Barreto et al. 2008). High-level violacein-producing novel strain of *Chromobacterium* sp. NIIST (MTCC 5522) was isolated from the clay mine acidic sediments of Thiruvananthapuram district in Kerala, India (8°55′53″ N, 76° 85′77″ E) (Sasidharan et al. 2015). Violacein-producing *Chromobacterium violaceum* UTM5 was isolated from soil sample collected from the vicinity of a wastewater treatment plant in one oil refinery premise in Negeri Sembilan, Malaysia (Chidambaram et al. 2016; Aruldass et al. 2015). Ten *Chromobacterium* spp. were isolated from soil and water samples collected from different localities at Trivandrum and Nedumangad, Kerala India (Vishnu and Palaniswamy 2016). *Chromobacterium violaceum* was isolated from the soil surface at Universiti Putra Malaysia (Loke et al. 2022). Various locations have been found to serve as a source of violacein-producing bacteria. Table 11.2 summarizes the violacein-producing bacterial genera with their source of isolation.

11.5 Factors Affecting Violacein Production

Violacein production is produced by a different path which depends on the strain type, the culture medium, and growth conditions. Microorganisms produce the purple pigment to overcome environmental stresses and a defense mechanism against external harmful factors.

11.5.1 Carbon Source

Biomass and the violacein production are affected by the carbon source. Glucose inhibits production, and glycerol is an enhancer (Pantanella et al. 2007). Glucose, xylose, and glycerol are inexpensive carbon sources. Fermentation of lignocellulosic biomass and biodiesel waste can be useful to obtain these sources (Kumar et al. 2009; Yang et al. 2012). When these sugars were added to the culture media, there was an increase in biomass but a decrease in violacein production. This justifies that metabolic pathway that guides violacein production is regulated by carbon source. Glycerol aids the production of violacein, LB supports production but to a lesser extent, and glucose completely inhibits production (Pantanella et al. 2007). When the antibiotic was examined as a stress inducer, cell growth was not affected, but ampicillin dose of around 0.1–0.4 mg/mL showed an increase in violacein production by 36% (Pantanella et al. 2007).

Less expensive carbon sources like glucose, glycerol, and xylose are taken as carbon sources for nutrition in microorganisms. Biomass production increased

Table 11.2 List of bacteria-producing violacein pigment with their source of isolation

Bacteria-producing violacein	Location isolated from
Chromobacterium marinum	Open ocean water
Alteromonas luteo-violaceus	Mediterranean coastal water
Alteromonas species	Kinko Bay, Kagoshima Prefecture, Japan
Pseudoalteromonas luteoviolacea	From the marine sponge *Acanthella cavernosa*
Pseudoalteromonas luteoviolacea *Pseudoalteromonas tunicate* *Pseudoalteromonas* sp.520P1	Marine environment
Pseudoalteromonas sp. 520P1	Pacific coast, Japan
Pseudoalteromonas ulvae TC14	Mediterranean Sea
Pseudoalteromonas amylolytica sp. nov.	Surface water of Arabian Sea (E67° N24°)
Pseudoalteromonas luteoviolacea	Coral reef regions of Minicoy Lagoon in the Lakshadweep Sea and Palk Bay, off Olaikuda village in Tamil Nadu, India
Collimonas species	Sea surface microlayer at the coast of Trondelag, Norway
Collimonas species	Submarine ikaite columns in Greenland
Massilia sp. BS-1	Soil samples collected from Akita Prefecture, Japan
Duganella violaceinigra str. NI28	Temperate forest soil sample from Ulsan, South Korea
Iodobacter sp. BJB	Hudson Valley freshwater sources
Iodobacter sp. 7MAnt	King George Island near Eduardo Frei Montalva, Chilean Air Base (62°12′0″S58°57′51″W), Antarctic Peninsula
Rugamonas violacea sp. nov. type strain P5900^{T}	Streams and lakes located in a deglaciated northern part of James Ross Island, Antarctica
Massilia antarctica sp. nov. P8398^{T}	Freshwater samples collected in deglaciated parts of James Ross Island and Eagle Island, Antarctica
Janthinobacterium lividum	Wet silk threads
Janthinobacterium lividum CM37	Mesotrophic lake
Janthinobacterium lividum	Skin of the salamander *Hemidactylium scutatum* from the George Washington National Forest, Rockingham Co., VA, USA
Janthinobacterium lividum	Soil core obtained from a site on an island in the Tanana River in the Bonanza Creek Long-Term Ecological Research site approximately 30 km southwest of Fairbanks, Alaska, USA (64° 40.24′N, 148°14.46′W)
Janthinobacterium lividum SMN 33.6	Soil sample in Fildes Peninsula (62°12′S, 58° 57′W) located in King George Island, Antarctica
Janthinobacterium lividum MMPP4	Manimahesh Lake of Chamba district, Himachal Pradesh, India
Janthinobacterium lividum-ROICE173	Antarctic snow

(continued)

Table 11.2 (continued)

Bacteria-producing violacein	Location isolated from
Janthinobacterium lividum	Rainbow trout fish *Oncorhynchus mykiss* farm located in Chungbuk Province, Korea
Janthinobacterium lividum strains EIF1 and EIF2	Environmental oligotrophic water ponds in Gottingen
Janthinobacterium lividum	Tropical environment of Trinidad
Chromobacterium violaceum ATCC 31532	Fresh water
Chromobacterium violaceum ATCC 12472^T	Fresh water from sites closed to Manaus city, state of Amazon, Brazil
Chromobacterium sp. NIIST (MTCC 5522)	Clay mine acidic sediments from Thiruvananthapuram district in Kerala, India (8°55′53″ N, 76° 85′77″ E)
Chromobacterium violaceum UTM5	Soil samples collected from the vicinity of a wastewater treatment plant in one oil refinery premise in Negeri Sembilan, Malaysia
Chromobacterium sp.	Soil and water samples collected from different localities at Trivandrum and Nedumangad, Kerala, India
Chromobacterium violaceum	Soil surface at Universiti Putra Malaysia

positively in NB-G, NB-GY, and NB-X. NB-G showed the highest biomass concentration of 2.25 mg/mL, but it inhibited violacein production compared to basic NB as a medium. NB-GY and NB-X also showed higher biomass production (1.60 and 1.55 mg mL, respectively) less than NB-G. NB showed maximum violacein production followed by NB-GY. An inverse relation was observed between glucose and violacein yields. The addition of limited amounts of glucose to the culture proved advantageous. Applying this fact increased the dry cell weight from 7.5 to 21 g/L, and the violacein yields from 170 to 430 mg/L.

To produce violacein, it is necessary to select suitable medium. When three different media were chosen, nutrient broth showed the highest pigment production (1000 μg/L) followed by peptone broth (500 μg/L), whereas brain heart infusion (BHI) showed the least production of pigment (400 μg/L) among the three (Dike-Ndudim et al. 2022).

For violacein production, glucose was the best carbon source which resulted in the highest amount of violacein (2300 μg/L). This was followed by dextrose and maltose. Starch as a carbon source gave the least production (Dike-Ndudim et al. 2022). Tryptophan, sodium butyrate, or Tween 80, when added, didn't show an increase in production of violacein. When glycerol or glucose was added to the medium, violacein production increased (Dike-Ndudim et al. 2022). The role of glucose is very crucial for violacein production. It also retards the violacein production by inhibiting even distribution of dissolved oxygen and resulting in anoxia. This is observed in *C. violaceum*, which is aerobic (Cheng et al. 2022).

11.5.2 pH

At pH 7.0 and 28 °C for 24 h, in a liquid medium and the absence of light, highest yields of violacein were obtained (Riveros et al. 1989). When violacein production was observed at different pH (4.0, 4.5, 5.0, 5.5, 6.0, 6.5, 7.0, 7.5 8.0, 8.5, 9.0), pH 4 had a distinct effect. The pH of 6.5 was the maximum pH for violacein production (2200 μg/L). The production of violacein decreased as the pH increased above 6.5 (Dike-Ndudim et al. 2022).

Violacein became colorless at a high pH of 11 after 24 h in the presence and absence of light. This resulted in 90% of the decomposition of pigment. At a highly acidic pH of 2, it was observed that 65% of residual pigment remained in solution after 30 days, reflecting moderate degree of decomposition. It acquires a dark blue shade at pH 2 and green at pH 13. In the pH range of 5–9, low effect was observed on the color intensity of the pigment. When pH was increased to alkaline side, excess OH− ions from NaOH deprotonate the phenolic group of the hydroxyindol and amine group of oxoindol and pyrrolidone, causing the formation of anion and destruction in the conjugated structure of the pigment (Wan et al. 2012).

In nutrient broth, the pH was varied between 3 and 11 and the effect was studied. pH between 7 and 9 gave the highest yields of the pigment. On both sides of the pH scale, acidic and alkaline low yields were observed indicating an unsuitable environment for the bacterium. *Chromobacterium violaceum* produces good violacein production at pH 7, indicating it is ideal for violacein production (Palukurty et al. 2016). *Y. lipolytica grows rapidly at pH 7.5, but violacein production doesn't increase at the same rate. When pH was adjusted between* 6.0 and 7.0, violacein production was seen to decrease. This occurred possibly due to the acidophilic nature of the violacein pathway (Tong et al. 2019).

11.5.3 Temperature

A psychotropic strain of *Chromobacterium*, *Chromobacterium lividum*, produced pigment, but the pigment production was inhibited at 0 °C, though their growth continued at that temperature. Temperatures of 15, 20, and 25 °C supported normal pigment formation (Cardona-Cardona et al. 2010; Inniss and Mayfield 1979). At 20 ° C, highest biomass and violacein production was obtained (Diego et al. 2020). At 30 °C, much less biomass was observed and there was no pigment production.

For *J. lividum*, temperature in the range of 20–25 °C is found to be the most favorable, whereas temperatures between 30 and 35 °C inhibited growth. An average concentration of 0.79 mg/mL was produced at 25 °C, which also showed maximum pigment production. When exposure was given to a temperature between 30 and 35 ° C, no growth or pigment production was observed. A biomass concentration of 0.38 mg/mL was observed at 20 °C with second-best violacein production. These observations justify the psychotropic character of *J. lividum* (Alem et al. 2022).

Janthinobacterium sp. B9-8 grows well in the temperature range of 4–37 °C. This organism was first isolated from low-temperature sewage (5–10 °C). When the temperature was increased to 25 °C, pigment production increased, beyond which it started decreasing (Alem et al. 2022). Completely pigmented colonies were observed at 20 °C. On the contrary, colonies when grown at 0 °C did not show pigment production. It was also observed that when growth occurred at 0 °C, pigment production could only occur when they were incubated at 20 °C (Rokade and Pethe 2017). Violacein is stable when heated at a temperature range of 40–100 °C for 2 h. (Nathan et al. 2019). This shows that, unlike pH, temperature has no effect on the stability of pigment (Wan et al. 2012).

Cultures show pigment production at 4 °C but not at 30 °C. As recent research suggests, violacein production is favored at low temperatures and it may have cryoprotectant properties. Pigment-producing bacteria showed increased resistance to freeze-thaw cycles than the nonpigmented ones. This indicates that pigments will likely play a significant role in maintaining membrane fluidity. It also protects the membranes from freezing damage and helps them to adapt to cold (Cortes-Osorio et al. 2017).

11.5.4 Rate of Agitation

Many different pigments were produced by *Pseudoalteromonas luteoviolacea. These* ranged from dark purple (0 rev min) to pale purple (50 rev min) and then to yellow (100–200 rev min). Bacterial extracts from low agitation speeds (0 and 50 rev min) absorbed different wavelengths from that at high agitation intensity (100–200 rev min), when exposed to UV-visible spectra. These bacterial cultures showed similar morphology under all treatments. An interesting observation was different cell arrangements under different agitation speeds. Bacterial clusters were formed when the bacterium was grown at zero or lower agitation (0, 50, and 100 rev min) (Yang et al. 2007a, b). In case of *C. violaceum*, crude violacein production was not affected by the intensity of agitation. At the same time, bacterial biomass production was affected.

11.5.5 Incubation Time

Microorganisms produce violacein as a secondary metabolite. Hence, when the cells reach the late log or stationary phase of growth, violacein production becomes apparent.

When glycerol is added as a carbon source to the medium, growth occurs till 96 h, whereas glycerol consumption continues till 120 h. It's to be noted that when glycerol is added, the C/N ratio increases and hence violacein production is boosted as a secondary metabolite (Kanelli et al. 2018). Cultures were exposed to different periods of incubation (12, 24, 36, 48, 72, 60, 84, 96 h), and violacein production was observed (Demoss and Evans 1959). It was maximum at 72 h (1800 μg/L). Beyond this, the production deteriorated (Kanelli et al. 2018).

11.6 Pharmaceutical Properties of Violacein

11.6.1 Antibacterial

Violacein has powerful antibacterial properties making it as effective as an antibiotic. So, it's a promising candidate in the current age of the emergence of antibiotic-resistant microorganisms. Furthermore, when it is used along with other antibiotics, the impact is very high in fighting bacteria than using antibiotics alone (Choi et al. 2015a, b, Subramaniam et al. 2014). The antimicrobial effects of violacein on various microorganisms affecting humans and other animals make violacein an appealing drug to battle resistant pathogens (Durán et al. 2021). This is particularly interesting considering antibiotic-resistant strains of infective bacteria, such as MRSA.

Violacein in the crude state obtained from *Duganella violaceinigra* str. NI28 indicated MIC of 15 μM for five multidrug-tolerant strains of *S. aureus* (Choi et al. 2015a, b). *S. aureus* ATCC 3359 and *S. aureus* CCBH 5330 resistive to oxacillin were inhibited by violacein at the MIC of 20 and 5 μg/mL, respectively. Violacein demonstrated a repressive effect on the biofilm formation of *S. aureus* at the concentration of 1.25 μg/mL (Batista et al. 2017). Ethanolic extracts of violacein from *Janthinobacterium* sp. Exhibited antibacterial action against carbapenemase-producing *Acinetobacter baumannii* strain, *Pseudomonas aeruginosa* strain, extended-spectrum β-lactamases producing two strains of *E. coli*, two strains of *Klebsiella pneumoniae*, and a chromosomal Amp C beta-lactamase producing two multiresistant strains of *Serratia marcescens* with MIC values varying between 0.5 and 16 μg/mL (Asencio et al. 2014). Ethanolic extracts of violacein from *Chromobacterium violaceum* (MTRI7) revealed antibacterial activity against *B. subtilis*, *S. aureus*, *E. coli*, *S. typhi*, and *K. pneumoniae* (Siva 2007).

Violacein from *C. violaceum* revealed synergistic action with penicillin G procaine against *S. aureus* isolates MBSA 24, 35, and 63, isolated from bovine mastitis (Malik et al. 2012). Antibacterial mode of action of violacein from *C. violaceum* UTM5 was examined against *Staphylococcus aureus* and methicillin-resistant *S. aureus* (MRSA) strains with minimum inhibitory concentration (MIC) of 3.9 μg/mL (Aruldass et al. 2018). The synergistic antibacterial action of violacein and silver nanoparticles (AgNPs) against ATCC bacteria, *Staphylococcus aureus* and *Escherichia coli*, was investigated. The bactericidal effect of violacein against *S. aureus* was higher when combined with AgNPs (synergistic). *Staphylococcus aureus* exhibited sensitivity to violacein with MIC of 200 μM, and the MIC with AgNPs between 250 and 125 μM (Nakazato et al. 2019).

The antimicrobial activity of violacein against attenuated *Mycobacterium tuberculosis* (H37Ra) with MIC of 64 μg/mL and MBC of 128 μg/mL was described in vitro (Duran et al. 2012). Violacein demonstrated antimicrobial activity on 11 plant pathogens including *Erwinia carotovora*, *Pseudomonas cichorii*, *Colletotrichum dematium*, and *Rosellinia necatrix* (Sen et al. 2019).

Violacein also exhibited additive or synergistic effects when mixed in 1:1 ratio with different antibiotics against infective organisms *Salmonella typhi*, *Vibrio cholerae*, *Pseudomonas aeruginosa*, *Klebsiella pneumoniae*, and *Staphylococcus aureus* (Sinha et al. 2017). A smaller number of antibiotics were reported to exhibit antagonistic activity when applied with violacein. The blend of violacein with aztreonam, cefuroxime, meropenem, and ampicillin was observed to be synergistic, while with Ceftazidime, it showed an additive effect against uropathogenic *E. coli* (Devi et al. 2018). Violacein from *Chromobacterium violaceum* revealed bactericidal and bacteriostatic action against *S. epidermidis* ATCC 12228 and *S. epidermidis* ATCC 35984 with MIC of 10 and 20 μg/mL, respectively. It also exhibited synergistic and additive effects with a decrease in MIC from 4- to 16-fold of antibiotics (Dodou et al. 2017).

Violacein is a quorum sensing metabolite and an effective antimicrobial agent. It works synergistically with most commercial antibiotics and could be used as medicine in combination with other antimicrobial agents. The MIC for violacein was noticed to be 5.7 μg/mL (*Staphylococcus aureus*) and 5.7 μg/mL (*Salmonella typhi*). Violacein-gentamicin and violacein-cefadroxil combinations had MIC of 1.0 μg/mL versus *S. aureus*. Violacein-macrolide and violacein-aminoglycoside class combinations revealed fractional inhibitory concentration indices (FICI) of <0.5, demonstrating synergism. Violacein-azithromycin and violacein-kanamycin combinations exhibited substantial synergy (FICI – 0.3) against *S. typhi* (Subramaniam et al. 2014).

The antibacterial activity of violacein was examined against *Staphylococcus aureus* isolated from bovine mastitis (Martins et al. 2011). Violacein presented a synergistic effect when combined with penicillin in the three selected isolates of *S. aureus* (Cazoto et al. 2011). *Janthinobacterium* is a Gram-negative bacterium in various ecosystems, including Antarctica (Asencio et al. 2014). Some members of this genus are violet-colored, a characteristic observed in other bacterial genera (e.g., *Chromobacterium*, *Collimonas*, *Duganella*, *Pseudoalteromonas*, etc.) Violacein has distinct biological activities, including a strong antibacterial effect against many important human pathogens (Martins et al. 2010; Cazoto et al. 2011; Berni et al. 2013). An inhibitory effect was observed against MRSA, MSSA, Enterococci, and *Enterobacteriaceae* isolates.

11.6.2 Antiviral

Violacein revealed antiviral activity against herpes simplex and polio viruses. Sixty-two percent HSV replication was inhibited by 0.25 μg/mL violacein, and 0.063 μg/mL violacein inhibited 56% of polio virus replication in HeLa cells (Duran et al. 2016). These results have indicated substantial violacein antiviral activity. Opposite to these results, violacein demonstrated weak inhibitory activity on HSV-1 (strains KOS and ATCC/VR733) and PV-2 replication. Cytotoxic effect of violacein on various cell lines can be obvious with the trypan blue dye exclusion method and MTT assay. Violacein-containing liposomes showed heightened antiviral activity against HSV-1 (Andrighetti-Frohner et al. 2003). Violacein revealed no cytopathic inhibition of herpes simplex virus type 1 (HSV-1) strain 29-R/acyclovir resistant, hepatitis A virus (HM175 and HAF-203), and adenovirus type 5. However, violacein did show a weak inhibition of viral replication (Andrighetti-Frohner et al. 2003). The MTT assay has shown 1-weak repression of HSV-1 (KOS-1.25 μM [22%]), ATCC/VR-733 strains (0.625 μM [9%]), poliovirus type 2 (1.25 μM [9%]), and simian rotavirus SA11 (1.25 μM [24%]) replication in cell cultures.

Cyclodextrin-violacein nanostructure (Duran et al. 1998) and cyclodextrin-Au-thiol-derivative/hydrophobic nanoparticles (Alves et al. 2005) were described as antiviral drugs. Cosmetic formulation based on the association between violacein and lipophilic materials was established to help protect an individual's skin and (semi)mucous membrane against viruses (Kallmayer et al. 2005). Indirect action by the genus *Chromobacterium* on dengue, chikungunya, and Zika virus vectors during *Aedes aegypti* treatment was described in the literature. Violacein could be the innovative therapeutic agent used to treat SARS-CoV-2. Violacein is expected to act as a protease inhibitor at ACE-2 receptor level and as an immunotherapeutic drug against Covid-19 (Duran et al. 2021).

11.6.3 Antifungal

Pure violacein obtained from *Chromobacterium* sp. NIIST (MTCC 5522) was highly efficient against human and plant pathogenic fungi. The eight fungi examined were *A. flavus*, *C. albicans*, *C. tropicalis*, *C. gastricus*, *T. rubrum*, *R. solani*, *F. oxysporum*, and *P. expansum*. Out of these, violacein was discovered to be more efficient against *A. flavus*, *R. solani*, *F. oxysporum*, and *P. expansum* as compared to the common antifungal agents bavistin and amphotericin B (Santhosh and Palaniswamy 2016a, b). Crude violacein extracts retrieved from *J. lividum* exhibited antifungal activity against phytopathogenic fungi *Colletotrichum dematium* and *Rosellinia necatrix* causing anthracnose and white root rot diseases of mulberry, respectively (Duran et al. 2012). Brazilian strains of *C. violaceum* were observed to be useful against seed-pathogenic fungi *Fusarium* sp., *Phomopsis* sp., and *Cercospora kikuchii*. The strains also demonstrated antifungal activity against *Corynespora* sp., *Aspergillus* sp., and *Colletotrichum* sp. The antifungal activity is attributed to the pigment violacein, chitinases, and cyanide production by these strains (Barreto et al. 2008).

Violacein from *J. lividum* displayed antifungal activity against *Batrachochytrium dendrobatidis* (Bd) which causes chytridiomycosis, a major factor responsible for the extinction of amphibian species (Harris et al. 2009; Tabor-Godwin et al. 2009). Out of the six *R. muscosa* juveniles' frogs examined, only one bioaugmented with *J. lividum* survived in the study. Five juveniles not treated with *J. lividum* developed chytridiomycosis, a skin infection, and died. These findings indicate a higher population of *J. lividum* leading to violacein production, which showed antifungal activity (Harris et al. 2009). Violacein was found to be effective against several plant and human pathogenic fungi and yeast species such as *Cryptococcus gastricus*, *Trichophyton rubrum*, *Fusarium oxysporum*, *Rhizoctonia solani*, *Aspergillus flavus*, *Penicillium expansum*, and *Candida albicans*. The greatest activity was recorded against *Trichophyton rubrum* (2 g/mL), a human pathogen responsible for causing athlete's foot disease. This is the first report on the antifungal activity of purified violacein against pathogenic fungi and yeast (Sasidharan et al. 2015).

11.6.4 Antiparasitic

Moderate activity of violacein against *Trypanosoma cruzi* and *Leishmania amazonensis* promastigotes was noted (Duran et al. 1989; Leon et al. 2001). The antiprotozoal properties of violacein could be used to cure diseases such as malaria and leishmaniasis (Leon et al. 2001; Lopes et al. 2009). In vitro and in vivo impact of violacein on human and murine blood stage forms of *Plasmodium* parasites was assessed. Antimalarial activity of violacein was exhibited by showing inhibition of the growth of human and mouse-derived *Plasmodium* parasites. Violacein was seen to be effective against young and mature forms of the human parasite, and its action extended to chloroquine-sensitive and chloroquine-resistant strains of *P. falciparum* (Lopes et al. 2009). The antiprotozoal activity was discovered to correlate positively with the violacein content of the bacterial strains (Matz et al. 2004). Anti-plasmodial activity of purified violacein and deoxy-violacein was investigated. Violacein was active against chloroquine-sensitive and chloroquine-resistant *Plasmodium* strains (IC50 ~0.4 μM against 3D7 and ~0.5 μM against W2 parasites). Deoxy-violacein demonstrated moderate activity against *Plasmodium* strains when compared with violacein (IC50 ~ 11 μM against 3D7 and ~14 μM against W2 parasites) (Bilsland et al. 2018).

Violacein demonstrated IC_{50} of 0.85 ± 0.11 μM against *P. falciparum* strain 3D7. The lowest possible tested concentration of violacein of 0.06 μM effectively inhibited parasite development after 48 h of incubation. C57BL/6 mice infected with a nonlethal (AS) *Plasmodium chabaudi chabaudi* used for in vivo assay showed 39% inhibition in parasitemia on day 7 post-infection, even at a low dose of 0.75 mg/kg/day. Similarly, animals infected with the lethal AJ strain of *P. chabaudi chabaudi* treated with 7.5 mg/kg/day violacein showed 80% survival on day 16 post-infection, significantly demonstrating a protective effect of violacein (Lee et al. 2019). Violacein glucoside derivatives showed anti-nematode activity against pine wood nematode *Bursaphelenchus xylophilus* (Kumar et al. 2015).

11.6.5 Immunomodulatory and Immunostimulatory

Violacein, an indole-derived, purple-colored natural pigment purified from *Chromobacterium violaceum*, has displayed multiple biological activities. It was stated that violacein shows immunosuppressive, analgesic, and antipyretic properties in mice (Ahmed et al. 2021; Antonisamy and Ignacimuthu 2010). The effect of violacein on various immune cell lines, namely, THP-1, MonoMac 6, ANA-1, Raw 264.7 cells, and human peripheral blood mononuclear cells (PBMCs), was studied. TNF-α production was stimulated in murine macrophages (ANA-1 and Raw 264.7), and in PBMCs, IL-6 and IL-1β secretion was detected. The molecular mechanism of activation was confirmed by determining the mRNA expression pattern upon therapy with violacein in Raw 264.7 cells. Incubation with violacein caused stimulation of pathways related to an immune and inflammatory response. Violacein causes TNF-α activation at noncytotoxic concentrations in two murine cell lines (Raw 264.7 and ANA-1) developed by retroviral infection, and cell death was observed at the highest concentration examined (12 μmol/L). Activation of PBMCs by violacein was found at higher concentrations than used for murine cells and was stopped by CU-CPT9a, a specific hTLR8 antagonist (Venegas et al. 2019).

The data utilizing TLR-transfected HEK-293 cells indicate that violacein activates the human TLR8 (hTLR8) receptor signaling pathway, not human TLR7 (hTLR7). It was noticed that the specific hTLR8 antagonist, CU-CPT9a, could suppress the immunostimulatory effect of violacein in PBMCs. The interaction of hTLR8 with violacein in silico was investigated, and it was obvious that violacein could comparably bind to hTLR8 to imidazoquinoline compounds. Therefore, it shows that violacein can contribute to immune therapy strategies. Mice simultaneously injected with 1 μg of LPS and 3.5 mg/kg violacein showed a significant reduction in the production of inflammatory cytokines. Transfer of violacein elicited regulatory T cells and significantly reduced experimental autoimmune Encephalomyelitis (Verinaud et al. 2015).

11.6.6 Antioxidant

Violacein, a violet pigment produced by *Chromobacterium violaceum*, has fascinated much attention due to its pharmacological properties. The antioxidant properties of violacein were examined. The reactivity with oxygen and nitrogen reactive species and 1,1-diphenyl-2-picryl-hydrazyl (DPPH), a stable free radical, was assessed. EPR findings were carried out to measure the reactivity with the hydroxyl radical (Konzen et al. 2006). The compound responded with DPPH (IC (Durán et al. 2021) =30 μM), nitric oxide (IC (Durán et al. 2021) =21 μM), and superoxide radicals (IC 50) = 125 μM) and reduced the hydroxyl radical EPR signal. The compound protected the membranes against peroxidation induced by reactive species in the micromolar range. The reconstitution of violacein into the membranes enhanced its antioxidant effect. The compound violacein has powerful antioxidant potential. Violacein plays a significant role in defense against oxidative stress with the microorganism membrane.

The antioxidant efficiency of the pigment was examined using DPPH as a free radical. In DPPH free radical scavenging assay, the percentage of inhibition also enhanced as the concentration of the pigment boosted. According to the previous study, it has been described that violacein exhibited efficient antioxidant activity (Konzen et al. 2006). The free radical scavenging capability of the pigment was determined against ascorbic acid as a positive control, where the percentage of inhibition was found to be higher with increasing concentration. Hence, the study concluded that the violacein pigment scavenges reactive oxygen species even at a lesser concentration (Gunasekaran et al. 2020).

11.7 Anticancer Properties of Violacein

11.7.1 Development of Cancer

11.7.1.1 Cell Cycle Regulation

Cellular components are doubled and accurately distributed into the daughter cells through a series of events called the cell cycle. Usually, progression through the cell cycle is tightly regulated. Sometimes errors arise in one or more of the cell cycle regulating components which cause cells to divide at inappropriate times or rates, leading to cancer. Actively dividing cells move to the M phase after passing through the interphase's G1, S, and G2 phases. Several internal and external signals influence the headway from one stage of the cell cycle to another and are regulated at key points called checkpoints (Figs. 11.7 and 11.8).

Crucial events of the cell cycle are controlled by cyclin-dependent kinases (CDKs), which activate or deactivate proteins by adding a phosphate group when associated with cyclin protein. Cyclin specifies which protein the CDK will

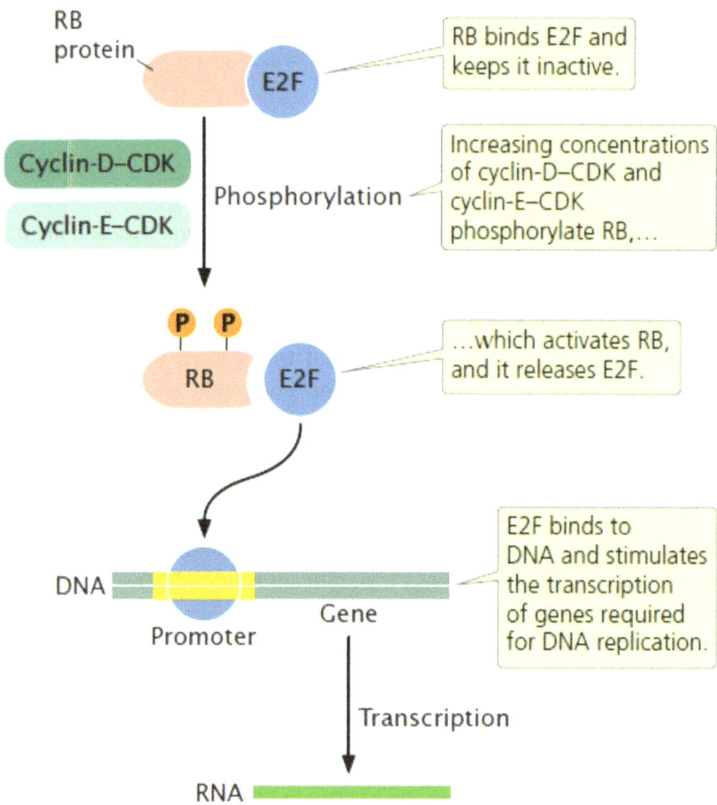

Fig. 11.7 Control by RB protein in the progression through the G_1/S checkpoint (Pierce, B. A., Genetics A Conceptual Approach, 4th Edition, W. H. Freeman and Company, 2012)

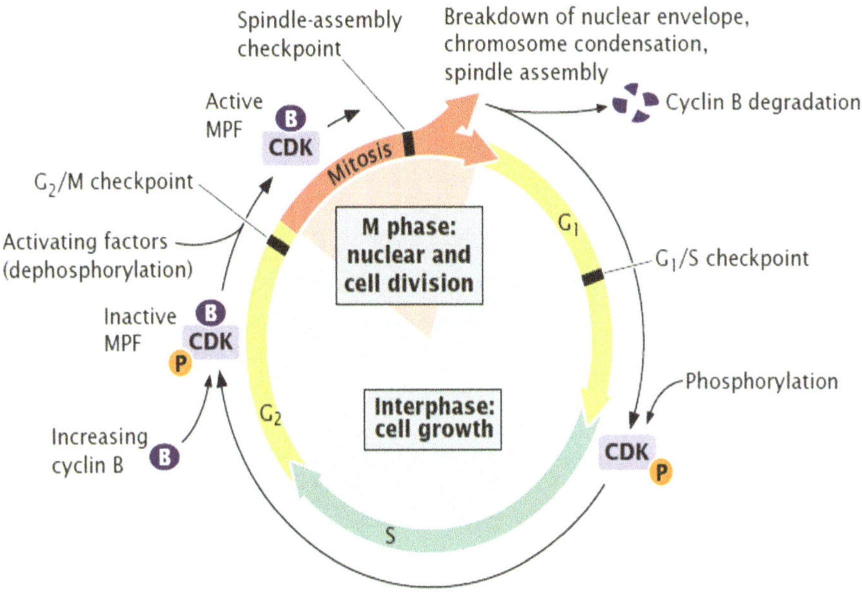

Fig. 11.8 Regulation of progression through G_2/M checkpoint by cyclin B. (Pierce, B.A., Genetics A Conceptual Approach, 4th Edition, W. H. Freeman and Company, 2012)

phosphorylate. Cyclin D-CDK and cyclin E-CDK act in the G_1-S transition with a molecule called retinoblastoma (RB) protein as a regulator. In contrast, cyclin B-CDK acts as a regulator in the G2-M transition with mitosis-promoting factor (MPF).

Many cancers result because of defects in the cell cycle regulatory machinery. Mutations in the genes encoding RB protein are associated with cancers like retinoblastoma. Fifty percent of all breast cancers and some esophageal and skin cancers are observed due to overexpression of genes encoding cyclin D (Pierce 2012).

11.7.1.2 Tumor Suppressor Proteins

Tumor suppressor proteins are the product of tumor suppressor genes that regulates cell proliferation. These proteins act variously to arrest the cell having abnormalities. They control cell division; they repair the damaged DNA, and if the cells' DNA damage cannot be fixed, then proteins like p53 trigger apoptosis in such cells. Tumor suppressor proteins effectively inhibit cell division by causing repression of genes essential for the continuation of the cell cycle. Some tumor suppressor proteins are involved in cell adhesion and are vital in inhibiting metastasis (Fig. 11.9).

DCC, *NF1*, *NF2*, *VHL*, *MLH1*, *MSH2*, *p53*, and *p21* are some of the tumor suppressor genes discovered. Loss of function of the tumor suppressor genes leads to cancer development. However, according to the widely accepted Knudson's two-hit model of tumorigenesis for both hereditary and sporadic cancers, cells must be

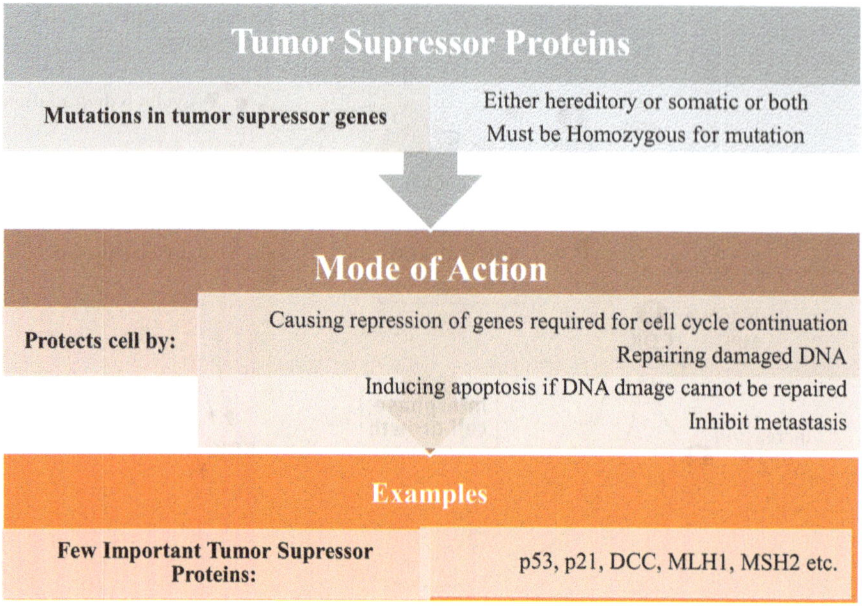

Fig. 11.9 A brief account of tumor suppressor proteins

homozygous for the loss-of-function mutations in a tumor suppressor gene to give rise to a tumor. Allelic loss or reduced expression of DCC protein is detected in 70% of colorectal carcinomas with high frequency in ovarian adenocarcinomas along with gastric, esophageal, pancreatic, breast, hematological, and glial malignancies. Fifteen percent of colon cancers showed a defect in MLH1 MSH2 genes. Seventy-five percent of all colon cancers are observed with a mutation in p53 genes, which regulates a potent inhibitor of CDK activity (Gamudi and Blundell 2010; Pierce 2012).

11.7.2 Mechanism of Violacein Action

11.7.2.1 By Inducing Apoptosis in Cancerous Cells

Violacein, a bisindole secondary metabolite, is known for its anticancer activity. Violacein caused a dose-dependent increase in apoptosis of HCT116 colon cancer cells and enhanced 5-fluorouracil-induced apoptosis (Fig.11.10). Western blot analysis of HCT116 colon cancer cells showed increased levels of pro-apoptotic Bad proteins, a decrease in the phosphorylated Bad^{ser136}, and reduced levels of FADD. This suggests the induction of an intrinsic apoptotic pathway than an extrinsic one with violacein treatment. The PI3 kinase/Akt pathway governs the apoptotic response, an important regulatory pathway. Phosphorylation regulates the activity of Akt. Time-dependent downregulation of phosphorylated Akt^{Thr308} was evident with violacein treatment, approving that violacein-induced apoptosis is mediated

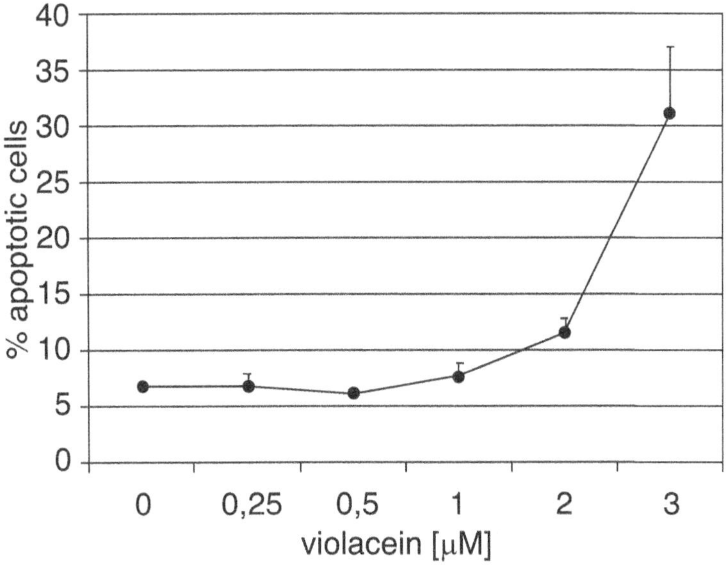

Fig. 11.10 Violacein treatment results in dose-dependent augmentation of apoptosis (Kodach et al. 2006)

through the inhibition of the Akt pathway. MTT assay and Annexin V measurements of HCT116 cells transfected with pmaxGFP plasmid and treated with violacein (0.5–3 μM) showed induction of apoptosis, indicating inhibition of the Akt pathway. Activation of NFκB blocks apoptosis and promotes cell proliferation. Western blot analysis of violacein-treated nuclear extracts of HCT116 cells showed a decrease in nuclear translocation of NFκB as IκBα phosphorylation is blocked by violacein. A time-dependent decrease of p50 and p65 subunits of NFκB expression was observed with violacein treatment. Time-dependent upregulation of p21, p27, and p53 protein expression, a significant reduction of cyclin D1 expression, and downregulation of pRb were also exhibited by violacein (Kodach et al. 2006).

Production of reactive oxygen species (ROS) in HT29 and Caco-2 human colon cancer cells with the treatment of violacein indicates its ability to induce programmed cell death (PCD). Excessive ROS leads to cell death by elevating calcium levels. This further impairs mitochondrial function, which may activate enzymes like endonucleases, proteases, and phospholipases, leading to irreversible damage to organelles, membranes, and chromatin, thereby causing cell death (Daniela et al. 2006; Carvalho et al. 2006). Direct and specific activation of violacein on TNF receptor I signaling in HL60 cells makes it a member of a novel class of cytotoxic drugs mediating apoptosis. Violacein-induced morphological changes are characteristic of apoptosis in human myeloid leukemia cells (Duran et al. 2007; Fakhr et al. 2012).Violacein negatively modulated kinases like CDK, Rock, Axl, and AurkA, which promote cell cycle progression and survival as evidenced by kinome profiling. Violacein moderated two kinases linked with reticulum stress and cell death by apoptosis. It increases the autophosphorylation of DAPK1 at its inhibitory site and leads to expressive activation of PKA by causing higher phosphorylation of its substrate, CREB1 (Queiroz et al. 2012). Violacein induce programmed cell death with an IC_{50} of 2 μmole/dm^3 in PCD-resistant, less differentiated TF1 leukemia cells as evident by MTT reduction assay (Duran et al. 2016).

11.7.2.2 By Inhibiting Proliferation of Tumors

Violacein showed decreased human leukemia cell proliferation with IC_{50} value of less than 1 μmol/L. One of the important components of cell signaling control is protein phosphatases. Twenty percent activation of total protein phosphatase activity was observed in human lymphocytes after 20 min of pre-incubation with 10 μmol/L of violacein. The inhibitory effects attributed to modifying sulfhydryl groups of protein tyrosine phosphatase (Bromberg et al. 2006). Significant antiproliferative activities in a dose-dependent manner of crude violacein extracts of ethyl acetate (AcCE) and ethanol (EtCE) of five *Chromobacterium* isolates CBMAI 307, 308, 310, 592, and 534 revealed on a panel of nine human cancer cell lines MCF-7 (breast), NCI/ADR-RES (ovarian cancer with a phenotype of multiple drug resistance), UACC-62 (melanoma), NCIH460 (non-small-cell lung), PC-3 (prostate), HT29 (colon), OVCAR-3 (ovarian), 786-0 (renal), and U251 (central nervous system) and one normal VERO cell line (monkey kidney normal cells). The most potent extract was AcCE 307, with mean total growth inhibition (TGI) value of 2.50 μg/mL. EtCE 307 and AcCE 592 presented moderate antiproliferative activity, while AcCE 308, AcCE 310, EtCE 310, and AcCE 534 showed weak antiproliferative activity (Menezes et al. 2013). Living cells convert MTT to formazan. MTT proliferation assay displayed antiproliferative activity of violacein with 0–10 μM concentration against A549 non-small-cell lung cancer, A431 melanoma, MCF-7 breast cancer, PC3 prostate cancer, HT29 colon cancer, and HeLa cervical cancer. It was found effective against HCT116 colon and HN5 head and neck cancers (Hashimi et al. 2015). Violacein extracts from *Chromobacterium vaccinii* CV5 showed a dose-dependent decrease in the proliferation of cervical (HeLa) and lung (A549) cancer cells after 48 h of incubation (Santhosh and Palaniswamy, 2017). Violacein showed dose-dependent inhibition of the human urinary bladder carcinoma T24 and 5637 cells, with IC_{50} values of 0.1135 ± 0.5 μM and 0.1129 ± 0.1 μM, respectively (Fig. 11.11). The effect of violacein was significant on the second-stage bladder cancer cell line (5637) (Neroni et al. 2022).

11.7.2.3 By Caspase Activity

Caspase is an enzyme that cleaves the protein substrate, leading to cell dismantling in the early stages of apoptosis. Caspase 3, a fundamental mediator of apoptosis, also known as death protease, cuts the precise protein substrate and initiates the process of apoptosis. Violacein exhibited a time-dependent increase in the caspase 3 activity with the concentration of 5 μM in human colon cancer cells Caco-2 and HT-29 (Carvalho et al. 2006) (Fig. 11.12). Western blot analysis of violacein-treated human colorectal HCT116 cancer cells displayed the activity of caspase 3 after 45 min of the

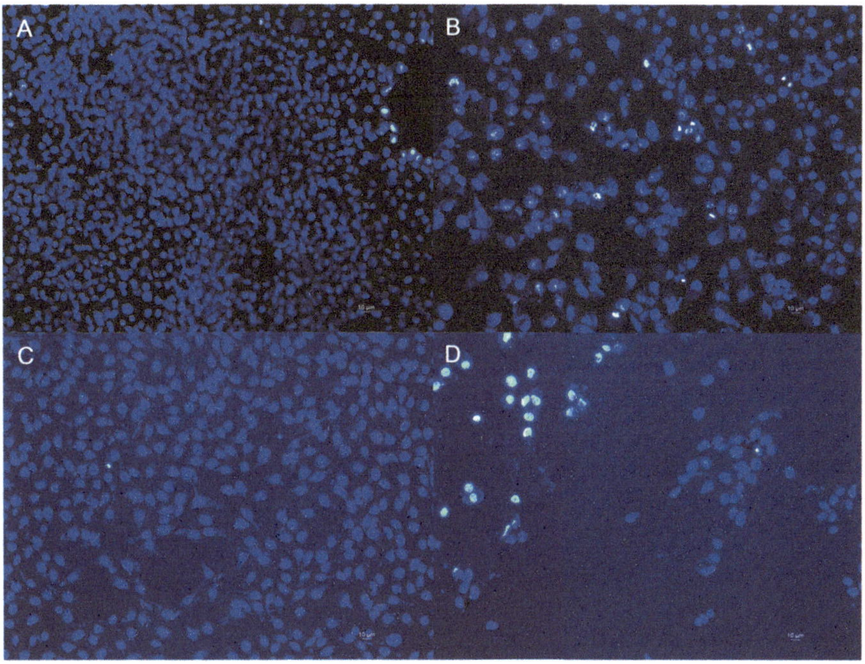

Fig. 11.11 Violacein exposure to T24 and 5637 urinary bladder carcinoma cells (left, untreated cells; right, treated cells; top, T24 cells; bottom, 5637 cells) (Neroni et al. 2022)

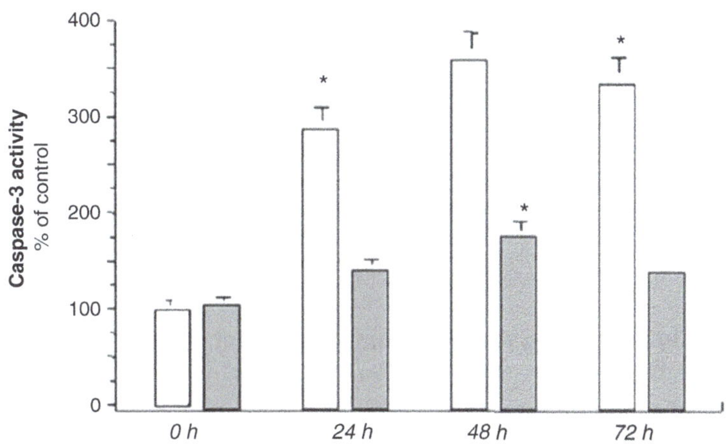

Fig. 11.12 Caspase 3 activity after 24, 48, and 72 h of incubation with 5 lM of violacein in Caco-2 (white) and HT-29 (gray) cancer cells. *$P < 0.001$ compared to control caspase 3 (Carvalho et al. 2006)

treatment. Further, MTT assay with specific caspase 8 and 9 inhibitors showed reduced violacein activity on HCT116 cells, indicating the activation of caspase 8 and 9 with violacein treatment (Kodach et al. 2006). Flow cytometric analysis of the human acute lymphoblastic leukemia Jurkat E6.1 cell lines displayed an increase in the caspase 3, 7, and 8 activities by 3.5, 4.5, and 6 times, respectively, with 7 μM of violacein and violacein coupled with dendrimers (Fakhr et al. 2012). Caspase-Glo 3/7 assay system exhibited induction of caspase 3/7 activity in A549 non-small-cell lung cancer cells (Hashimi et al. 2015). Caspase 3 fluorometric assay showed maximum expression of active caspase 3 after 3 h of violacein treatment in urinary bladder carcinoma T24 and 5637 cells, later higher (Neroni et al. 2022). All the data indicates violacein induces apoptosis in cancer cells in a caspase-dependent pathway.

11.7.2.4 By Exhibiting Cytotoxic Effects

MTT assay of Caco-2 and HT29 cells grown in the presence of violacein showed 50% reduction in the viability of Caco-2 cells after 24 h of incubation with IC_{50} value of 2 μM, while HT29 cells showed a reduction in viability after 72 h of exposure to violacein with IC_{50} values of more than 10 μM. These observed differences in the IC_{50} values of the two cell lines indicate violacein's cell type-specific mode of action (Carvalho et al. 2006). Violacein inhibits the growth of Caco-2, HCT116, SW480, and DLD-1 cells after 48 h of incubation and displayed an IC_{50} value of ~1–2 μM. Further, when these cells were grown with 5-fluorouracil (5-FU) alone and 5-FU with violacein for 48 h, violacein enhanced the effect of 5-FU synergistically (Kodach et al. 2006). Cytotoxic assay of human acute lymphoblastic leukemia Jurkat E6.1 cells was carried out with violacein alone and the formation of nanocomplex by coupling violacein with nanoparticle dendrimer-ascorbic acid (Den-AA-Vio). The results revealed two times more activity of Den-AA-Vio nanocomplex than free violacein with IC_{50} values of 4 and 7 μM, respectively (Fakhr et al. 2012). Violacein exhibited a cytotoxic effect on HeLa (cervical cancer) and A549 (lung cancer) cell lines with a better effect on A549 cells (Vishnu and Palaniswamy 2017). Cytotoxic effect of violacein on T24 and 5637 human urinary bladder carcinoma cells was evaluated by measuring cell viability with MTT assay and trypan blue dye. The MTT assay values indicate the induction of a permanent cytotoxic effect with violacein treatment. Cell viability in both cell lines was inhibited with violacein treatment in a dose-dependent manner (Neroni et al. 2022).

11.7.2.5 By Displaying Anti-inflammatory Effect

Matrix metalloproteinases (MMP) play a vital role in cancer metastasis. MCF-7 breast cancer cells treated with 1 μM violacein in DMSO downmodulated expression of MMP-2 from 156 ± 22 to 42 ± 10 (densitometric units), $p < 0.05$. This is credited to the anti-inflammatory effect of violacein. There is a correlation between cancer metastasis and chemokine CXCL12 and its interaction with receptor CXCR4. CXCL12/CXCR4 is required for angiogenesis. Violacein downregulated expression of chemokine/receptor CXCL12/CXCR4, which leads to hypoxic conditions in the absence of angiogenesis. Direct inhibition of pro-inflammatory cytokine-induced

MMP-2-mediated CXCL12 secretion was displayed in violacein-treated MCF-7 breast cancer cells. Violacein was also demonstrated to inhibit MMP-9 expression and activity efficiently (Platt et al. 2014). When administered orally, violacein decreases inflammatory cytokines, particularly TNF-α, thereby inhibiting inflammation and angiogenesis, maintaining a balance of cytokines and promoting healing (Choi et al. 2021).

11.7.2.6 By Persuading Stress

Violacein has been reported to cause programmed cell death in HT29 and Caco-2 human colon cancer cells by producing reactive oxygen species (ROS). Excessive ROS leads to cell death by elevating calcium levels, further impairing mitochondrial function, which may activate enzymes like endonucleases, proteases, and phospholipases, leading to irreversible damage to organelles, membranes, and chromatin, thereby causing cell death (Carvalho et al. 2006).

11.8 Conclusion

Violacein is a versatile bacterial secondary metabolite. It is produced by condensation of two tryptophan molecules. Its production is governed by *vioABCDE* operon. Violacein biosynthesis is influenced by a small molecule responsible for quorum sensing called acyl-homoserine lactone. Hence, violacein production is subjected to the attainment of a specific cell population. Different factors like carbon source, pH, temperature, and agitation rate affect violacein production.

Different bacteria from the proteobacteria group are known to produce violacein, a purple-violet colored pigment. They include genera like *Chromobacterium*, *Janthinobacterium*, *Duganella*, *Collimonas*, *Pseudoalteromonas*, *Iodobacter*, *Rugamonas*, and *Massilia* isolated from diverse habitats including simple habitats like soil, freshwater to extreme environments like Antarctic regions, and coral and marine sponges.

This bisindole compound is catching the attraction of the pharmaceutical industry for its versatile biological activities. The biological activities of violacein include antiviral activity with the exhibition of potential antiviral agent against SARS-2 coronavirus and antibacterial, antifungal, antiparasitic, antiprotozoal, antimalarial, and antinematodal activities. Among its biological activities, its anticancer activities make it a potential therapeutic candidate for various cancer treatments. Violacein showed good antitumor activity against various resistant cell lines. It induces apoptosis by creating oxidative stress, activating caspases, regulating cell cycle proteins, modulating different cellular signaling pathways, and controlling inflammatory cytokines. Violacein also showed synergistic antitumor activity with other chemotherapeutic drugs and nanoparticles. Most importantly, the low concentrations of violacein show promising effects on various cancerous cells with no or minimal effect on normal cells, hoping its use in cancer treatment is without any side effects. Further clinical trials must be conducted to make this multipurpose metabolite a promising therapeutic agent.

Acknowledgments Authors are thankful to Haribhai V. Desai College of Arts, Science and Commerce, Pune, Maharashtra, India, affiliated to Savitribai Phule Pune University (SPPU), for constant support in the completion of the present Springer book chapter.

Conflict of Interest Statements The authors declare that there is no conflict of interests regarding the publication of this article.

References

Agematu H, Suzuki K, Tsuya H (2011) *Massilia* sp. BS-1, a novel violacein-producing bacterium isolated from soil. Biosci Biotechnol Biochem 75(10):2008–2010

Ahmed A, Ahmad A, Li R, Al-Ansi W, Fatima M, Mushtaq BS, Basharat S, Li Y, Bai Z (2021) Recent advances in synthetic, industrial and biological applications of violacein and its heterologous production. J Microbiol Biotechnol 31(11):1465–1480. https://doi.org/10.4014/jmb.2107.07045

Alem D, Canclini L, Castro-Sowinski S, Martínez-López W (2022) Chemosensitizer effect of violacein on cisplatin-treated bladder cancer cells. Clin Complementary Med Pharma 2 (2):100036. https://doi.org/10.1016/j.ccmp.2022.100036

Alves OL, Gimenez IF, De Azevedo MMM, Duran N, Melo PS (2005) Pharmacological use of cyclodextrine-Au-thiol-derivative/hydrophobic compound nanoparticles as antih1moral, antibacterial, antiviral and/or antiparasites, its obtention process and formulation. Brazil Pat; PIBr 0502657-1

Andrighetti-Frohner CR, Antonio RV, Creczynski-Pasa TB, Barardi CRM, Simoes CMO (2003) Cytotoxicity and potential antiviral evaluation of violacein produced by *Chromobacterium violaceum*. Mem Inst Oswaldo Cruz 98(6):843–848

Antonio RV, Creczynski-Pasa TB (2004) Genetic analysis of violacein biosynthesis by *Chromobacterium violaceum*. Genet Mol Res 3(1):85–91

Antonisamy P, Ignacimuthu S (2010) Immunomodulatory, analgesic and antipyretic effects of violacein isolated from *Chromobacterium violaceum*. Phytomedicine 17:300–304. https://doi.org/10.1016/j.phymed.2009.05.018

Aruldass CA, Rubiyatno CKV, Wan AA (2015) Violet pigment production from liquid pineapple waste by *Chromobacterium violaceum* UTM5 and evaluation of its bioactivity. R Soc Chem Adv 5:51524–51536

Aruldass CA, Raj S, Masalamany L, Venil CK, Ahmad WA (2018) Antibacterial mode of action of violacein from *Chromobacterium violaceum* UTM5 against *Staphylococcus aureus* and methicillin-resistant *Staphylococcus aureus* (MRSA). Environ Sci Pollut Res 25:5164. https://doi.org/10.1007/s11356-017-8855-2

Asencio G, Lavin P, Karen A, Dominguez M, Bello H, Gonzalez-Rocha G, Gonzalez-Aravena M (2014) Antibacterial activity of the Antarctic bacterium *Janthinobacterium* sp. SMN 33.6 against multi-resistant gram-negative bacteria. Electron J Biotechnol 17:1–5

Atalah J, Blamey L, Munoz-Ibacache S, Gutierrez F, Urzua M, Encinas MV, Paez M, Sun J, Blamey JM (2020) Isolation and characterization of violacein from an Antarctic *Iodobacter*: a non-pathogenic psychrotolerant microorganism. Extremophiles 24:43–52

August PR, Grossman TH, Minor C, Draper MP, MacNeil IA, Pemberton JM, Call KM, Holt D, Osburne MS (2000) Sequence analysis and functional characterization of the violacein biosynthetic pathway from *Chromobacterium violaceum*. J Mol Microbiol Biotechnol 2(4):513–519

Aye AM, Jusserand MB, Jaisson FB, Magne AO, Culioli G, Nevry RK, Rabah N, Blache Y, Molmeret M (2015) Modulation of violacein production and phenotypes associated with biofilm by exogenous quorum sensing N-acylhomoserine lactones in the marine bacterium *Pseudoalteromonas ulvae* TC14. Microbiology 161:2039–2052

Baricz A, Teban A, Chiriac CM, Szekeres E, Farkas A, Nica M, Dascalu A, Oprişan C, Lavin P, Coman C (2018) Investigating the potential use of an Antarctic variant of *Janthinobacterium lividum* for tackling antimicrobial resistance in a one health approach. Sci Rep 8:15272

Barreto ES, Torres AR, Barreto MR, Vasconcelos ATR, Filho SA, Hungria M (2008) Diversity in antifungal activity of strains of *Chromobacterium violaceum* from the Brazilian Amazon. J Ind Microbiol Biotechnol 35:783–790

Batista AHM, Moreira ACD, Carvalho RM, Sales GWP, Nogueira PCN, Grangeiro TB, Medeiros SC, Silveira ER, Nogueira NAP (2017) Antimicrobial effects of violacein against planktonic cells and biofilms of *Staphylococcus aureus*. Molecules 22:1534

Berni E, Marcato PD, Nakazato G, Kobayashi RKT, Vacchi FI, Umbuzeiro GA, Durán N (2013) Violacein/poly(ε-caprolactone)/chitosan nanoparticles against bovine mastitis: antibacterial and ecotoxicity evaluation. J Phys 429:012030

Bilsland E, Tavella TA, Krogh R, Stokes JE, Roberts A, Spring JD, Andricopulo AD, Costa FTM, Oliver SG (2018) Antiplasmodial and trypanocidal activity of violacein and deoxyviolacein produced from synthetic operons. BMC Biotechnol 18(1):22

Bromberg N, Justo GZ, Haun M, Duran N, Ferreira CV (2006) Violacein cytotoxicity on human blood lymphocytes and effect on phosphatases. J Enzyme Inhib Med Chem 20(5):449–454

Brotosudarmo THP (2021) Exploring bioactive pigments from marine bacterial isolate from the Indonesian Seas. Earth Environ Sci 944:12016. https://doi.org/10.1088/1755-1315/944/1/012016

Cardona-Cardona V, Arroyo D, Scellekens J, Rios-Velazquez C (2010) Characterisation of blue pigmented bacteria isolated from Puerto Rico. Curr Res Technol Educ Topics Appl Microbiol Microbial Biotechnol 1:117–123

Cazoto LL, Martins D, Ribeiro MG, Duran N, Nakazato G (2011) Antibacterial activity of violacein against *Staphylococcus aureus* isolated from Bovine mastitis. J Antibiot 64:395–397

Chatragadda R, Nambali VV, Ramalingam K (2019a) Marine pigmented bacteria: a prospective source of antibacterial compounds. J Nat Sci Biol Med 10(2):104–113

Chatragadda R, Nambali VV, Ramalingam K, Chidambaram KV, Laurent D (2019b) Multifaceted applications of microbial pigments: current knowledge, challenges and future directions for public health implications. Microorganisms 7:186

Cheng KC, Hsiao HC, Hou YC, Hsieh CW, Hsu HY, Chen HY, Lin SP (2022) Improvement in violacein production by utilizing formic acid to induce quorum sensing in *Chromobacterium violaceum*. Antioxidants 11:849. https://doi.org/10.3390/antiox11050849

Chidambaram KV, Nur ZY, Claira AA, Wan AA (2016) Application of violet pigment from *Chromobacterium violaceum* UTM5 in textile dyeing. Biologia 71:121. https://doi.org/10.1515/biolog-2016-0031

Choi SY, Kim S, Lyuck S, Kim SB, Mitchell RJ (2015a) High-level production of violacein by the newly isolated *Duganellaviolaceinigra* str. NI28 and its impact on *Staphylococcus aureus*. Sci Rep 5:15598

Choi SY, Yoon K, Lee JI, Mitchell RJ (2015b) Violacein: properties and production of a versatile bacterial pigment. BioMed Res Int 2015:465056, 8p

Cortes-Osorio N, Cardoso-Cardoso M, Chavarro-Anzola V, Prada-Salcedo LD, Reyes-Cardenas A, Robles-Camargo J, Franco-Correa M (2017) Influence of environmental factors on the production of violacein synthesized by *Janthinobacterium lividum*. Int J Eng Sci 6(1):76–83

Dang HT, Yotsumoto K, Enomoto K (2014) Draft genome sequence of violacein-producing marine bacterium *Pseudoalteromonas* sp. 520P1. Genome Announc 2(6):e01346

Daniela D, Carvalho D, Costa FTM, Duran N, Haun M (2006) Cytotoxic activity of violacein in human colon cancer cells. Toxicol In Vitro 20:1514–1521

de Carvalho DD, Costa FT, Duran N, Haun M (2006) Cytotoxic activity of violacein in human colon cancer cells. Toxicol in Vitro 20(8):1514–1521. https://doi.org/10.1016/j.tiv.2006.06.007

Demoss RD, Evans NR (1959) Physiological aspects of violacein biosynthesis in nonproliferating cells. J Bacteriol 78:583–588

Devi P, Kannan SS, R and Mariappan T. K. (2018) Production of violacein pigment from *Chromobacterium violaceum* and its antibacterial activity and synergism on *E. coli* isolated from UTI samples. Int J Recent Sci Res 9(2):24479–24484

Diego AJ, Marizcurrena V, Saravia D, Lopez WM, Sowinski SC (2020) Production and antiproliferative effect of violacein, a purple pigment produced by an Antarctic bacterial isolate. World J Microbiol Biotechnol 36:120

Dike-Ndudim JN et al (2022) Studies on production and *in vitro* antibacterial potentials of violacein from *Chromobacterium violaceum* isolated from Otamiri River in Owerri, Imo State. J Microbiol Biotechnol 7(1):000215

Dodou HV, de Morais Batista AH, Sales GWP, de Medeiros SC, Rodrigues ML, Nogueira PCN, Silveira ER, Nogueira NAP (2017) Violacein antimicrobial activity on *Staphylococcus epidermidis* and synergistic effect on commercially available antibiotics. J Appl Microbiol 123:853–860

Doing G, Perron GG, Brooke AJ (2018) Draft genome sequence of a violacein-producing *Iodobacter* sp. from the Hudson Valley watershed. Am Soc Microbiol 6(1):e01428

Duran N, Campos V, Riveros R, Joyas A, Pereira MF, Haun M (1989) Bacterial chemistry. III. Preliminary studies on trypanosomal activities of Chromobacterium violaceum products. An Acad Bras Cienc 61:31–36

Duran N, De Azevedo MBM, Alderete J (1998) Formulation process of cyclodextrin/violacein for using as antibacterial, antitumoral, antiviral and trypanocide. Brazil Patent 1998; PIBr 9801307

Duran N, Justo GZ, Ferreira CV, Melo PS, Cordi L, Martins D (2007) Violacein: properties and biological activities. Biotechnol Appl Biochem 48:127–133

Duran M, Ponezi AN, Faljoni-Alario A, Teixeira MFS, Justo GZ, Duran N (2012) Potential applications of violacein: a microbial pigment. Med Chem Res 21:1524–1532

Duran N, Justo GZ, Justo DM, Brocchi M, Cordi L, Tasic L, Castro GR, Nakazato G (2016) Advances in *Chromobacterium violaceum* and properties of violacein-its main secondary metabolite: a review. Biotechnol Adv 34:1030–1045

Duran N, Justo GZ, Nakazato G, Fávaro WJ (2021). Violacein, a microbial antiviral product: does play key role as active agent against SARS-CoV-2?. AIJR preprints, 315, version 1

Durán N, Nakazato G, Durán M et al (2021) Multi-target drug with potential applications: violacein in the spotlight. World J Microbiol Biotechnol 37:151. https://doi.org/10.1007/s11274-021-03120-4

Fakhr FA, Khanafari A, Baserisalehi M, Yaghoobi R, Shahghasempour S (2012) An investigation of antileukemia activity of violacein-loaded dendrimer in Jurkat cell lines. Af J Microbiol Res 6(33):6235–6242

Friedrich I, Hollensteiner J, Schneider D, Poehlein A, Hertel R, Daniel R (2020) First complete genome sequences of *Janthinobacterium lividum* EIF1 and EIF2 and their comparative genome analysis. Genome Biol Evol 12(10):1782–1788

Fuller JJ, Rene R, Joern K, Rennhack KE, Daniel NP, Wulf B, Stefan S, Dieter J, Jurgen M (2016) Biosynthesis of violacein, structure and function of L-Tryptophan oxidase VioA from *Chromobacterium violaceum*. J Biol Chem 291(38):20068–20084. https://doi.org/10.37231/jab.2022.13.1.262

Gamudi D, Blundell R (2010) Tumor suppressor genes. Res J Med Sci 4(4):280–284. https://doi.org/10.3923/rjmsci.2010.280.284

Gunasekaran R, Janarthanam H, Selvaraj V (2020) Antagonistic and antioxidant efficiency of bacterial pigment isolated from Chromobacterium violaceum (BRT-GR2). Int J Curr Microbiol Appl Sci 9(7):4050–4059

Hakvag S, Fjærvik E, Klinkenberg G, Borgos SEF, Josefsen KD, Ellingsen TE, Zotchev SB (2009) Violacein-producing *Collimonas* sp. from the sea surface microlayer of coastal waters in Trondelag, Norway. Mar Drugs 7:576–588

Harris RN, Brucker RM, Walke JB, Becker MH, Schwantes CR, Flaherty DV, Lam BA, Woodhams DC, Briggs CJ, Vredenburg VT, Minbiole KPC (2009) Skin microbes on frogs prevent morbidity and mortality caused by a lethal skin fungus. Int Soc Microbiol Ecol J 3:818–824

Hashimi SM, Xu T, Wei MQ (2015) Violacein anticancer activity is enhanced under hypoxia. Oncol Rep 33:1731–1736

Inniss WE, Mayfield CI (1979) Effect of temperature on violacein production in a psychrotrophic*Chromobacterium* from Lake Ontario sediment. Microb Ecol 5:51–56

Jayasree VS, Sobhana KS, Priyankaa P, Keerthia KR, Jasmine S, Ranjith L, Ramkumar S, Saravanan R, Kingsley HJ, Sreenath KR, George RM, Joshi KK, Gopalakrishnan A (2021) Characterization and antibacterial activity of violacein producing deep purple pigmented bacterium *Pseudoalteromonas luteoviolacea* (Gauthier, 1982) isolated from coral reef ecosystems. Ind J Geo Marine Sci 50(8):620–634

Kallmayer V, Lanzendoerfer G, Meiring U, Mocigemba N, Reidel H, Schaefer J et al (2005) Cosmetic preparation, useful e.g. for the protection of skin and (semi)mucous membrane against bacteria and/or virus, comprises violacein dye in combination with lipophilic and/or hydrophilic substances. Germany Patent 2005; DE102005051869 Al

Kanade YB, Waman M, Patwardhan RB (2022) Violacein: a promising bacterial secondary metabolite. Res J Chem Environ 26(6):165–177

Kanelli M, Mandic M, Kalakona M, Vasilakos S, Kekos D, Nikodinovic-Runic J, Topakas E (2018) Microbial production of violacein and process optimization for dyeing polyamide fabrics with acquired antimicrobial properties. Front Microbiol 9:1495. https://doi.org/10.3389/fmicb.2018.01495

Kodach LL, Bos CL, Duran N, Peppelenbosch MP, Ferreira CV, Hardwick JCH (2006) Violacein synergistically increases 5-fluorouracil cytotoxicity, induces apoptosis and inhibits Akt-mediated signal transduction in human colorectal cancer cells. Carcinogenesis 27(3):508–516

Konzen M, Marco DD, Cordova CAS, Vieira TO, Antonio RV, Creczynski-Pasa TB (2006) Antioxidant properties of violacein: possible relation on its biological function. Bioorg Med Chem 14:8307–8313

Kothari V, Sharma S, Padia D (2017) Recent research advances on *Chromobacterium violaceum*. Asian Pac J Trop Med 10(8):744–752

Kumar P, Kumar P, Barrett DM, Barrett DM, Delwiche MJ, Delwiche MJ et al (2009) Methods for pretreatment of lignocellulosic biomass for efficient hydrolysis and biofuel production. Ind Eng Chem 48:3713–3729

Kumar A, Vishwakarma HS, Singh J, Dwivedi S, Kumar M (2015) Microbial pigments: production and their applications in various industries. Int J Pharm Chem Biol Sci 5(1):203–212

Lee YJ, Bashyal P, Pandey RP, Sohng JK (2019) Enzymatic and microbial biosynthesis of novel violacein glycosides with enhanced water solubility and improved anti-nematode activity. Biotechnol Bioprocess Eng 24:366–374

Leon LL, Miranda CC, De Souza AO, Duran N (2001) Antileishmanial activity of the violacein extracted from Chromobacterium violaceum. J Antimicrob Chemother 48:449–450

Loke WK, Saud HM (2022) The identification of *Chromobacterium violaceum* from soil based on the production of violacein. J Agrobiotech 13(1):33–39

Lopes SCP, Blanco YC, Justo GZ, Nogueira PA, Rodrigues FLS, Goelnitz U, Wunderlich G, Facchini G, Brocchi M, Duran N, Costa FTM (2009) Violacein extracted from *Chromobacterium violaceum* inhibits *Plasmodium* growth in vitro and in vivo. Antimicrob Agents Chemother 53(3):2149–2152

Malik K, Tokkas J, Goyal S (2012) Microbial pigments: a review. Int J Microbial Resour Technol 1(4):361–365

Martins D, Costa F, Brocchi TMM, Durán N (2010) Evaluation of the antibacterial activity of poly-(d,l-lactide-co-glycolide) nanoparticles containing violacein. J Nanopart Res 13:355–363

Martins D, Ribeiro MG, Duran N, Gerson N (2011) Antibacterial activity of violacein against *Staphylococcus aureus* isolated from Bovine mastitis. J Antibiotics 64:395–397

Matz C, Deines P, Boenigk J, Arndt H, Eberl L, Kjelleberg S, Jürgens K (2004) Impact of violacein-producing bacteria on survival and feeding of bacterivorous nanoflagellates. Appl Environ Microbiol 70(3):1593–1599. https://doi.org/10.1128/AEM.70.3.1593-1599.2004. PMID: 15006783; PMCID: PMC368400

May G, Brummer B, Ott H (1991) Treatment of prophylaxis of polio and herpes virus infections comprises admin of 3-(1,2-dihydro-5-(5-hydroxy-1H- indol-3-yl)-2-oxo3H-pyrrole-3-ylidene)-1,3-dihydro-2H-indol-2-one. GerOffenDE:3935066

McClean KH, Winson MK, Fish L, Taylor A, Chhabra SR, Camara M, Daykin M, Lamb JH, Swift S, Bycroft BW, Stewart GSAB, Williams P (1997) Quorum sensing and *Chromobacterium violaceum*: exploitation of violacein production and inhibition for the detection of N-acyl homoserine lactones. Microbiology 143:3703–3711

Menezes CBA, Silva BP, Sousa IMO, Ruiz ALTG, Spindola HM, Cabral E, Eberlin MN, Tinti SV, Carvalho JE, Foglio MA, Fantinatti-Garboggini F (2013) *In vitro* and *in vivo* antitumor activity of crude extracts obtained from Brazilian *Chromobacterium* sp isolates. Braz J Med Biores 46: 65–70

Momen ZM, Hoshino T (2000) Biosynthesis of violacein: intact incorporation of the tryptophan molecule on the oxindole side, with intramolecular rearrangement of the indole ring on the 5-hydroxyindole side. Biosci Biotechnol Biochem 64(3):539–549

Morohoshi T, Fukamachi K, Kato M, Kato N, Ikeda T (2010) Regulation of the violacein biosynthetic gene cluster by acylhomoserine lactone-mediated quorum sensing in *Chromobacterium violaceum* ATCC 12472. Biosci Biotechnol Biochem 74(10):2116–2119

Nakazato G, Gonçalves MC, da Silva das Neves M, RKT K, Brocchi M, Durán N (2019) Violacein@Biogenic Ag system: synergistic antibacterial activity against *Staphylococcus aureus*. Biotechnol Lett 41(12):1433–1437. https://doi.org/10.1007/s10529-019-02745-8. Epub 2019 Oct 24. PMID: 31650420

Nathan VK, Subha Rajam K, Rani ME, Rathinasamy G, Kannan ND (2019) Surface culturing of Chromobacterium violaceum MTCC 2656 for violacein production and prospecting its bio-activities, chapter 3, Current Research in Micro

Neroni B, Antonella ZM, Radocchia G, Ciardi MR, Mosca L, Pantanella F, Schippa S (2022) Evaluation of the anti-proliferative activity of violacein, a natural pigment of bacterial origin, in urinary bladder cancer cell lines. Oncol Lett 23:132. https://doi.org/10.3892/ol.2022.13252

Oh WT, Giri SS, Yun S, Kim HJ, Kim SG, Kim SW, Kang JW, Han SJ, Kwon J, Jun JW, Park SC (2019) *Janthinobacterium lividum* as an emerging pathogenic bacterium affecting rainbow trout (*Oncorhynchus mykiss*) fisheries in Korea. Pathogens 8:146. https://doi.org/10.3390/pathogens8030146

Palukurty MA, Pyla SP, Silarapu S, Subba RS (2016) Analysing alternative nutrient supplements and optimization of production parameters for violacein using central composite design. Int J Sci Engi Res 7(7):294

Pantanella F, Berlutti F, Passariello C, Sarli S, Morea C, Schippa S (2007) Violacein and biofilm production in *Janthinobacterium lividum*. J Appl Microbiol 102:992–999. https://doi.org/10.1111/j.1365-2672.2006.03155

Pawar R, Mohandass C, Rajasabapathy R, Meena RM (2018) Molecular diversity of marine pigmented bacteria in the central Arabian Sea with special reference to antioxidant properties. Cah Biol Mar 59:409–420

Pierce BA (2012) Genetics a conceptual approach, 4th edn. W. H. Freeman, New York

Platt D, Amara S, Mehta T, Vercuyssee K, Myles EL, Johnson T, Tiriveedhi V (2014) Violacein inhibits matrix metalloproteinase mediated CXCR4 expression: potential anti-tumor effect in cancer invasion and metastasis. Biochem Biophys Res Commun 5(455):107–112

Queiroz KCS, Milani R, Ruela-de-Sousa RR, Fuhler GM, Justo GZ, Zambuzzi WF, Duran N, Diks HS, Spek CA, Ferreira CV, Peppelenbosch MP (2012) Violacein induces death of resistant leukaemia cells via kinome reprogramming, endoplasmic reticulum stress and golgi apparatus collapse. PLoS One 7(10):e45362

Ramdass AC, Rampersad SN (2021) Molecular signatures of *Janthinobacterium lividum* from Trinidad support high potential for crude oil metabolism. BMC Microbiol 21:287. https://doi.org/10.1186/s12866-021-02346-4

Riveros R, Haun M, Durán N (1989) Effect of growth conditions on production of violacein by *Chromobacterium violaceum* (BB-78 strain). Braz J Med Biol Res 22(5):569–577

Rokade MT, Pethe AS (2017) Isolation, identification, extraction and production of antibacterial violacein pigment by psychrotrophic bacterium MTRI7 strain. J Global Biosci 6(6):5077–5083

Santhosh VT, Palaniswamy M (2016a) Isolation and identification of *Chromobacterium* sp. from different ecosystems. Asian J Pharm Clin Res 9(suppl 3):253–257

Santhosh VT, Palaniswamy M (2016b) Isolation and identification of *Chromobacterium* sp. from different ecosystems. Asian J Pharma Clin Res 9(suppl 3):253–257

Santhosh VT, Palaniswamy M (2017) Evaluation of *in vitro* cytotoxic effect of violacein produced by novel isolate *Chromobacterium vaccinii* CV5 against the cervical and lung cancer cell. Asian J Pharm Clin 10(10):227–229

Sasidharan A, Sasidharan NK, Amma DBNS et al (2015) Antifungal activity of violacein purified from a novel strain of *Chromobacterium* sp. NIIST (MTCC 5522). J Microbiol 53:694–701. https://doi.org/10.1007/s12275-015-5173-6

Schloss PD, Allen HK, Klimowicz AK, Christine M, Gross JA, Sarah S, Jennifer ME, Jon C, Ruess RW, Jo H (2010) Psychrotrophic strain Janthinobacterium lividum from a cold Alaskan soil produces prodigiosine. DNA Cell Biol 29(9):533–541

Sedlacek I, Holochova P, Sobotka R, Busse H-J, Svec P, Kralova S, Sedo O, Pilny J, Stankova E, Koublova V, Sedlar K (2021) Classification of a violacein-producing psychrophilic group of isolates associated with freshwater in Antarctica and description of *Rugamonas violacea* sp. nov. Microbiol Spectr 9:e00452–e00421. https://doi.org/10.1128/Spectrum.00452-21

Sedlacek I, Holochova P, Busse H-J, Koublova V, Kralova S, Svec P, Sobotka R, Stankova E, Pilny J, Sedo O, Smolikova J, Sedlar K (2022) Characterisation of waterborne psychrophilic *Massilia* isolates with violacein production and description of *Massilia antarctica* sp. nov. Microorganisms 10:704. https://doi.org/10.3390/microorganisms10040704

Sen T, Barrow CJ, Deshmukh SK (2019) Microbial pigments in the food industry - challenges and the way forward. Front Nutr 6:7

Shirata A, Tsukamo T, Yasuj H, Hata T, Hayasaka SS, Kojima AA, Kato H (2000) Isolation of bacteria producing bluish-purple pigment and use for dyeing. Japan Agric Res Q 34:131–140

Sinha S, Choubey S, Ajay KA, Bhosale P (2017) Identification, characterization of pigment producing bacteria from soil and water and testing of antimicrobial activity of bacterial pigments. Int J Pharm Sci Rev Res 42(2):119–124

Siva R (2007) Status of natural dyes and dye yielding plants in India. Curr Sci 92(7):916–925

Soliev AB, Hosokawa K, Enomoto K (2011) Bioactive pigments from marine bacteria: applications and physiological roles. Evid Based Complement Alternat Med 2011:670349, 1–17. https://doi.org/10.1155/2011/670349

Srinivasan R, Kannappan A, Shi C, Lin X (2021) Marine bacterial secondary metabolites: a treasure house for structurally unique and effective antimicrobial compounds. Mar Drugs 19:530. https://doi.org/10.3390/md19100530

Subramaniam S, Ravi V, Sivasubramanian A (2014) Synergistic antimicrobial profiling of violacein with commercial antibiotics against pathogenic microorganisms. Pharm Biol 52(1):86–90

Suman R, Sharma P, Gupta S, Sourirajan A, Dev K (2015) A novel psychrophilic *Janthinobacterium lividum* MMPP4 isolated from Manimahesh Lake of Chamba District of Himachal Pradesh, India. J Biochem Technol 6(1):846–851

Tabor-Godwin J, Stuart R, Leon Zayas RI, Rajakuberan C (2009) What you didn't know about *Janthinobacterium*. Physiol Genet, Teachers Corner

Tong Y, Zhou J, Zhang L, Xu P (2019) Engineering oleaginous yeast *Yarrowia lipolytica* for violacein production: extraction, quantitative measurement and culture optimization. BioRxiv. https://doi.org/10.1101/687012

Venegas FA, Köllisch G, Kerstin M, Diederich WE, Kaufmann A, Bauer S, Chavarria M, Araya JJ, García-Pineres AG (2019) The bacterial product violacein exerts an immunostimulatory effect via TLR8. Sci Rep 9(1):13661

Verinaud L, Lopes SCP, Naranjo Prado IC, Alves T, da Costa FZ, Gangi RD, Issayama LK, Carvalho AC, Bonfanti AP, Niederauer GF, Duran N, Maranhao Costa FT, Oliveira ALR, da Cruz Hofling MA, Stach Machado DR, Thome R (2015) Violacein treatment modulates acute and chronic inflammation through the suppression of cytokine production and induction of regulatory T cells. PLoS One 10:e0125409

Vishnu TS, Palaniswamy M (2016) Isolation and identification of Chromobacterium SP. from different ecosystems. Asian J Pharm Clin Res 9(suppl 3):253–257

Vishnu TS, Palaniswamy M (2017) Impact of various fermentation conditions on the production of violacein by the novel isolate Chromobacterium vaccinii cv5. Int J Pharm Bio Sci 8(3):514–522

Vishnu TS, Palaniswamy M (2018) Systematic approach on evaluating the *in vitro* antioxidant activity of violacein; novel isolate *Chromobacterium Vaccinii* CV5. Biomed Pharmacol J 11(2): 703–709

Wan AA, Nur ZY, Nordin N, Zakaria ZA, Rezali MF (2012) Production and characterization of violacein by locally isolated *Chromobacterium violaceum* grown in agricultural wastes. Appl Biochem Biotechnol 167(5):1220–1234. https://doi.org/10.1007/s12010-012-9553-7

Wu YH, Cheng H, Xu L, Jin XB, Wang CS (2017) Physiological and genomic features of a novel violacein-producing bacterium isolated from surface seawater. PLoS One 12(6):e0179997. https://doi.org/10.1371/journal.pone.0179997

Yang LH, Xiong H, Lee OO, Qi S-H, Qian P-Y (2007a) Effect of agitation on violacein production in *Pseudoalteromonas luteoviolacea* isolated from a marine sponge. Lett Appl Microbiol 44: 625–630

Yang LH, Xiong H, Lee OO, Qi SH, Qian PY (2007b) Effect of agitation on violacein production in *Pseudoalteromonas luteoviolacea* isolated from a marine sponge. Lett Appl Microbiol 44:625–630

Yang F, Hanna MA, Sun R (2012) Value-added uses for crude glycerol–a byproduct of biodiesel production. Biotechnol Biofuels 5(13):1–10

Exploitation of Marine-Derived Multifunctional Biomaterials in Biomedical Engineering and Drug Delivery

12

Madhubalaji Chegu Krishnamurthi, Sonal Tiwari,
Pallaval Veera Bramhachari, and G. V. Swarnalatha

Abstract

Most of Earth's resources are found in the oceans, making marine-derived biomaterials a viable, sustainable energy source. Marine-derived materials include porifera, marine sponges, sea urchins, marine fauna, marine algae, coral, sea urchin, foraminifera, and red sea algae, and their byproducts can be employed in biomedical applications. Natural polymeric substances (alginate, carrageenan, sulfate polysaccharides, chitosan, agar, ulvan, fucoidan, and collagen) and bioactive compounds (fatty acids, carotenoid sterol, phlorotannins, squalene, MMA, minerals, and vitamins) have been found in marine-derived biomaterial, which has a wide range of properties including anti-acne, antioxidant, antibacterial, antitumor, anti-inflammatory, and anti-obesity properties. As a result of recent rapidly evolving technological advancements, various marine-derived biomaterials have major biomedical applications such as drug delivery agents (liposome-based drug delivery), tissue engineering (3D printing, wound healing, and bone tissue engineering), dentistry, cosmetics (bioactive metabolite properties), and diabetes.

Keywords

Marine resources · Biomaterials · Bioactive · Polymers · Tissue engineering

M. Chegu Krishnamurthi · S. Tiwari
Plant Cell Biotechnology Department, CSIR-Central Food Technological Research Institute, Mysore, Karnataka, India

P. Veera Bramhachari
Department of Biotechnology, Krishna University, Machilipatnam, Andhra Pradesh, India

G. V. Swarnalatha (✉)
Department of Biochemistry, Rayalaseema University, Kurnool, Andhra Pradesh, India

P. Veera Bramhachari, C. V. Berde (eds.), *Marine Bioactive Molecules for Biomedical and Pharmacotherapeutic Applications*,
https://doi.org/10.1007/978-981-99-6770-4_12

12.1 Introduction

Oceans cover about 71% of the earth surface; consequently, the marine resources and solutions developed from them are quite potent. The ocean is not only a supply of water but also a habitat for a wide variety of microbes and a significant source of biological material. More than 25,000 physiologically active chemicals have been identified in these ecosystems. Shellfish, macroalgae, fungi, microalgae, and corals create polysaccharides with a wide range of structures and biological functions that are not seen in terrestrial animals. Biodiversity is mostly due to the maritime environment's saline conditions, which include high pressures and low temperatures. After extensive research, most marine biomaterials have yet to be identified and defined (91% are still unidentified). Humans have relied on biological materials derived from water since time began because of the abundance of resources available. Further, Human infections are generally found to be absent from marine-derived biological material.

Researchers have focused much on marine biomaterials for various biomedical uses in the last two decades. The development of marine biotechnology relies heavily on the isolation, characterization, and application of marine biomaterials. Bioactive elements can now be efficiently extracted from the ocean through a new technical innovation. Rhodophyceae, Phaeophyceae, and Chlorophyceae are the three main types of polysaccharides based on the photosynthetic pigments supplied. There are numerous bioactive chemicals found in marine algae that have a wide range of health benefits. Because of this, the biomedical, pharmaceutical, cosmetic, and food industries are paying increased attention to the use of algae. The biomaterials' market offers considerable prospects for entrepreneurs, but much research is needed to generate new and improved products at a reasonable price. These bioactive chemicals can have pharmacological effects because of their non-cytotoxicity, biodegradability, and biocompatibility, which form the foundation for biomedical applications.

Algal biomass is rich in marine resources and contains diverse bioactive molecules with significant potential for use in biomedicine. In recent decades, significant advances have been made in using natural polymeric biomaterials in new biomedical applications. There are several different natural biopolymers whose anti-inflammatory and antibacterial characteristics have attracted the interest of biomedical scientists and surgeons. These include polysaccharides, Carrageenan, alginate, and agar are a few algae-based polysaccharides that have been widely used in medicinal applications in recent years. Biochemical processes can be exploited in designing materials for drug delivery systems to alter their affinity for certain medicines. In addition, chemical reactions or interactions with other natural polymers can enhance the ability to integrate medications and increase their release effectiveness. They come in various sizes, shapes, liquefaction, chemical stability, and cross-linking abilities. With their softness and ability to expand when exposed to stimuli, algal polysaccharides are ideal for usage in the biomedical industry.

12.2 List of Marine Biomaterials

The biomedical field continues to demand new biomaterials with innovative properties. Natural polymers are the materials of choice for this goal because of their biocompatibility and biodegradability. Marine biomaterial extraction methods and their yields with different methods are presented in Table 12.1.

12.2.1 Alginate

Controlled pharmacological release, cell encapsulation, ligament scaffolding, tissue engineering, and regeneration of nearly all human tissue or dental impressions are among the most common uses for alginates derived from macroalgae (Lee and Mooney 2012). In the cell walls of brown algae such as *Ascophyllum*, *Laminaria hyperborea*, *Laminaria digitata*, and *Macrocystis pyrifera*, anionic polysaccharides known as alginates are found (Rahmati et al. 2019). As the name suggests, the polysaccharide is made of D-mannuronic acid (M), which is linked to guluronic acid by a chain of amino acids called -(1-4)-linked D-mannuronic acid (M), and -L-guluronic acid (G). These considerations must be considered because alginate extraction methods, seasons, and regions affect the relative abundance and sequential distribution of M and G. The G:M ratio, the kind of ionic crosslinker, and the concentration and viscosity of the beginning alginate solution all have a major role in alginate-based biomaterials' mechanical strength and porosity. Such polysaccharides are found in marine brown algae and soil bacteria as capsular polysaccharides. Biodegradable alginates may be capable of binding with other biomolecules that have a controlled amount of porosity.

Alginate beads can improve cell growth and survival when cells are encased within them. Patients can be treated for wounds using hydrogels or sponges made of alginates and their salts, which have hemostatic qualities. As a skeletal component, alginate is incorporated into algae's cell wall and intercellular matrix to enhance its ability to increase. A wide range of biomedical applications is possible because of alginate's biocompatibility, chelating ability, and non-antigenicity. Some studies have suggested using these materials as excipients to stabilize or modulate the rate of distribution of drugs. It is soluble even at high pH because carboxyl groups are charged at pH > 3–4. The drug's protective action is enhanced and absorption in the stomach is facilitated because of this pH sensitivity. Because of its solubility and pH sensitivity, this polysaccharide is an excellent candidate for use in the production of drug delivery systems. To make tablets, especially for insoluble pharmaceuticals, sodium alginate is one of the most important alginates. As a hydrogel, the polysaccharides are commonly employed for follicles (Tagler et al. 2012), islets of Langerhans (Johnson et al. 2011), and stem cells in neural tissue engineering (Hamblin 2016), bone and skin tissues, and regeneration of the osteochondral interface (Khanarian et al. 2012), as well as in skeletal muscle regeneration (Liu et al. 2013), capture endothelial cells (Khanarian et al. 2012), and capture fibroblasts (Khanarian et al. 2012; Hatch et al. 2011). Low to moderately oozing wounds benefit

Table 12.1 Extraction methods used and their yields for marine biomaterials

Marine biomaterial	Source	Extraction method	Yield (%)	References
Alginate	*Colpomenia peregrina*	Enzyme-assisted extraction	Alcalase – 3.8%	Rostami et al. (2017)
			Cellulase – 6.6%	
	L. digitata	Conventional extraction with HCl	51.8%	Fertah et al. (2017)
	Sargassum binderi	Ultrasound-assisted extraction	54%	Youssouf et al. (2017)
Carrageenan	*Solieria chordalis*	Microwave-assisted extraction	29.3%	Boulho et al. (2017)
	F. lumbricalis, C. truncatus	Conventional extraction (NaOH pre-treatment)	72.6%	Tuvikene et al. (2006)
	Mastocarpus stellatus	Enzyme-assisted extraction	28.65%	Blanco-Pascual et al. (2014)
	Eucheuma denticulatum and *Kappaphycus alvarezii*	Ultrasound-assisted extraction	27% and 44.46%	Youssouf et al. (2017) and Azis (2019)
Agar	*Gracilaria vermiculophylla*	Microwave-assisted extraction	14.4%	Sousa et al. (2010)
	Gracilaria lemaneiformis	Conventional extraction (alkali pre-treatment with hot water)	25.8%	Li et al. (2008)
	Gracilaria lemaneiformis	Enzyme-assisted extraction	12.52%	Xiao et al. (2019)
Ulvan	*Ulva pertusa*	Microwave-assisted extraction	41.91%	Le et al. (2019)
	Ulva intestinalis	Ultrasound-assisted extraction	8.30%	Rahimi et al. (2016)
	Ulva sp.	Enzyme-assisted extraction	42.5%	Fournière et al. (2019)
Fucoidan	*Nizamuddinia zanardinii*	Ultrasound-assisted extraction	3.51%	Alboofetileh et al. (2019b)
	S. henslowianum	Conventional extraction 95% EtOH; 2 × 12 h	5.1%	Sun et al. (2020)
	N. zanardinii	Enzyme-assisted extraction	Alcalase – 5.58%	

			Cellulase – 4.80%	Alboofetileh et al. (2019a)
			Flavourzyme – 4.36%	
			Viscozyme – 4.28%	
			Hot water – 5.20%	
Laminarin	*S. latissima*	Conventional extraction (hot H_2SO_4)	20%	Devillé et al. (2004)
	L. hyperborea, *A. nodosum*	Ultrasound-assisted extraction	5.29–6.24%	Kadam et al. (2015)
	U. pinnatifida	Enzyme-assisted extraction	3.2 ± 0.9%	Je et al. (2009)
Algal polysaccharide-based 3D bioprinting for tissue engineering applications				
Source	**Application**	**3D bioprinting technology**	**Bioink composites**	**References**
Alginate	Bone tissue	Jet bioprinting	Collagen/alginate/fibrin/hMSCs	Ojansivu et al. (2019)
	Cartilage tissue	Inkjet printing	Cellulose/alginate/chondrocyte	Thayer et al. (2018)
	Cardiac tissue	Coaxial extrusion bioprinting	Alginate/PEG-fibrinogen/HUVECs/iPSC-CMs	Maiullari et al. (2018)
	Dental tissue	Extrusion bioprinting	Alginate/gelatin/human dental pulp stem cells	Yu et al. (2019)
	Liver tissue	Extrusion bioprinting	Alginate/mouse-induced hepatocyte-like cells	Kang et al. (2018)
	Neural tissue	Coaxial extrusion bioprinting	Alginate/gelatin/Schwann cell RSC96s	Li et al. (2018)
	Vessel system	Coaxial extrusion bioprinting	Alginate/fibroblasts/smooth muscle cells	Gao et al. (2017)
Carrageenan	Cartilage tissue	Extrusion bioprinting	MA-κ-carrageenan/2D nanosilicates/hMSC	Thakur et al. (2016)
	Tissue scaffolds	Extrusion bioprinting	Carrageenan/alginate/hADMSCs	Kim et al. (2019)
Algal polysaccharide-based drug delivery				
Source	**Composition**	**Drug delivery route**	**Application**	**References**
Alginate	Polyvinyl alcohol-alginate hydrogel	Oral	Metformin was delivered as an antidiabetic with a 78% entrapment efficiency	Kumar et al. (2017)

(continued)

Table 12.1 (continued)

Marine biomaterial	Source	Extraction method	Yield (%)	References
	Alginate nano-emulsion	Skin	Lidocaine is delivered transdermally with a 97% entrapment efficiency	Sarheed et al. (2021)
Fucoidan	Zein/fucoidan-based nanoparticles	Oral	Pterostilbene was delivered with a 95% entrapment efficiency	Liu et al. (2020)
	Fucoidan/chitosan nanocarriers	Mucosal	Proteins were delivered with a 100% entrapment efficiency	Ferreira (2012)
Carrageenan	Carrageenan/chitosan nanocarrier	Topical	Naproxen was delivered with 97.55% encapsulation efficiency	Noreen et al. (2021)
	Ovalbumin/κ-carrageenan complex	Oral	Curcumin was delivered with 91% encapsulation efficiency	Xie et al. (2019)
Agar	Agar/alginate hydrogel beads	Oral	Delivered biomedicine with 94% loading efficiency	Yin et al. (2018)

from alginate dressings, which enhance hemostasis and manage exudates. Proliferation and adhesion are two of the most common cellular functions that can be aided by calcium alginate. As a result of its hydrophilic nature, alginate is known to be left behind since biological fluids and water weigh more than 20 times their weight. The presence of a moist environment will accelerate the wound-healing process.

Biodegradation of alginic acid and its salts occurs within the body. It has the potential to boost cell growth and viability. Wounded patients benefit from alginate biomaterials because they maintain a moist wound environment, reduce bacterial burden, and speed up wound healing. Calcium alginates can boost the proliferation and adhesion of cells. Injectable hydrogels, hydrogel sponges, hydrofiber films and membranes, dressings and films are some of the forms in which it is employed (Hassan et al. 2014; Abdel-Mohsen et al. 2012).

12.2.2 Carrageenan

The galactan family has a common name: carrageenan. Carrageenan is the most common and abundant component of red algal cell walls. Linear polysaccharide chains with sulfate monoesters connected to sugar units are known as carrageenans. Carrageenans can dissolve entirely in water due to their chemical characteristics, which results in a viscous solution that is stable throughout a wide pH range. Premium carrageenan, which accounts for around 90% of the world market value, is mostly utilized for human consumption; the rest is used in the same way as animal food. Red algal cell walls comprise carrageenan, accounting for 30–75% of their dry weight. Hydrophilic, high molecular weight sulfated carrageenans are made of D-galacto-pyrannosaccharides joined by alternating 1,3,4-linkages or linked together by a 3,6-anhydropyran hemi-lactose unit. Antiviral and anticancer effects of carrageenan have been reported. The industry's most frequent forms of carrageenan are kappa, iota, and lambda, depending on their degree of sulfation. Structural and gelling qualities are the fundamental differences between carrageenan types. Because of its capacity to generate hydrogels, carrageenan is widely utilized in the pharmaceutical sector as a stabilizer, gelling agent, and emulsifier. Most of the polysaccharide is derived from the *Kappaphycus* and *Chondrus crispus* species. Moreover, extracts from *Hypnea*, *Gigartina*, *Furcellaria*, and *Iridae* genera have also recently been introduced to extract carrageenan from those species.

C. crispus and *Mastocarpus stellatus* carrageenans, which date back to the 1830s, offer a wide range of medical benefits. Traditional medical teas and cough suppressants for Ireland's common cold and chronic bronchitis still contain carrageenan. Carrageenan possesses antiviral and mucus-loosening effects (Skoler-Karpoff et al. 2008; Vlieghe et al. 2002). Other medications, ointments, and lotions use these chemicals as stabilizers and suspending agents. Some other therapeutic uses include the treatment of dysentery, diarrhea, and constipation in the intestines, among other things. Internal medicine can induce stomach ulcers, which are also used to treat them. Using *C. crispus* carrageenan gel may help prevent the spread of

HIV and other STDs, including genital warts, gonorrhea, and herpes simplex virus (HSV), according to recent studies.

Algae species, and strains within a single genus can produce sulfate polysaccharides that are both complex and diverse. Chemical structures are typically difficult to detect; thus, their biological functions are unknown. Few scientists have taken on the challenge of examining the potential link between polysaccharide chemistry and activity. Depending on the bacterial species, locale, growth conditions, extraction procedures, and analytical methodologies, the glycosidic linkages and sulfate groups in various sulfate polysaccharides might vary greatly. Structural factors such as the level of sulfation and the placement of sulfate groups along the polysaccharide backbone, molecular weight, content of sugar residues, and stereochemistry all play a role in sulfate polysaccharides' biological and pharmacological actions. Anticoagulant properties of fucoidan can be explained by its sugar composition and molecular weight, its degree of sulfation, and the position of sulfate groups on the sugar backbone.

12.2.3 Chitosan

In nature, chitosan is the second most common polymer after cellulose and is a cationic polysaccharide (Kumar 2000; Peniche et al. 2008; Lizardi-Mendoza et al. 2016). It is a deacetylated derivative of chitin. In marine polysaccharides, chitosan is one of the most common and widely used. The exoskeletons of various marine invertebrates depend on it heavily (Silva et al. 2012). 70% to 90% of chitosan comprises D-glucosamine, whereas the remaining 10–30% comprises *N*-acetyl-D-glucosamine (Kumar 2000). Chitosan's deacetylation depends on the ratio of the two units. Deacetylated chitosan is more easily dissolved in an acidic aqueous solution than chitin, which is more difficult to dissolve (Chang et al. 1997). Protonation of the free amino groups in dilute acid makes chitosan completely soluble at pH 5 (Suh and Matthew 2000). Because of this, chitosan-based hydrogels are sensitive to pH changes. Adding NaOH or KOH coagulant solutions may improve viscosity and be used in various neural tissue engineering applications at high pH settings (Kumar 2000). The charge interaction with bacteria membranes gives chitosan additional antimicrobial capabilities. These chitosan-based hydrogels for bone tissue engineering demonstrate the practicality and printability of the chitosan bioink and the chitosan-based hydrogels. This mineralization and osteogenic differentiation process occurs after 21 days of culture in chitosan. For specialized applications, the amino group of chitosan can be chemically changed to generate additional derivatives with suitable solubility and mechanical and biological properties. However, because chitosan is not thermoplastic, thermomechanical processing does not work well with it. Chitosan degradation can occur if the processing temperature is too high, therefore temperature control is necessary.

12.2.4 Agar

The rheological qualities of red algal galactans make them particularly valuable in the food business. Agarans and carrageenans are the most common types of polysaccharides found in nature. Alpha-galactose galactan with four connected L- and D-series, *Polysiphonia* species, *such as P. strictissima, P. abscissoides*, and *P. nigrescens,* produce sulfated agars. Adding various O-linked substitutions and sulfate groups has been reported to disrupt traditional agar backbones. Sulfated xylose residues and 4,6-pyruvate xylose were found in agar recovered from *Robinia pseudoacacia*, as well as substantial levels of sulfuration at the C-2 position of the galactose unit. Biomedical engineers can employ this polysaccharide for drug delivery. Polysaccharides such as agarose and carrageenan are both made from agar.

12.2.5 Collagen

Fibroblasts can be attracted to wounds by collagen, its breakdown products, and metabolic byproducts, resulting in wound healing and fracture enhancement. Collagen-based formulations have advantages over cell- and growth factor-based therapies in various wound repair applications. It is suggested to use collagen-based bandages derived from swine, chicken, or bovine sources to treat a wide range of wounds. For example, collagen is contraindicated for use on inquisition skin burns. With a three-dimensional collagen 1 and 3 scaffolding, a dermal transplant mimics the dermis and has demonstrated remarkable wound-healing capacity in people with diabetes. In diabetic wounds, a collagen-based skin graft was also found to have strong healing potential. The in vivo toxicity and inflammation caused by collagen are negligible, as it is biodegradable. Cellular biomaterial interactions, ECM deposition, and cell adhesion are stimulated by collagen. To ensure the survival of cells, it promotes the movement of nutrients and regulatory substances and enables gaseous exchange. It's employed in 3D scaffolds, films, and nanoparticles embedded in hydrogels and sponges, among other forms (Dhandayuthapani et al. 2011). Non-immunogenic gelatin is biodegradable, biocompatible, and biodegradable. The hydrolysis of collagen produces gelatin, which preserves a significant peptide that increases signal transmission and cell adhesion in a well-organized wound-healing process. 3D scaffolds, hydrogels, dressings, and micro cryogel sponges are all examples of applications for this material (Hsu et al. 2019).

12.2.6 Hyaluronic Acid (HA)

Hyaluronic acid is included in the family of glycosaminoglycans (GAGs). *N*-Acetylated hexosamine and uronic acid repeating disaccharide units constitute GAGs (except keratan sulfate), which are linear, negatively charged heteropolysaccharides (Afratis et al. 2012). Synovial fluid, vitreous body, and cartilage all contain significant amounts of HA, making it an important extracellular matrix component. Hyaluronic acid is a great biomedical material because of its high viscosity, making it useful for various bodily functions. Because of its involvement in cell proliferation, differentiation and migration could be used in various medical applications (Kogan et al. 2007). Hyaluronic acid can be utilized as a biomarker for rheumatoid arthritis because it is found in the synovial fluid of affected joints (Lee et al. 2008). HA is also slated for tissue engineering, specifically in structures for wound healing and as a supplement for arthritis patients (Chen 2002). Hyaluronic acid's high water sorption and retention capacity is one of its many unique features. It is nonallergenic and degrades in the environment. HA plays only a few roles in wound healing: inflammation, reepithelization, granulation tissue development, proliferation, and remodeling. Sponge, film, and membrane are just a few of the many biomaterial shapes it can take (Humbert et al. 2013).

12.2.7 Ulvan

Polysaccharides derived from *Ulva* sp., a member of the Ulvales, are now referred to as *ulvan*. Polyelectrolytes are the basic building blocks of ulvans. Xylose, uronic acid, and rhamnose are the three most common monomeric sugars in them. Alduronic acid, a frequent component disaccharide, is also found in them. Iduronic acid is another sugar component. Ulvans have an average molecular mass of 189–8200 kDa (Lahaye 1998). Cell wall polysaccharides comprise 38% and 54% of ulvans algal dry weight (Lahaye 1998). Ulvan comes in various forms, the most prevalent of which are insoluble and water soluble. An important part of green algal cell walls, ulvans comprise between 8 and 29% of the dry weight (Lahaye and Robic 2007). Sulfated polysaccharides are green algal sulfurized polysaccharides in nature. A central disaccharide backbone unit is linked to L-rhamnose-3-sulfate via D-glucuronic acid (ulvabiouronic acid A), L-iduronic acid (ulvabiouronic acid B), D-wood sugar-4-sulfate (ulvabios unit A), or D-xylose residues, and they are then attached to L-rhamnose (ulvabios unit B).

The viscosity of ulvan solutions is on par with gum arabic, which means they are extremely thick. The presence of additional functional groups, such as Arabian, in green algal sulfurized polysaccharides has yet to be determined. pH, divalent cations, and boric acid influence ulvans' gelling capabilities. When thermal hysteresis is absent, ulvans can be reversible in the hot or cold. Under specific conditions, ulvan's gelling properties are critical for releasing trapped particles or molecules.

12.2.8 Fucoidan

A polysaccharide group known as fucoidans is made up of sulfated L-fucose. More than 90% of it is composed of other monosaccharides. Fucoidans are crucial chemicals for reinforcing the cell wall. When algae are exposed to the air at low tide, this prevents them from drying out. Brown algae are the only organisms to have fucoidan, which has a variety of intriguing biological functions. Cell wall-enhancing chemical fucoidan is essential. During low tide, they help to keep the algae from drying out due to exposure to the air. Nearly a century ago, Kylin discovered that fucoidan possesses fascinating biological properties (Kylin 1913). Fucoidan levels in the *Fucus vesiculosus* species are the highest (up to 20% of its dry weight). Conversely, fucoidan accounts for over 40% of the cell wall's dry weight.

Additionally, acidic solutions or hot water can efficiently extract fucoidan for further study (Berteau and Mulloy 2003). As their concentration drops, their viscosity increases. A mixture of alkalis and acids can also quickly destroy them, making them vulnerable to corrosion. Fucoidan, a polymer generated by algae, aids in developing intercellular mucus. Fucoidans have varying degrees of linkage, substituents, branching, and sulfation in their structural complexity, depending on the source of polysaccharides for finer structure.

12.2.9 Fibrin and Fibrinogen

Fibrin and fibrinogen are biocompatible and biodegradable, and they have a low level of toxicity. The fibrinogen component is responsible for its role in the hemostasis phase of the wound-healing cascade. The fibrin matrix promotes cell migration, attraction, proliferation, and adhesion during the wound-healing process. This helps the wound to heal more quickly. It will be utilized as scaffolds, dressings, bandages, nanoparticles, and 3D-printed structures (Mosesson et al. 2001; Medved et al. 2001).

12.2.10 Keratin

Keratin helps to improve aeration and facilitates the release of wound fluids and exudates. Because it stimulates the activity of Schwann cells, keratin is a material that can greatly benefit the nerve regeneration process. In addition, it results in the suppression of microbial load in the environs of the wound. It will be utilized as dermal patches, membranes, films, scaffolds, and dressings (Nayak and Gupta 2017; Mosesson et al. 2001).

12.2.11 Carboxymethyl Cellulose

Carboxymethyl cellulose has higher compatibility with mucous membranes and skin and has a greater capacity for absorbing water. It comes at a reasonable cost and doesn't pose any health risks. Scaffolds made out of sodium carboxymethyl cellulose stimulate early wound healing in rats and considerably affect the regulation of transepidermal water loss. The CMC scaffold can prevent excessive moisture loss at the wound site while retaining some of it. It can take the shape of hydrogel, sponges, hydrofibers, films, or membranes when it is put to use.

It can be difficult to explain all of the specific marine biomaterials and the applications for those biomaterials because they contribute to the formation of so many different compounds. This chapter focuses on marine biomaterials, which are essential for a wide range of medical applications.

12.3 Applications of Marine Biomaterials

12.3.1 Drug Delivery Systems

Marine polysaccharides as drug delivery systems are gaining attention due to their potential to enable less invasive and more individualized treatment methods. Polysaccharides derived from marine organisms are commonly employed in medicine delivery devices. The biodegradability and stimulus responsiveness of these polymers make them ideal for designing complicated loading devices that can be gradually released over time. These devices can be made in various ways and formats, such as hydrogels, particles, and capsules, and they can all be protected by them. Research shows that various biomaterials allow lower drug dose loading, which can reduce side effects caused by drugs significantly. In addition to being diagnostic tools, these materials can be used as signaling markers to direct the carrier and deliver to specified places. The health risks associated with viral vectors limit the usefulness of these systems in gene therapy. On the other hand, biomaterials provide many advantages for encapsulating genetic material and other therapeutic substances, including stabilization and protection, increased solubility, permitting prolonged release, and biocompatibility, in certain cases, biodegradability.

12.3.1.1 Liposome-Based Drug Delivery

When combined with proteins and medications, liposomes can treat various ailments, including cancer. Genes and medicines of varied sizes and shapes can be efficiently encapsulated and immobilized using these nanoparticles. Liposomes can adjust the concentration of hydrogen ions, pH, and other ions without altering the liposome core. Encapsulated biological molecules, such as peptides, are likewise protected from degradation and harm by liposomes. In addition, they have a larger loading capacity for water-soluble additives than microemulsions. In cells, liposome stimuli are formed from cells and closely match the borders and defined vesicles of the cells in which they are present.

Marine Biomaterials in Tissue Engineering

To create fictitious materials for humans, such as curing or replacing infected organs, tissue engineering is an expanding field of research. Pore distribution, adsorption, swelling studies, in vitro degradation, biomineralization, and biocompatibility should all be considered when developing ocean-derived biological materials for bone tissue engineering. Chitosan is a common ingredient in marine-derived biomaterials. Different forms, such as chitosan composites with different bioactive substances, chitin microspheres, silica nanoparticles, bioactive glass, and chitin nanogels, can achieve desired results. Tissue engineering products comprised of chitosan can take on any shape.

12.3.1.2 3D Printing

There has been a surge in interest in bioprinting applications in recent years. Bibliotheca of biopolymers with a wide range of biological signals can be found in marine species. A lack of efficient use of marine life remains, though. Marine biomaterials are employed in 3D bioprinting because of the wide variety of marine biopolymers and the ease with which they may be separated. Chemical structures and functionalities of marine-derived chemicals make them suitable for a wide range of potential uses. Alginate, collagen, hyaluronic acid, carrageenan, chitosan, and gelatin are marine-derived macromolecules commonly used in 3D bioprinting for regenerative medicine applications. Printability, mechanical integrity, and biocompatibility are all met by biomaterials in 3D bioprinting. Bioinks generated from marine sources for 3D bioprinting have the qualities and advantages needed for tissue engineering. Unlike mammalian resources, natural marine resources are free of infectious disease hazards and religious limitations. As a result, marine resources are now a popular supply of biomaterials for 3D bioprinting due to their high yields and low cost. Because bioinks contain living cells, biomaterials are essential in 3D bioprinting. In addition to their printability, these biomaterials have good mechanical and functional qualities.

Digital data in CAD files guide the layer-by-layer direct production of items via three-dimensional (3D) printing. Different 3D printing technologies have emerged in the recent decade and transformed the notion of directly producing things for many purposes. Biomedical devices and tissue engineering benefit greatly from 3D printing technology's ability to produce small quantities or distinctive items. Patient's requirements are considered while allocating portions at no additional cost to the rest of the community. The design can be tailored to the needs of each patient while allowing for a wide range of start-up materials. Despite this, there are still many worries concerning the widespread usage of 3D-printed biological materials, such as challenges with regulation, sterile manufacturing settings, and achieving desirable material properties with desired architecture.

3D bioprinted construction relies heavily on cells, biomaterials, and other biological variables. To better understand how diseases emerge and how organs and tissues are replaced, 3D bioprinting offers great promise as a tool for creating biomimetic systems. Printability, mechanical integrity, and biocompatibility are all characteristics of materials that are ideal for 3D bioprinting. Poly-L-lactic acid (PLLA), polycaprolactone (PCL), and poly(lactic-co-glycolic) acid are commonly used synthetic polymers with great mechanical strength; however, they lack cell attachment sites. Cell encapsulation is problematic because of these materials' high printing temperatures and organic solvents. Although natural materials are more likely to interact with cells, their mechanical strength and deterioration cannot be controlled.

Polysaccharides with sulfated sugars, such as carrageenan, fucoidan, and green algae sulfated polysaccharides, have a variety of useful properties, such as scaffolding, hydrogel formation, alkaline phosphatase activity, extracellular matrix imitation, and improved differentiation and biomineralization of sheared cells.

12.3.1.3 Dentistry

Collagen, alginate, and chitosan are marine biomaterials that have all been used in dentistry. Freshwater fish such as tilapia have been utilized to rebuild dentin and pulp using collagen obtained from the fish. Chitosan is another important element for dental applications because of its biocompatibility, bioactivity, and antibacterial characteristics. Chronic periodontitis bone defects can be filled using chitosan hydrogels. A good choice for integration into toothpaste and mouthwash is chitosan due to its antibacterial activity against dental biofilms and its potential to lower *Streptococcus mutans* levels. Dentin and pulp can also be regenerated using the scaffolding technique.

It helps transfer active chemicals and cells to pulp compartments via a scaffold. Both pulp and dentin regeneration can be accomplished using chitosan-based scaffolds. The expression of mineralization indicators like osteopontin and alkaline phosphatase is increased when these two supplements are taken together with beta-tricalcium phosphate. In vitro and In vivo investigations have shown that they increase the proliferation, migration, and differentiation of dental pulp and mesenchymal stem cells. Human periodontal ligament cells can produce more dentin due to exposure to these substances.

12.3.1.4 Cartilage Tissue Engineering

Because of its lack of vascularity, cartilage has a very limited potential to repair. Using deep-sea fish skin collagen, adipose stromal cells can become more chondrogenic. In human mesenchymal stem cells, they induced chondrogenesis by increasing chondrogenic and decreasing osteogenic markers. To regenerate cartilage tissue, coral materials can be used as new scaffolds for growth. For the regeneration of cartilage, their outstanding features, such as osteogenic capacity, biodegradability, biocompatibility, and stimulatory effects on cell invasion, adhesion, and proliferation, make them an excellent choice. The regeneration of cartilage is another popular application for chitosan. Glycosaminoglycans in articular cartilage are structurally identical to those mentioned above. To heal cartilage abnormalities, chitosan can be combined with additional components such as silk fibroin and gelatin, which are excellent transporters for stem cells, to increase the attachment and proliferation of chondrocytes.

12.3.1.5 Wound Healing

Oxygen is a critical component in cellular metabolism, communication, and a wide range of repair processes during wound healing. It is important to note that vascular rupture and vasoconstriction can lead to tissue hypoxia and hinder the healing process. Wound healing can now be aided by boosting oxygenation to hypoxic areas in the body. Wound oxygenation may benefit from regenerative therapy in this case. A hydrogel made from *S. platensis* and the polymer carboxymethyl chitosan was produced to keep the algae's ability to produce oxygen. Laser-irradiated

Spirulina platensis gel is also capable of photodynamically killing bacteria in the affected area by creating ROS from chlorophyll generated by algae.

As a result, wound healing and skin remodeling are enhanced by the use of marine collagen. These proteins also regulate epidermal and fibroblast growth factors. As a result, they're an essential part of the scaffolding-building process. It's an ideal dressing for wound care because it promotes healing and acts as an antimicrobial. Alginate is an effective tool for transplantation into severe wounds, which can rebuild skeletal muscle cells. Biodegradability, non-toxicity, antibacterial property, and biocompatibility are just a few of chitosan's many advantages in wound healing. The use of these two supplements achieves blood clotting and immune system stimulation. Hydrogel dressings that promote granulation, tissue development, and vascularization can be made from these marine-derived materials when combined with other marine-derived materials. Polysaccharide molecules with amino groups on their backbone make up the chitosan structure.

The negatively charged molecules on the surface of red blood cells interact electrostatically with these positively charged molecules. Increased blood viscosity and platelet adhesion and aggregation are caused by this interaction, which also boosts platelet transport to the vascular wall, ultimately resulting in hemostasis. The quantity of positively charged amino groups in chitosan determines its hemostatic characteristics. Chitosan's antibacterial impact on bacteria, yeast, fungi, and algae is also due to the cationic character of the substance. Chitin and alginate films from marine sources are good wound dressing materials because of their flexibility, transparency, cytocompatibility, and adhesive characteristics. The composition, growth, and function of biomaterials can constantly adapt to dynamically changing environmental conditions. When adaptations fail, they also show self-healing capacities.

12.3.2 Diabetes

Wound-healing times for diabetics are often longer and more susceptible to infection and other problems, which considerably impacts healthcare expenses. The average cost of treating a diabetic wound is $50,000. Chronic wounds that do not heal place a heavy burden on the healthcare system and the patients themselves. Chronic wounds, especially those caused by diabetes, can impair mobility, necessitate amputation, and possibly result in death. Diabetes-related sores and ulcers of the limbs are still major health issues today. As a result of rising medical expenses, patients face an increasingly onerous clinical burden.

The highly coordinated sequence of events is disrupted during the wound closure process when diabetes is present. Tissue and organ replacement and improvement can be achieved with biomaterials, which are nondrug materials. They can remain in contact with bodily fluids and tissues for a long time without causing any harm: bioactive dressings and marine biopolymers (polymers) promote wound healing. Advances in diabetic wound pathophysiology have led to the development of medication delivery devices that can promote wound healing by targeting various phases of the damage process. A physiological cascade involved in tissue regeneration necessitates effective delivery mechanisms in wound environments that frequently contain degrading enzymes and high pH.

12.3.3 Bioactive Properties

Polysaccharides, carotenoids, sterols, tannins, fatty acids, minerals, and vitamins found in marine biomaterials have been shown to have antibacterial, anticancer, and anti-oxidative characteristics. These substances are promising candidates for biomedical research. This has led to a growing interest in the use of algae in the pharmaceutical, cosmetic, and food industries, among others. Algae have been utilized in diet and nutritional therapy for a long time because of their immunomodulatory and disease-fighting properties. There is a growing interest in using seaweed as an ingredient in pharmaceutical and cosmetic products. In particular, seaweed's anti-melanoma activities are promising for treating skin problems. Chemicals and metabolites from marine algae are used in dermatology to treat acne, block ultraviolet radiation, lighten skin, and fight melanoma. Polysaccharides, carotenoids, and sterols have all been studied for use in cosmeceutical preparations, as have several other substances. According to research, bioactive derived from seaweed could significantly impact the skin care sector if they are proven helpful in treating dermatological diseases and routine skin care. Acne therapy, sun protection, skin whitening, and melanoma prevention can all benefit from marine algae-derived bioactive chemicals and metabolites. In vitro and in vivo investigations looked at bone repair and regeneration. Algae have long been employed in food diets and nutritional therapies for their immune-adjusting and disease-defending properties. The potential for marine algae in the pharmaceutical and cosmetics industries has recently gained great acclaim. Dermatology, in particular, could benefit from the use of marine algae. Algae-derived chemicals have been found to perform various skin health and care functions.

12.4 Conclusion

Biopolymers are abundant in the ocean. As a result, marine raw materials will be a valuable new resource for biomedical research. Biopolymers are raw materials for biomedical applications as they withstand various environmental stressors and stimuli; marine biopolymers are ideal. Marine biomimetic materials' invention and widespread application in tissue regeneration technologies have occurred over the last two decades. Biomedical applications can benefit from marine biomaterials and their derivatives since each polymer has unique chemical and biological characteristics. The ultimate goal of 3D bioprinting is to replace diseased or damaged organs with artificial tissues or organs. For artificial tissue or organ regeneration, biomaterials that allow for the controlled breakdown of the materials, such as those that may be injected into the body, still need to be developed. In vivo.

However, In vivo testing is still required to optimize the integration of additional materials with marine biomaterial based technologies. Clinical applications of marine biomaterials for regenerative medicine are very limited at this stage. Future advancements in industrial control and insulating systems will be necessary.

References

Abdel-Mohsen AM, Abdel-Rahman RM, Hrdina R, Imramovský A, Burgert L, Aly AS (2012) Antibacterial cotton fabrics treated with core–shell nanoparticles. Int J Biol Macromol 50(5): 1245–1253

Afratis N, Gialeli C, Nikitovic D, Tsegenidis T, Karousou E, Theocharis AD, Pavão MS, Tzanakakis GN, Karamanos NK (2012) Glycosaminoglycans: key players in cancer cell biology and treatment. FEBS J 279(7):1177–1197

Alboofetileh M, Rezaei M, Tabarsa M (2019a) Enzyme-assisted extraction of Nizamuddinia zanardinii for the recovery of sulfated polysaccharides with anticancer and immune-enhancing activities. J Appl Phycol 31(2):1391–1402

Alboofetileh M, Rezaei M, Tabarsa M, You SG (2019b) Ultrasound-assisted extraction of sulfated polysaccharide from Nizamuddinia zanardinii: process optimization, structural characterization, and biological properties. J Food Proc Eng 42(2):e12979

Azis A (2019) Optimization of temperature and time in carrageenan extraction of seaweed (Kappaphycus Alvarezii) using ultrasonic wave extraction methods. IOP Conf Ser Earth Environ Sci 370:012076

Berteau O, Mulloy B (2003) Sulfated fucans, fresh perspectives: structures, functions, and biological properties of sulfated fucans and an overview of enzymes active toward this class of polysaccharide. Glycobiology 13(6):29R–40R

Blanco-Pascual N, Alemán A, Gómez-Guillén MC, Montero MP (2014) Enzyme-assisted extraction of κ/ι-hybrid carrageenan from Mastocarpus stellatus for obtaining bioactive ingredients and their application for edible active film development. Food Funct 5(2):319–329

Boulho R, Marty C, Freile-Pelegrín Y, Robledo D, Bourgougnon N, Bedoux G (2017) Antiherpetic (HSV-1) activity of carrageenans from the red seaweed Solieria chordalis (Rhodophyta, Gigartinales) extracted by microwave-assisted extraction (MAE). J Appl Phycol 29(5): 2219–2228

Chang KLB, Tsai G, Lee J, Fu W-R (1997) Heterogeneous N-deacetylation of chitin in alkaline solution. Carbohydr Res 303(3):327–332

Chen WYJ (2002) Functions of hyaluronan in wound repair. In: Hyaluronan. Elsevier, Amsterdam, pp 147–156

Devillé C, Damas J, Forget P, Dandrifosse G, Peulen O (2004) Laminarin in the dietary fibre concept. J Sci Food Agric 84(9):1030–1038

Dhandayuthapani B, Yoshida Y, Maekawa T, Kumar DS (2011) Polymeric scaffolds in tissue engineering application: a review. Int J Polym Sci. 2011.1-19, doi:10.1155/2011/290602

Ferreira SA (2012) Development of fucoidan/chitosan nanoparticulate systems for protein administration through mucosal routes. PhD Thesis

Fertah M, Belfkira A, Taourirte M, Brouillette F (2017) Extraction and characterization of sodium alginate from Moroccan Laminaria digitata brown seaweed. Arab J Chem 10:S3707–S3714

Fournière M, Latire T, Lang M, Terme N, Bourgougnon N, Bedoux G (2019) Production of active poly-and oligosaccharidic fractions from Ulva sp. by combining enzyme-assisted extraction (EAE) and depolymerization. Metabolites 9(9):182

Gao Q, Liu Z, Lin Z, Qiu J, Yu L, Liu A, Wang Y, Xiang M, Chen B, Jianzhong F (2017) 3D bioprinting of vessel-like structures with multilevel fluidic channels. ACS Biomater Sci Eng 3(3):399–408

Hamblin MR (2016) Shining light on the head: photobiomodulation for brain disorders. BBA Clin 6:113–124

Hassan AF, Abdel-Mohsen AM, Elhadidy H (2014) Adsorption of arsenic by activated carbon, calcium alginate and their composite beads. Int J Biol Macromol 68:125–130

Hatch A, Hansmann G, Murthy SK (2011) Engineered alginate hydrogels for effective microfluidic capture and release of endothelial progenitor cells from whole blood. Langmuir 27(7): 4257–4264

Hsu Y-Y, Liu K-L, Yeh H-H, Lin H-R, Hua-Lin W, Tsai J-C (2019) Sustained release of recombinant Thrombomodulin from cross-linked gelatin/hyaluronic acid hydrogels potentiate wound healing in diabetic mice. Eur J Pharm Biopharm 135:61–71

Humbert P, Mikosinki J, Benchikhi H, Allaert F-A (2013) Efficacy and safety of a gauze pad containing hyaluronic acid in treatment of leg ulcers of venous or mixed origin: a double-blind, randomised, controlled trial. Int Wound J 10(2):159–166

Je J-Y, Park P-J, Kim E-K, Park J-S, Yoon H-D, Kim K-R, Ahn C-B (2009) Antioxidant activity of enzymatic extracts from the brown seaweed Undaria pinnatifida by electron spin resonance spectroscopy. LWT Food Sci Technol 42(4):874–878

Johnson AS, O'Sullivan E, D'Aoust LN, Omer A, Bonner-Weir S, Fisher RJ, Weir GC, Colton CK (2011) Quantitative assessment of islets of Langerhans encapsulated in alginate. Tissue Eng Part C Methods 17(4):435–449

Kadam SU, O'Donnell CP, Rai DK, Hossain MB, Burgess CM, Walsh D, Tiwari BK (2015) Laminarin from Irish Brown seaweeds Ascophyllum nodosum and Laminaria hyperborea: ultrasound assisted extraction, characterization and bioactivity. Mar Drugs 13(7):4270–4280

Kang K, Kim Y, Jeon H, Lee SB, Kim JS, Park SA, Kim WD, Yang HM, Kim SJ, Jeong J (2018) Three-dimensional bioprinting of hepatic structures with directly converted hepatocyte-like cells. Tissue Eng A 24(7–8):576–583

Khanarian NT, Jiang J, Wan LQ, Mow VC, Helen HL (2012) A hydrogel-mineral composite scaffold for osteochondral interface tissue engineering. Tissue Eng A 18(5–6):533–545

Kim MH, Lee YW, Jung W-K, Junghwan O, Nam SY (2019) Enhanced rheological behaviors of alginate hydrogels with carrageenan for extrusion-based bioprinting. J Mech Behav Biomed Mater 98:187–194

Kogan G, Šoltés L, Stern R, Gemeiner P (2007) Hyaluronic acid: a natural biopolymer with a broad range of biomedical and industrial applications. Biotechnol Lett 29(1):17–25

Kumar MNVR (2000) A review of chitin and chitosan applications. React Funct Polym 46(1):1–27

Kumar S, Bhanjana G, Verma RK, Dhingra D, Dilbaghi N, Kim K-H (2017) Metformin-loaded alginate nanoparticles as an effective antidiabetic agent for controlled drug release. J Pharmacy Pharmacol 69(2):143–150

Kylin H (1913) Zur Biochemie Der Meeresalgen, vol 83. Walter de Gruyter, Berlin, p 171

Lahaye M (1998) NMR spectroscopic characterisation of oligosaccharides from two Ulva rigida ulvan samples (Ulvales, Chlorophyta) degraded by a lyase. Carbohydr Res 314(1–2):1–12

Lahaye M, Robic A (2007) Structure and functional properties of ulvan, a polysaccharide from green seaweeds. Biomacromolecules 8(6):1765–1774

Le B, Golokhvast KS, Yang SH, Sun S (2019) Optimization of microwave-assisted extraction of polysaccharides from Ulva pertusa and evaluation of their antioxidant activity. Antioxidants 8(5):129

Lee H, Lee K, Kim IK, Park TG (2008) Synthesis, characterization, and in vivo diagnostic applications of hyaluronic acid immobilized gold nanoprobes. Biomaterials 29(35):4709–4718

Lee KY, Mooney DJ (2012) Alginate: properties and biomedical applications. Progress in polymer science 37(1):106–126

Li H, Xingju Y, Jin Y, Zhang W, Liu Y (2008) Development of an eco-friendly agar extraction technique from the red seaweed Gracilaria lemaneiformis. Bioresour Technol 99(8):3301–3305

Li X, Wang X, Wang X, Chen H, Zhang X, Zhou L, Tao X (2018) 3D bioprinted rat Schwann cell-laden structures with shape flexibility and enhanced nerve growth factor expression. 3 Biotech 8(8):1–10

Liu J, Xu HHK, Zhou H, Weir MD, Chen Q, Trotman CA (2013) Human umbilical cord stem cell encapsulation in novel macroporous and injectable fibrin for muscle tissue engineering. Acta Biomater 9(1):4688–4697
Liu Q, Chen J, Qin Y, Jiang B, Zhang T (2020) Zein/fucoidan-based composite nanoparticles for the encapsulation of Pterostilbene: preparation, characterization, physicochemical stability, and formation mechanism. Int J Biol Macromol 158:461–470
Lizardi-Mendoza J, Argüelles Monal WM, Goycoolea Valencia FM (2016) Chemical characteristics and functional properties of chitosan. In: Chitosan in the preservation of agricultural commodities. Elsevier, Amsterdam, pp 3–31
Maiullari F, Costantini M, Milan M, Pace V, Chirivì M, Maiullari S, Rainer A, Baci D, Marei HE-S, Seliktar D (2018) A multi-cellular 3D bioprinting approach for vascularized heart tissue engineering based on HUVECs and IPSC-derived cardiomyocytes. Sci Rep 8(1):1–15
Medved L, Tsurupa G, Yakovlev S (2001) Conformational changes upon conversion of fibrinogen into fibrin: the mechanisms of exposure of cryptic sites. Ann N Y Acad Sci 936(1):185–204
Mosesson MW, Siebenlist KR, Meh DA (2001) The structure and biological features of fibrinogen and fibrin. Ann N Y Acad Sci 936(1):11–30
Nayak KK, Gupta P (2017) Study of the keratin-based therapeutic dermal patches for the delivery of bioactive molecules for wound treatment. Mater Sci Eng C 77:1088–1097
Noreen S, Pervaiz F, Ashames A, Buabeid M, Fahelelbom K, Shoukat H, Maqbool I, Murtaza G (2021) Optimization of novel naproxen-loaded chitosan/carrageenan nanocarrier-based gel for topical delivery: ex vivo, histopathological, and in vivo evaluation. Pharmaceuticals 14(6):557
Ojansivu M, Rashad A, Ahlinder A, Massera J, Mishra A, Syverud K, Finne-Wistrand A, Miettinen S, Mustafa K (2019) Wood-based nanocellulose and bioactive glass modified gelatin–alginate bioinks for 3D bioprinting of bone cells. Biofabrication 11(3):035010
Peniche C, Argüelles-Monal W, Goycoolea FM (2008) Chitin and chitosan: major sources, properties and applications. In: Monomers, polymers and composites from renewable resources. Elsevier, Amsterdam, pp 517–542
Rahimi F, Tabarsa M, Rezaei M (2016) Ulvan from green algae Ulva intestinalis: optimization of ultrasound-assisted extraction and antioxidant activity. J Appl Phycol 28(5):2979–2990
Rahmati M, Alipanahi Z, Mozafari M (2019) Emerging biomedical applications of algal polysaccharides. Curr Pharm Des 25(11):1335–1344
Rostami Z, Tabarsa M, You SG, Rezaei M (2017) Relationship between molecular weights and biological properties of alginates extracted under different methods from Colpomenia peregrina. Process Biochem 58:289–297
Sarheed O, Dibi M, Ramesh KVRNS, Drechsler M (2021) Fabrication of alginate-based O/W nanoemulsions for transdermal drug delivery of lidocaine: influence of the oil phase and surfactant. Molecules 26(9):2556
Silva TH, Alves A, Ferreira BM, Oliveira JM, Reys LL, Ferreira RJF, Sousa RA, Silva SS, Mano JF, Reis RL (2012) Materials of marine origin: a review on polymers and ceramics of biomedical interest. Int Mater Rev 57(5):276–306
Skoler-Karpoff S, Ramjee G, Ahmed K, Altini L, Plagianos MG, Friedland B, Govender S, De Kock A, Cassim N, Palanee T (2008) Efficacy of carraguard for prevention of HIV infection in women in South Africa: a randomised, double-blind, placebo-controlled trial. Lancet 372(9654):1977–1987
Sousa AMM, Alves VD, Morais S, Delerue-Matos C, Gonçalves MP (2010) Agar extraction from integrated multitrophic aquacultured Gracilaria vermiculophylla: evaluation of a microwave-assisted process using response surface methodology. Bioresour Technol 101(9):3258–3267
Suh J-KF, Matthew HWT (2000) Application of chitosan-based polysaccharide biomaterials in cartilage tissue engineering: a review. Biomaterials 21(24):2589–2598
Sun Q-L, Li Y, Ni L-Q, Li Y-X, Cui Y-S, Jiang S-L, Xie E-Y, Juan D, Deng F, Dong C-X (2020) Structural characterization and antiviral activity of two fucoidans from the brown algae Sargassum henslowianum. Carbohydr Polym 229:115487

Tagler D, Tao T, Smith RM, Anderson NR, Tingen CM, Woodruff TK, Shea LD (2012) Embryonic fibroblasts enable the culture of primary ovarian follicles within alginate hydrogels. Tissue Eng A 18(11–12):1229–1238

Thakur A, Jaiswal MK, Peak CW, Carrow JK, Gentry J, Dolatshahi-Pirouz A, Gaharwar AK (2016) Injectable shear-thinning nanoengineered hydrogels for stem cell delivery. Nanoscale 8(24): 12362–12372

Thayer PS, Orrhult LS, Martínez H (2018) Bioprinting of cartilage and skin tissue analogs utilizing a novel passive mixing unit technique for bioink precellularization. J Vis Exp 131:e56372

Tuvikene R, Truus K, Vaher M, Kailas T, Martin G, Kersen P (2006) Extraction and quantification of hybrid carrageenans from the biomass of the red algae Furcellaria lumbricalis and Coccotylus truncatus. Proc Estonian Acad Sci Chem 55:40

Vlieghe P, Clerc T, Pannecouque C, Witvrouw M, De Clercq E, Salles J-P, Kraus J-L (2002) Synthesis of new covalently bound κ-Carrageenan-AZT conjugates with improved anti-HIV activities. J Med Chem 45(6):1275–1283

Xiao Q, Weng H, Ni H, Hong Q, Lin K, Xiao A (2019) Physicochemical and gel properties of agar extracted by enzyme and enzyme-assisted methods. Food Hydrocoll 87:530–540

Xie H, Xiang C, Li Y, Wang L, Zhang Y, Song Z, Ma X, Xiaoxing L, Lei Q, Fang W (2019) Fabrication of ovalbumin/κ-carrageenan complex nanoparticles as a novel carrier for curcumin delivery. Food Hydrocoll 89:111–121

Yin Z-C, Wang Y-L, Wang K (2018) A PH-responsive composite hydrogel beads based on agar and alginate for oral drug delivery. J Drug Deliv Sci Technol 43:12–18

Youssouf L, Lallemand L, Giraud P, Soulé F, Bhaw-Luximon A, Meilhac O, D'Hellencourt CL, Jhurry D, Couprie J (2017) Ultrasound-assisted extraction and structural characterization by NMR of alginates and carrageenans from seaweeds. Carbohydr Polym 166:55–63

Yu H, Zhang X, Song W, Pan T, Wang H, Ning T, Wei Q, Xu HHK, Buling W, Ma D (2019) Effects of 3-dimensional bioprinting alginate/gelatin hydrogel scaffold extract on proliferation and differentiation of human dental pulp stem cells. J Endod 45(6):706–715

Marine Phytoplankton: Bioactive Compounds and Their Applications in Medicine

13

Rajalakshmy Menon, Rincy Thomas, Neeru Sood, and Trupti Gokhale

Abstract

Phytoplanktons include cyanobacteria, dinoflagellates, and diatoms, which form the foundation of marine and freshwater food webs while effectively maintaining the oxygen and carbon levels globally. They also contribute to 50–80% of the total oxygen produced on Earth. Phytoplanktons produce structurally and functionally diverse bioactive metabolites with many applications in medicine and industry. Phytoplankton metabolites (secondary or non-primary metabolites) such as fatty acids, phycobiliproteins, fucoidans, fucoxanthin, laminarin, mannitol, alginic acids, carotenoids, vitamins, and mineral salts are shown to prevent various human diseases through their application as antidiabetic, antioxidant, antibacterial, anti-inflammatory, anticancer, antifungal, antimalarial, antiviral, and prevention of other metabolic diseases. Some of the metabolites offer an extended role in defense (allelopathic, antipredator, and antibacterial compounds) and/or cell-to-cell communication (e.g., polyunsaturated aldehydes (PUAs)). Until recently, marine microbes were almost ignored and remained unexplored as a potential depot of unique biologically active metabolites. Hence, researchers have now highlighted the prominence of marine microorganisms in producing unusual bioactive metabolites that are otherwise absent in terrestrial species. The presence of these unusual metabolites encourages the application of marine phytoplankton to produce industrially vital compounds. Like phytoplankton, seaweeds or macroalgae are also known for their bioactive metabolites and phytochemicals, demonstrating promising applications in nutraceuticals,

R. Menon · R. Thomas · N. Sood · T. Gokhale (✉)
Department of Biotechnology, Birla Institute of Technology and Science, Pilani, Dubai Campus, Dubai International Academic City, Dubai, United Arab Emirates
e-mail: p20180914@dubai.bits-pilani.ac.in; p20190001@dubai.bits-pilani.ac.in; sood@dubai.bits-pilani.ac.in; trupti@dubai.bits-pilani.ac.in

P. Veera Bramhachari, C. V. Berde (eds.), *Marine Bioactive Molecules for Biomedical and Pharmacotherapeutic Applications*,
https://doi.org/10.1007/978-981-99-6770-4_13

pharmaceuticals, and edible seaweeds. This chapter gives a detailed discussion on synthesizing the bioactive metabolites in marine phytoplankton and macroalgae, their pharmacological activity, and their potential application in medicine.

Keywords

Fatty acids · Phycobiliproteins · Carotenoids · Antioxidants · Anticancer

13.1 Introduction

Algae are generically categorized as Rhodophyta (red algae), Phaeophyta (brown algae), and Chlorophyta (green algae), while based on their size, they are distinguished as macroalgae or microalgae. Microalgae are small single-celled organisms that may be prokaryotic, like cyanobacteria (Chloroxybacteria), or eukaryotic, like green algae (Chlorophyta). Macroalgae are multicellular and large, visible to the unaided eye (Khan et al. 2018). Phytoplankton includes cyanobacteria, dinoflagellates, and diatoms, which form the foundation of marine and freshwater food webs while serving an inevitable role in the global carbon cycle. They also contribute to 50–80% of the total oxygen produced on Earth. Phytoplankton produces structurally and functionally diverse bioactive metabolites with many applications in medicine and industry (Becker 1994; Das et al. 2011). Phytoplankton metabolites (secondary or non-primary metabolites) such as fatty acids, phycobiliproteins, fucoidans, laminarin, alginic acids, carotenoids, vitamins, and mineral salts are shown to prevent various human diseases through their application as antioxidants, antiviral, antibacterial, antifungal, anti-inflammatory, antimalarial, anticancer, antidiabetic, and prevention of other metabolic diseases (Del Campo et al. 2007; Rengasamy et al. 2020). Some of the metabolites offer an extended role in defense (allelopathic, antipredator, and antibacterial compounds) and/or cell-to-cell communication (e.g., polyunsaturated aldehydes (PUAs)) (Paul et al. 2012). Until recently, marine microbes were almost ignored and remained unexplored as a potential depot of unique biologically active metabolites. Hence, researchers have now highlighted the prominence of marine microorganisms in producing unusual bioactive metabolites that are otherwise absent in terrestrial species. The presence of these unusual metabolites encourages the application of marine phytoplankton to produce industrially vital compounds. Like phytoplankton, seaweeds or macroalgae are known for their bioactive metabolites and phytochemicals, demonstrating promising applications in nutraceuticals, pharmaceuticals, and edible seaweeds (Sanjeewa et al. 2018). This chapter gives a detailed discussion of the bioactive metabolites produced by marine phytoplankton and macroalgae, their pharmacological activity, and their potential application in medicine.

13.2 Nutraceuticals

Food products that offer medical benefits for better health can be broadly called nutraceuticals (Bhattacharya et al. 2021; Borowitzka 2013). There has been a commercialized introduction of various biochemical constituents of food in different forms, including gels, granules, capsules, etc. The contribution of these in the health sector is highly undeniable, and the products can be collectively categorized under nutraceuticals, as they involve a beautiful blend of nutrients and pharmaceuticals (Palthur et al. 2010). Several scientists have reported the enormous potential of microalgae as a significant source of nutraceuticals. Several biochemical components produced by microalgae, such as carbohydrates, pigments, PUFAs, vitamins, minerals, etc., can be used as nutraceuticals (Hudek et al. 2014). Algal cultivation offers several advantages such as fast growth, being autotrophs, and they do not compete for land with agriculture but are versatile and can be cultivated in seawater or wastewater too, which makes them a potential candidate to produce nutraceuticals (Abdelaziz et al. 2013). Figure 13.1 illustrates an outline of the applications of microalgae in nutraceuticals.

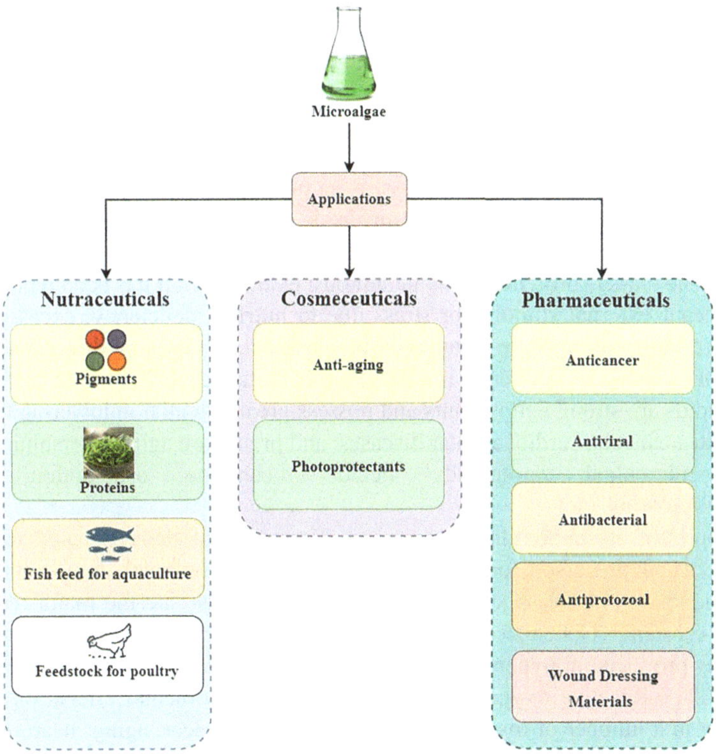

Fig. 13.1 An outline of the applications of microalgae in healthcare

The nutrient-loaded by-products are currently available as tablets, extracts, or capsules in their pure form and are also supplemented to numerous foods, like pasta, gums, candy bars, and beverages (Liang et al. 2004). Several microalgal strains, like *Arthrospira*, *Aphanizomenon flos-aquae*, and *Chlorella*, have been grown commercially for their proteins and an array of nutritionally valuable compounds (Sahin 2018; Soletto et al. 2005). Many of these by-products possess antihyperlipidemic and antihyperglycemic properties, which can be used in medicine for controlling obesity and diabetes, respectively.

13.2.1 Pigments

Photosynthetic pigments of cyanobacteria and microalgae vastly include carotenoids, chlorophyll, and phycobiliproteins (phycoerythrin, allophycocyanin, and phycocyanin), which impart yellow/orange, green, and blue/red to the algal species (Christaki et al. 2011; Siqueira et al. 2018). Phycoerythrin, a red-colored accessory pigment mainly seen in cryptophytes and red algae, is involved along with the major pigments in photosynthesis. Phycoerythrobilin is the chromophore of phycoerythrin, which imparts its characteristic red color to the pigment (Ficner and Huber 1993). Most of these naturally derived pigments obtained from the algal species find potent applications in food, pharmaceuticals, nutraceuticals, cosmetics, etc. In addition to photosynthesis, these pigments are important as antioxidants, neuroprotectants, and anticancer agents (Borowitzka 1995).

Carotenoids, which constitute 0.1–0.2% of algal dry matter, actively participate in photosynthesis and impart shades of yellow and red to the algal species (Vílchez et al. 2011). They fall into two classes: carotenes (possessing only carbon and hydrogen) and xanthophylls (possessing at least one oxygen). Pigments such as violaxanthin, lutein, and fucoxanthin are xanthophylls, while β-carotene belongs to the carotene class (Zorofchian Moghadamtousi et al. 2014). It has been reported that a significant external stimulus or stress due to nutrient deficiency, excessive salt, excessive light, or exposure to unfavorable physical factors may direct microalgae toward the enhanced production of carotenoids (Ahmed et al. 2014; Sun et al. 2018). Carotenoids are strong antioxidants and possess great potential in lowering the risks of arthritis, cancers, cardiovascular diseases, and premature aging (Vershinin 1999). Major microalgal carotenoids include β-carotene, astaxanthin, lutein, phycobiliproteins, etc.

Because of its antioxidant properties, β-carotene (a deoxygenated carotene) (Fig. 13.2), a precursor for vitamin A, remains a prominent carotenoid. Interthylakoid spaces of the specialized organelle, chloroplast, are the major reservoirs of beta-carotene (Gellenbeck 2012). It has been observed that the compound is primarily produced in hypersaline microalgae such as *Dunaliella salina* (Levasseur et al. 2020). Reactive oxygen species, which have detrimental effects on tissues, resulting in a number of diseases and conditions like cancer, aging, heart diseases, etc., are effectively trapped by beta-carotene (Muller-Feuga 2000), thereby enhancing the potential application of the pigment as a nutraceutical. Since it possesses

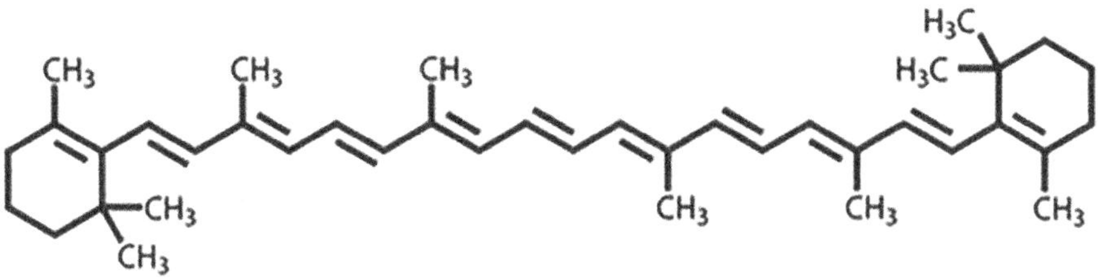

Fig. 13.2 Structure of β-carotene (Marino et al. 2020)

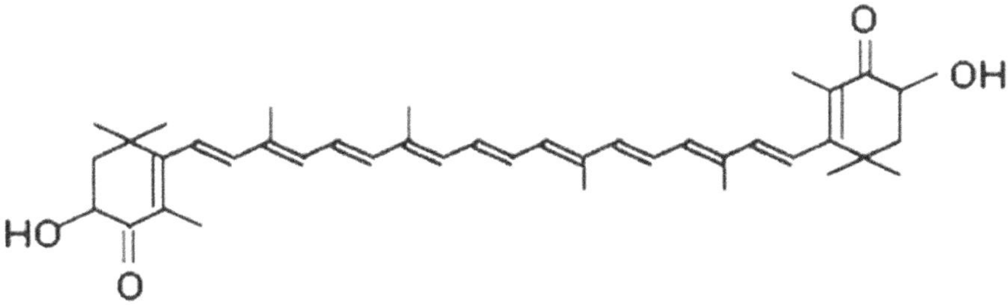

Fig. 13.3 Chemical structure of astaxanthin (Yuan et al. 2011)

remarkable antioxidant activity, many researchers have emphasized the efficacy of beta-carotene in preventing different types of cancer, including cancer affecting the lungs, breasts, colon, ovaries, etc. (Kim et al. 2015b) The pigment is also proven to have a major role in intracellular communication and elevating immune responses (Becker 2004). Since the pigment possesses high antioxidant activity, it can be used as a health food in nutraceuticals and animal fodder (Schubert et al. 2006). It has the ability to fight free radical formation, and ROS finds their extensive application as natural coloring agents in the food and pharmaceutical industry (Guedes et al. 2011).

Although *Dunaliella* species constitute a major source of beta-carotene, there are two other microalgal species, *Coelastrella multistriata* (Abe et al. 2007) and *Scenedesmus almeriensis* (Macías-Sánchez et al. 2010), which produce this pigment at elevated levels and can be exploited for commercial production.

Astaxanthin (oxygenated carotene) (Fig. 13.3) is a red nutraceutical that imparts pink/red color to the species in which it is abundant and is observed in *Haematococcus lacustris*. It is one of the largest constituents (81%) of algae-synthesized carotenoids and is considered another industrially vital pigment (Molino et al. 2018; Rammuni et al. 2019). The antioxidant property of astaxanthin is tenfold higher than beta-carotene and approximately 100-fold higher than α-tocopherol (Lorenz and Cysewski 2000; Miki 1991). Astaxanthin also plays pivotal roles in decreasing triglycerides in the plasma, maintaining blood pressure in the arteries, etc. (Robinson et al. 2005). Enhanced immunity, reproductive behavior, prevention of oxidative damage of PUFAs, etc. are some of the highly noticed metabolic functions of the pigment in humans (Ambati et al. 2014; Xiao et al. 2018). Alongside

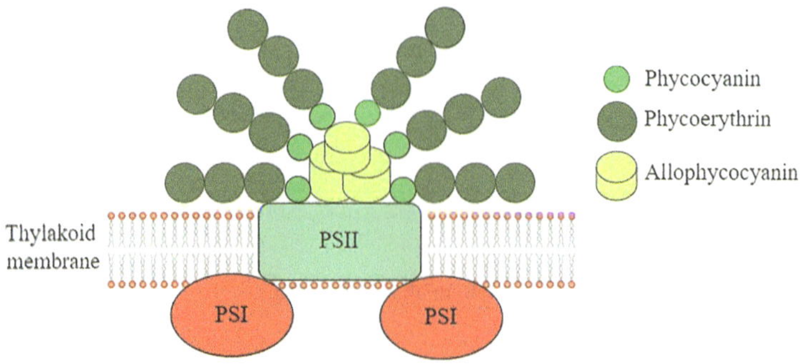

Fig. 13.4 Model of phycobilisome in microalgae

Haematococcus sp., other microalgae such as *Chlorella zofingiensis*, *Chlorococcum* sp., and *Scenedesmus* spp. have been commercially exploited for astaxanthin production. The antioxidant and dyeing properties of the pigment are highly noteworthy and are comparable to strong pigments, including lutein, tocopherols, lycopene, etc. (Cuellar-Bermudez et al. 2015). Astaxanthin finds major applications in pharmaceuticals and nutraceuticals since it is a strong antioxidant and exhibits antiaging, antihypertensive, and anti-inflammatory properties (Guedes et al. 2011). Although the primary source of astaxanthin is *Haematococcus pluvialis*, which yields 8.0 mg L^{-1} day^{-1}, *Chlorella zofingiensis* is equally competitive, yielding 10.3 mg L^{-1} (García-Malea et al. 2009; Ip and Chen 2005).

Bluish-red colored pigments, widely known as phycobiliproteins, are highly water-soluble and are present in phycobilisomes (light-harvesting antennae of PSII) (Fig. 13.4) in species such as Cryptophyta, Cyanophyta, Glaucophyta, and Rhodophyta (Green et al. 2003). Phycobiliproteins actively absorb light in the visible region (450–560 nm) and can be distinguished into four broad classes: phycocyanin (blue), phycoerythrin (red), phycoerthrocyanin (bluish red), and allophycocyanin (blue).

Phycocyanins (PCYs) are extensively constituted of blue-green phycobiliproteins and are primarily found in blue-green algae, cryptomonads, and red algae (Ashaolu et al. 2021). Phycocyanins are heterodimers of α- (17 kDa) and β-subunits (19.5 kDa). There are three classes of PCYs: R-PCY, R-PCY II, and C-PCY, which are extensively found in *Synechococcus* sp. and *Polysiphonia urceolata* (Wang et al. 2014). *Spirulina* sp. is a considerable reservoir of phycocyanin, which imparts bluish color to the alga (Paniagua-Michel 2015). Phycocyanins exhibit a strong antioxidant potential and are widely used in nutraceuticals against diseases involving ROS formation (Fratelli et al. 2021).

Porphyridium sp. (Rhodophyta), *Aphanizomenon flos-aquae* (Cyanobacteria), and *Spirulina* sp. are a few noteworthy microalgae that produce phycobiliproteins in abundance (Levasseur et al. 2020). A range of potential applications of phycobiliproteins include natural colorants in cosmetics, pharmaceuticals, dye, and food industries. It is interesting to learn when PCY is used as a food colorant in milk-based products, strong antioxidant activity is observed (de Amarante et al. 2020).

Phycoerythrin imparts red color to the microalgal species and are majorly seen in Cryptophyceae, red algae, and Cyanophyceae (Glazer 1994). They are also heterodimers of 160–180 amino acids of α- and β-subunits (Sonani et al. 2017). Phycoerythrin also possesses antioxidant and radical scavenging activities, finding major applications in nutraceuticals and pharmaceuticals. Phycoerythrin possesses fluorescent properties and is used to produce fluorescent immunological markers (Levasseur et al. 2020).

Lutein is another major pigment belonging to the category of carotenoids, which find major applications as nutraceuticals. In addition to zeaxanthin, lutein is the only pigment in the retina and lens (Hammond, and Frickj 2007). Lutein is most effective in preventing various types of cancer, enhanced vision (by imparting protection against macular degeneration), and cognitive impairment and nonproliferative diabetic retinopathy (Hu et al. 2011; Levasseur et al. 2020). *Muriellopsis* sp., *Chlorella protothecoides*, *Chlorella zofingiensis*, *Chlorococcum citriforme*, *Neospongiococcus gelatinosum*, and *Scenedesmus almeriensis* are prominent sources of lutein which can synthesize up to 0.5% lutein of total dry biomass (Bhalamur et al. 2018; Borowitzka 2013; Lin et al. 2015; Yaakob et al. 2014).

Astaxanthin and fucoxanthin are the two principal carotenoid pigments found in macroalgae. Less than 20% of the world's total carotenoids are produced from fucoxanthin production (Romay et al. 2003). It has been observed that fucoxanthin can protect the cells from the damage caused by free radicals, have antitumor and anti-inflammatory activity, and can also help control weight gain by enhancing metabolism (Gammone and D'Orazio 2015; Heo et al. 2009; Pádua et al. 2015; Zhang et al. 2015; Zheng et al. 2013).

Diverse techniques have been utilized to extract and release bioactive components from seaweeds; however, green technology using enzymes (Heo et al. 2005) and sub- and supercritical fluid (Plaza et al. 2008) are gaining attention.

13.2.2 Fatty Acids and Lipids

Microalgal lipids are functionally classified into two broad classes: storage lipids and structural lipids. Triglycerides are among the most significant lipids in algae which are produced through photosynthesis and stored in the cells (Kumari et al. 2013). On the other hand, phospholipids and sterols are present in plastids, cell membranes, etc., belonging to structural lipids (Hamed 2016). Although 20–50% of the total dry algal biomass consists of lipids, under unfavorable conditions such as increased salinity, nitrogen concentrations, etc., the amount may further enhance to 80% of the microalgal dry mass (Singh et al. 2005).

DHA (docosahexaenoic acid) and EPA (eicosapentaenoic acid) are two remarkable lipid-based health food supplements, categorized as polyunsaturated fatty acids (PUFAs) that possess 20 or more carbon atoms. Omega-3 and omega-6 PUFAs contribute toward tissue integrity and are pivotal in preventing arthritis, headache, asthma, heart problems, inflammatory diseases, etc. The human body does not produce these, so it is essential to supplement these through foods. DHA, eicosapentaenoic acid, arachidonic acid, linoleic acid, and gamma-linolenic acid have been demonstrated to reduce cholesterol, prevent cardiovascular diseases, have antiaging properties, and protect membrane integrity (Bannenberg et al. 2017; Hu et al. 2002). Microalgae species like *Odontella*, *Porphyridium cruentum*, *I. galbana*, and *Arthrospira platensis* have been explored for their ability to synthesize these commercially important lipids (Dammak et al. 2016). *Pavlova lutheri* has been reported to produce elevated quantities of PUFA (Kumar et al. 2019), while *A. platensis* has been a primary producer of sitosterol, stigmasterol, and γ-linolenic acid (Santhosh et al. 2016).

The amount and composition of fatty acids are highly species-dependent. For instance, *Chlorella*, which is remarkably popular for its abundant accumulation of DHA and EPA (3.24% and 8.9% of the fatty acid methyl esters, respectively) is also rich in saturated fatty acid (Hamed 2016). *Haematococcus pluvialis* produces short-chain fatty acids, which exhibit antimicrobial activity (García-Malea et al. 2009), while *Spirulina* sp. acts as a significant source of fatty acids such as palmitic acids, oleic acids, DHA, lauric acids, etc. (Tornabene et al. 1985). While EPA constitutes up to 15–30% of the total fatty acids in diatoms, dinoflagellates are a promising source of the production of DHA, which in turn constitutes 12–51% of the total fatty acids present in the species and find a major commercial application primarily in the pharmaceutical sector (Dammak et al. 2016; Hamed 2016).

Polyunsaturated fatty acids produced by microalgae are important commercial products of high therapeutic value for cardiac diseases, asthma, and arthritis. There exists sufficient evidence to substantiate the medicinal properties of fatty acids and lipids produced from microalgae, for instance, the cytotoxic, antiviral, and antitumor activities of lipopeptides produced from cyanobacteria as reported by Rocha et al. (2007) and Singh et al. (2005). It is also reported in this regard that *Skeletonema marinoi* is a primary source of producing monoacylglycerides that are functionally active against colon cancer cell lines (Miceli et al. 2019). Recently, several researchers have been trying to enhance the production of EPA and BHA by tailoring the metabolic pathways via genetic recombination, e.g., *P. tricornutum* has been genetically modified to enhance the production of EPA and DHA, etc. (Khan et al. 2018). Today, several commercially important microalgal products, such as eicosapentaenoic acid and DHA, have been made available in the market by numerous biotech and pharma companies.

13.2.3 Vitamins and Minerals

Vitamins and minerals are essential micronutrients for the human body as they contribute to fighting infection, wound healing, strengthening bones, and regulating hormones. Though required in smaller quantities, vitamins are essential for any living organism to survive. Both macroalgae and microalgae are regarded as a factory of vitamins and minerals. They are observed to be the best producers of fat-soluble vitamins such as vitamins A, D, E, and K, in addition to the production of water-soluble vitamins, including vitamin C (Galasso et al. 2019).

Few microalgal species have been reported to produce vitamins in a much larger quantity than conventional food products that are claimed to be vitamin-rich. Studies have revealed some microalgal species to be immensely rich in vitamin E (tocopherol) (*Dunaliella tertiolecta*), vitamin B12 (folic acid) (*Dunaliella tertiolecta*), and vitamin B1 (thiamine) (*Tetraselmis suecica*) as compared to vitamin-rich conventional foods (Fabregas and Herrero 1990). Vitamin B12 (folic acid) is found in around 9–18% of the identified *Chlorella* strains (Shim et al. 2008) and red seaweed *Porphyra, Enteromorpha* sp., and *Spirulina. Porphyridium cruentum* (red microalga) is demonstrated to be a prominent source of vitamin E (Durmaz et al. 2007), while vitamin C is found in *Laminaria digitata* and *U. pinnatifida* (Lordan et al. 2011).

Minerals are elements that are absolutely required by the human body and are assimilated either in elemental form or in the form of compounds. Some essential minerals include phosphorus, sodium, magnesium, manganese, selenium, iron, sulfur, calcium, cobalt, zinc, etc. (Marschner 1997). Although this area of research is not prominent, microalgae and macroalgae are reported as good accumulators of minerals such as calcium, sodium, potassium, iron, manganese, zinc, copper, etc. (Brasil et al. 2017). Some noticeable producers of minerals such as zinc, phosphorus, iron, manganese, sodium, calcium, potassium, and magnesium include *D. tertiolecta, Isochrysis galbana, Tetraselmis suecica, Chlorella vulgaris* (Tokuşoglu and Üunal 2003), and Rhodophyta such as *Chondrus crispus* and *Gracilariopsis* and brown algae such as *Sargassum* and *Ulva* spp. (Circuncisão et al. 2018; Rupérez et al. 2002). Several researchers have emphasized the consumption of these microalgae to meet the daily requirements of minerals in adults and children, as they are considered more bioavailable than traditional inorganic mineral supplements (Brasil et al. 2017). An extensive study has been performed on *Nannochloropsis oceanica* to produce inorganic selenium, which can be used to enhance the food products for aquaculture nutritionally. The study reported an enormous intracellular content of 0.131 g Se/kg algae biomass after 12 days (Guimarães et al. 2021). Macroalgae also contain antioxidant, anti-goiter, and anticancer effects (Eskin et al. 1995; Mohamed et al. 2012).

13.2.4 Proteins

Single-celled algae such as *Dunaliella, Spirulina*, or *Chlorella* contain up to 70% protein, similar to traditional protein sources like eggs and meat. Proteins derived from phytoplanktons contain all essential amino acids and are at par with other protein sources like soybean and egg (Bleakley 2017). Microalgae such as *Chlorella, Dunaliella, Spirulina, Chlamydomonas, Scenedesmus*, etc. produce numerous types of proteins in enormous quantities, making microalgae a sustainable and promising candidate for food and feed products (Becker 2007). Effective extraction of bioactive proteins is possible even in spent algal biomass where all the other biochemical components have been extracted (Dewapriya and Kim 2014). Amino acids lysine, valine, leucine, and isoleucine constitute 35% of the proteins in human muscles. These amino acids are abundant in microalgae, reflecting the potential consumption of microalgae, such as through proteinaceous products (Dewapriya and Kim 2014). The protein cluster, phycobiliproteins, seen in certain microalgae, including *Porphyridium* spp., *S. platensis*, etc., possess features that protect the liver, reduce inflammation, and have antioxidant properties (Ryu et al. 2012; Stengel et al. 2011). The peptides in microalgae also have antagonistic effects on increased cholesterol levels, hypertension, etc. (Borowitzka 1995; Morançais et al. 2018). This is substantiated by the DNA protective effects of peptides isolated from *Chlorella vulgaris* and its ability to prevent cardiac diseases and some types of cancer (Sheih et al. 2009). Very high antioxidant activity was exhibited by the polypeptides produced from *P. lutheri* (Ryu et al. 2012). The development of recombinant proteins remains an unexplored area of research, though with a huge scope (Agyei and Danquah 2011). This is evidenced by the production of recombinant therapeutic proteins produced from *Chlamydomonas reinhardtii*, because of its outstanding protein folding and assembly (Malla et al. 2021; Mayfield et al. 2007; Rasala et al. 2010). *Chlamydomonas* also acts as a host to produce other recombinant proteins, including HSV8-scFv (against herpes simplex virus) and human glutamic acid decarboxylase 65 that aids in diagnosing type 1 diabetes (Gong et al. 2011).

13.2.5 Polysaccharides

Notable marine phytoplankton carbohydrates include alginates, agarans, carrageenan, fucoidans, laminarin, and ulvans (Bilal and Iqbal 2019).

13.2.5.1 Fucoidan

Sargassum species belonging to the Phaeophyta class have a sulfated polysaccharide called fucoidan. The structure of fucoidan is composed of α-(1→3)-L-fucose residues or alternating α-(1→3) and α-(1→4)-linked L-fucose moieties. Depending on the algal species, fucoidan structure branches with an acetyl group and others with sulfate or galactose group (Fig. 13.5). The extracellular matrix enables these algae to deal with difficult situations caused by low tidal currents (Bilal and Iqbal 2019).

Just about 20%of the dry mass of algae is made up of fucoidan; however, these polysaccharides include many L-fucose and sulfate ester groups in addition to monosaccharides, including oligosaccharides and glucose. This proportion changes for each species and during the year (Rengasamy et al. 2014).

Most species of algae are not soluble in water and have a sticky consistency due to the different types of polysaccharides and parts of them present in their outer skeletal structure. As a result, hydrolysis with acid or enzyme (Fig. 13.6) is used to convert them into smaller derivative products, improving their absorption rate without compromising their physicochemical functions. The derivative products like fucoidan saccharides, laminarin saccharides, and alginate oligosaccharides are

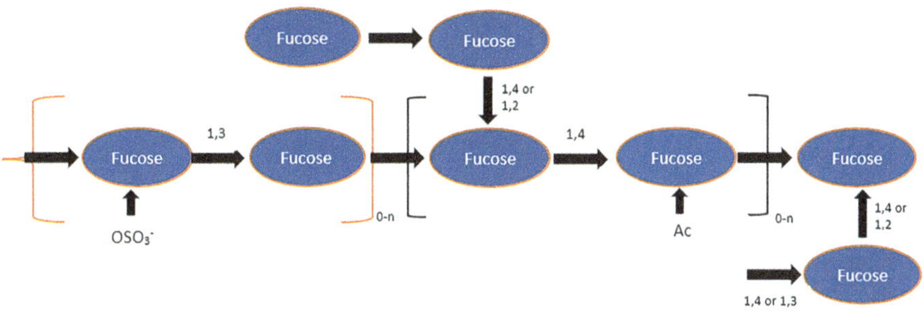

Fig. 13.5 Basic structure of fucoidan isolated from brown algae (Li et al. 2008; Sanjeewa et al. 2018)

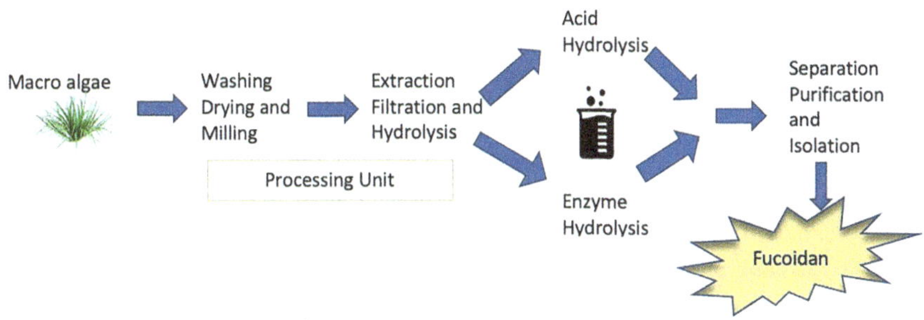

Fig. 13.6 Steps involved in the extraction of fucoidan from marine algae (Bilal and Iqbal 2019)

derived from their corresponding polysaccharide like fucoidan, laminarin, and alginate (Bilal and Iqbal 2019). Fucoidans display multiple pharmacological properties and effectively eliminate lifestyle diseases like diabetes and cancer. They also possess properties that act against viruses, and reactive oxygen species, reducing inflammation and being an anticoagulant. Latest research has also emphasized their use in improving gut health by increasing the number of good bacteria and initiating the lymphocytes and immune cells (Pradhan et al. 2020).

Galactofuran, a fucoidan from *F. evanescens* and *A. utricularis*, exhibits numerous pharmaceutical applications, such as antitumor and antithrombin effects, as an inhibitor of avian RT and HSV 1 and 2 (Ponce et al. 2003; Queiroz et al. 2008). Fucoidan polysaccharides have also been reported to demonstrate activities for tyrosine inhibition in Keln (Yu and Sun 2014), anti-atopic dermatitis in *Fucus vesiculosus* (Tian et al. 2019), and anticoagulant, anti-inflammatory, antiadhesive, and antiangiogenic activities in *Ascophyllum nodosum*, *Cladosiphon okamuranus*, and many species of *Fucus*, *Laminaria digitata*, and *L. saccharina* (Cumashi et al. 2007; Rupérez et al. 2002).

13.2.5.2 Alginic Acid

Alginic acid or alginate is a polymer made from various groups of carbohydrate moieties. They contain 1,4-linked β-D-mannuronic [M] and α-L-guluronic acid

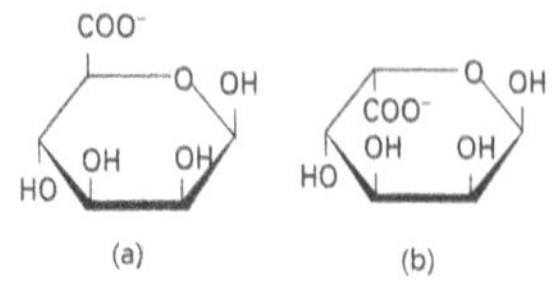

Fig. 13.7 Monomers of alginate; (**a**) β-D mannuronic acid; (**b**) α-L-guluronic acid (Rengasamy et al. 2014b)

[G] (Fig. 13.7), which are monomeric residues arranged in a non-regular, blockwise order along the chain.

Macroalgae belonging to the Phaeophyta group have an inherent capability to bind to metal atoms or molecules. This is mostly seen in sodium and calcium alginate, a common ingredient in the culinary and healthcare sectors since it can create highly viscous compounds. Alginates from *Sargassum vulgare C. Agardh* reportedly showed anticancer efficacy in vivo when implanted with Sarcoma 180 cells (Davis et al. 2003; De Sousa et al. 2007).

Scientific reports on *L. hyperborean* macroalgae have shown how calcium alginate can increase collagen production from the third to the fourth day. Alginates could also limit tissue injury by acting as blocks against intruding fibroblasts, thereby promoting tissue repair. These are excellent for venous, neuropathic, and ischemic lesions as they do not cause discomfort on their introduction. The medical industry can widely explore this property and tissue engineering fields (Tai-fang et al. 2007; Wang et al. 2015b). Other macroalgal species that produce alginate are *Ascophyllum*, *Durvillea*, *Lessonia*, *Macrocystis*, *Turbinaria*, *Laminaria*, *Ecklonia*, and *Sargassum*.

These, like fucoidans, are polysaccharides with sulfur demonstrating their significant medical value since they can fight diabetes, cancer, and virus and also in hematological uses (Zaharudin et al. 2018).

13.2.5.3 Laminarin

Schmiedeberg extracted and purified laminarin from Laminariaceae in 1885 for the first time, which is a source of food and a major part of these species. It is also known as laminarin or leucosin. Laminarins are low-molecular-weight storage β-glucans (Rioux et al. 2009), made of two types of chains, one ending with mannitol residue and the other with glucose. Laminarin composition varies according to algal species. Its basic structure comprises (1,3)-β-D-glucopyranose residues having 6-O branching in the main chain and β-(1,6)-intrachain links—two types of laminarin chains (M or G) with distinct reducing ends. M chains terminate with a mannitol residue, while G chains terminate with a glucose residue (Fig. 13.8) (Neyrinck et al. 2007).

Most laminarins are high in fibers and can be included in the diet due to their resistance to breakdown with water molecules in the top region of the intestine. They create complex molecules supported by cross-functional and cross-hydrogen bonding. Ambient elements such as temperature and the amount of salt and tidal currents in their surroundings are hypothesized to affect the morphology and bioactivity of laminarin and galactofucan (Kadam et al. 2015).

Antiapoptotic activity of laminarin polysaccharides extracted from *Laminaria japonica* was studied in the mice thymocyte cells. Species like *Laminaria*, *Eisenia*

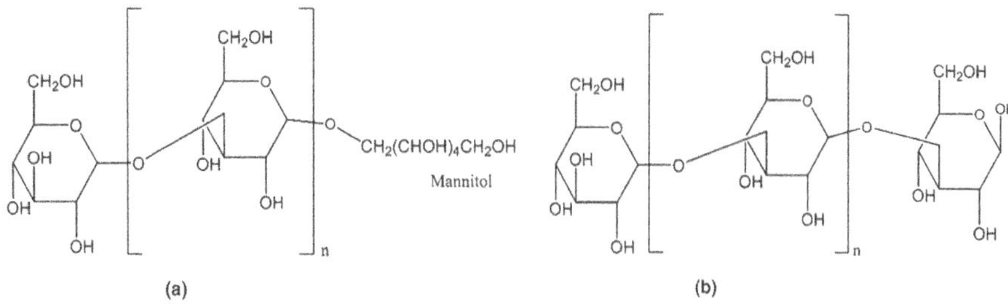

Fig. 13.8 Chemical structure of M and G unit of laminarin (Kadam et al. 2015). (**a**) M chain laminarin, (**b**) G chain laminarin

bicyclis, and *Laminaria digitata* also possess other biological properties like antibacterial, anti-inflammatory, hepatoprotective, immunostimulatory, and anticancer (Kuda et al. 2015; O'Shea et al. 2016; Park et al. 2013).

13.2.6 Phlorotannins

Phlorotannins are generally known as algal polyphenols or polymers of phloroglucinol (Pádua et al. 2015). Derived from tannins (Fig. 13.9), phlorotannin supports the macroalgae's outer structure and protects against the mechanical and osmotic stress in kelps, rockweeds, or sargassacean species and some red algae. It has been found that the content of phlorotannin in brown algae varies greatly across different genera of brown algae and between different regions of the globe (Targett and Arnold 1998). Large amounts of phlorotannin in *Ecklonia cava* species of algae have demonstrated antitumor and sunscreen protection properties. Phlorotannin dieckol showed a tyrosinase inhibition effect at doses significantly greater than the marketed drug kojic acid, suggesting it could be used in cosmetics (Heo et al. 2005). Based on their linkage, these can be classified into four groups (Singh and Bharate 2006):

1. Fuhalols and phlorethols (ether linkage)
2. Fucols (phenyl linkage)
3. Fucophlorethols (ether and phenyl linkage)
4. Eckols and carmalols (dibenzodioxin linkage)

Phlorotannins possess chemopreventive, antibacterial, antiproliferative, antioxidant, antitumor, and anti-HIV properties (Lordan et al. 2011; Rengasamy et al. 2013). Studies have shown that Phlorofucofuroeckol B could reduce allergy symptoms (Sugiura et al. 2007) and block the epithelial-mesenchymal transition, reducing the cancer cell's ability to spread over the body (Dembitsky and Maoka 2007), thus signifying their potential application in pharmaceuticals.

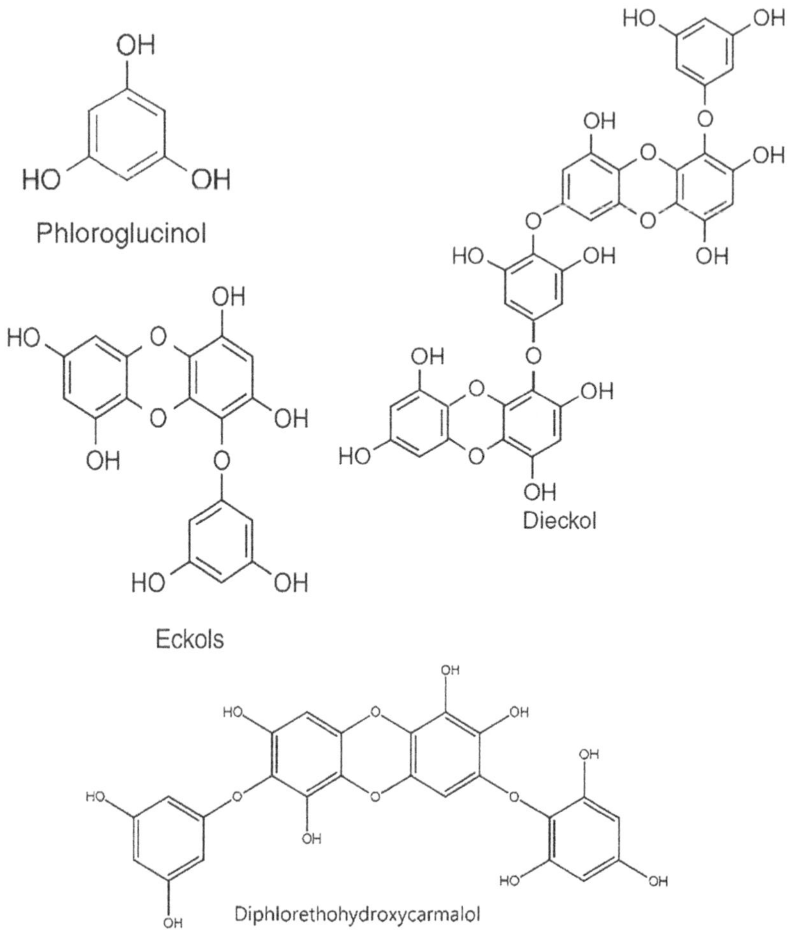

Fig. 13.9 Structure of different phlorotannins in marine algae (Heo et al. 2009)

13.2.7 Alkaloids

Alkaloids are nitrogenous organic compounds possessing halogens and sulfurin in their structure and exhibiting various biological activities. Alkaloids are present in different parts of the plant and are an important bioactive compound class exhibiting antioxidant and anti-inflammatory activities (Peng et al. 2019). There are three groups of alkaloids in macroalgal species.

The alkaloids found in marine algae are classified into three groups: phenylethylamine alkaloids [PEA], indole and halogenated indole alkaloids, and other alkaloids.

Alkaloids from brown marine algae and red algae have shown various pharmacological activities as neuromodulators and neurotransmitters in the brain (Güven et al. 2010). PEAs from marine algae have been demonstrated to act as hormones, stimulants, antidepressants, bronchodilators, hallucinogens, etc. (Güven et al. 2010). Crude extract of caulerpin from macroalga *Caulerpa racemose* exhibited cytotoxic activity against tumor cell lines (Queiroz et al. 2008; Lunagariya et al. 2019). Caulerpin is produced in different species of the genus *Caulerpa*, such as *Caulerpa peltata*, *C. racemosa*, *C. cupressoides*, etc., besides being found in the red algae *Chondria armata* (Nagappan and Vairappan 2014). Racemosin is another indolic alkaloid produced by the genus *Caulerpa* (Xiao et al. 2018). Besides exhibiting cytotoxicity, caulerpin also demonstrated antiviral, spasmolytic, anti-inflammatory, antibacterial, and antinociceptive activities (Macedo et al. 2012; Nagappan and Vairappan 2014; Schröder et al. 1998).

Other alkaloids like lophocladines A and B also demonstrated cytotoxic activity against tumor cells (Gross et al. 2006; Souza et al. 2020), while almazole isolated from *Haraldiophylum* species was reported for its antibacterial activity (Fresneda et al. 2007). Martefragin A from *Martensia fragilis* and azocinyl morpholinone from *Gracilaria opuntia* exhibited antioxidant and anti-inflammatory activities, respectively (Takahashi et al. 1998).

13.3 Pharmaceuticals

Cyanobacteria and microalgae find tremendous application in the pharmaceutical sector since diverse metabolites, including lipids, proteins, pigments, etc., can be extracted from these species (Lau et al. 2015; Koller et al. 2014). Recent studies have diverted attention toward the bioactive compounds produced by marine microalgae, which exhibit anticancer, antiaging, anti-obesity, antimicrobial, and antiviral activities and hence could be potential sources in pharmaceuticals (Khavari et al. 2021).

13.3.1 Anticancer Activity

Chemotherapy, though it has a few demerits, such as the development of drug-resistant cancer cells, anemia, secondary infections, hair loss, mucositis, etc., is still the commonly followed treatment for cancer (Love et al. 1989; Vijayakumar and Menakha 2015). Researchers have speculated that cancer treatment may improve considerably by using natural leads or biomolecules, such as metabolites produced by cyanobacteria or microalgae. Microalgae are found to produce biomolecules that possess anti-cancerous activity, such as grassystatin A–B for lung carcinoma, dolastatin 15 for breast cancer, and kempopeptin A for colon cancer (Khavari et al. 2021; Vijayakumar and Menakha 2015) (Table 13.1). It is also noteworthy to know that *Microcystis aeruginosa*, which forms HABs (harmful algal blooms), actively produces microviridin toxin BF-4, and siatoxin, which possess strong antibacterial and anticancer activities (Khavari et al. 2021). Other examples of molecules produced by microalgae include borophycin (obtained from *Nostoc spongiaeforme*) (Davidson 1995), apratoxin D (isolated from *L. sordida* and

Table 13.1 Some of the cyanobacterial species that effectively produce biomolecules that possess anti-cancerous activity (Singh et al. 2011)

Species	Biomolecules	Type of cancer
Nostoc sp. GSV 224	Cryptophycin-1	Breast and lung cancer
Tolypothrix nodosa	Tolyporphin	EMT-6 tumor cells
Lyngbya majuscula	Somocystinamide A	Neuro-2a neuroblastoma cells
Symploca sp.	Dolastatin 10	Affects microtubule assembly
Lyngbya majuscula	Pitipeptolides AeB	Colon cancer

L. majuscule), tolyporphin (*Tolypothrix nodosa*), and dolastatin 10 (*Symploca* sp.) (Orjala et al. 1995). Some microalgal species, like *U. pinnatifida*, have also been extensively studied in vitro for their anticancer, anti-obesity, and antidiabetic properties (Beppu et al. 2012; Hosokawa et al. 2010). Several studies have shown how fucoidan, fucoxanthin, and phloroglucinol obtained from macroalgae can be combined with conventional anticancer drugs like cisplatin, tamoxifen, or paclitaxel to induce better apoptosis and arrest cell cycle in breast carcinoma (Wu et al. 2013).

13.3.2 Antiviral Activity

Novel drugs are always in demand for better managing viral infections, including Ebola, SARS, H1N1, monkeypox, and many more (WHO 2022). Another persistent viral infection is AIDS (acquired immunodeficiency syndrome), where the human immunodeficiency virus prominently attacks the body's immune system, thereby reducing its capacity to fight other infections (Mastro and De Vincenzi 1996). ART (antiretroviral therapy) is one of the common modes for treating HIV, despite its limitations. The treatment results in development of strong viral resistance, which has shifted the research towards exploration of natural products with strong antiviral properties.

Cyanovirin, which effectively fights HIV, is hugely produced from *Nostoc* (Khan et al. 2018; Carbone et al. 2021). *Spirulina* sp. produces polysaccharides, spirulina, and Ca-spirulan that have significant antiviral properties against HIV-1 and HIV-2 and certain other viruses, including the influenza virus (Ayehunie et al. 1998). These products act by inhibiting HIV-1's reverse transcriptase activity along with the blockage for the attachment of viral cells to the host cells (Singh et al. 2011).

Herpes simplex virus 1 (HSV 1), though usually nonharmful, may result in serious health complications among immunocompromised individuals or people with lowered immune efficiency (Mahomoodally et al. 2019; WHO 2022). *Nostoc flagelliforme* produces nostoflan (acidic polysaccharide) with significant virucidal activity against HSV-1 (Kanekiyo et al. 2007) and may be considered for treating HSV-1.

Another major antiviral biomolecule isolated from *Nostoc ellipsosporum* is the polypeptide cyanovirin-N, which can also treat HIV and other lentiviruses. Cyanovirin-N blocks the attachment of gp120 proteins of HIV to the target cells'

CD4+ cells, thereby stopping HIV fusion with the CD4 cell membrane (Singh et al. 2011). Scytovirin, which binds to the gp120, gp41, and gp160 of HIV, thereby inactivating the virus, is largely produced from *Scytonema varium* (Bokesch et al. 2003).

13.3.3 Antibacterial Activity

The emergence of multidrug resistance (MDR) in a few bacteria has resulted in a major challenge faced by the healthcare sector across the globe. Bacteria such as *Staphylococcus aureus* have developed resistance against methicillin (Stapleton and Taylor 2002), *Enterococci* has exhibited resistance against vancomycin (Reinert et al. 2007), while the prevalence of *MDR in Pseudomonas aeruginosa* has increased over the last few decades (Darwesh et al. 2022; Horcajada et al. 2019; Khavari et al. 2021). This has resulted in an alarming situation where searching for alternative options or novel antibiotics is highly imperative. Since algae are reported to synthesize bioactive molecules, cyanobacteria were explored for their potential to produce antibacterial compounds (Biondi et al. 2008) (Table 13.2). An exciting finding is the sensitivity of vancomycin-resistant *Staphylococcus aureus* toward *Anabaena* extract, with a MIC of 32–64 μg/mL (Bhateja et al. 2006). *Nostoc commune* also produces noscomin 57, which actively fights diverse bacteria, including *Bacillus cereus* (MIC 32 ppm), *E. coli* (MIC 128 ppm), and *Staphylococcus epidermidis* (MIC 8 ppm) (Singh et al. 2011).

13.3.4 Antiprotozoal Activity

Protozoal infections caused by *Plasmodium* sp., *Schistosoma*, *Trypanosoma*, *Leishmania*, etc., remain to infect more than one billion people worldwide, as estimated by WHO (Simmons et al. 2008). Just like bacterial infections, protozoa also exhibit drug resistance, which is another challenge that the pharma sector largely faces. Resistance to new-generation drugs and high communicability create hurdles in treating protozoal infections like malaria (Lanzer and Rohrbach 2007) and leishmaniasis (Prioto et al. 2007). Some algae, like *Amphidinium*, produce karatungiols

Table 13.2 Antimicrobial compounds produced by some microalgal species

Species	Antimicrobial products	Type of bacterium/bacteria
Carbamidocylcophanes	*Nostoc* sp. CAVN 10	*S. aureus* (Bhateja et al. 2006)
Ambiguine-I—isonitrile	*Fischerella* sp.	*Bacillus subtilis and Staphylococcus albus* (Raveh and Carmeli 2007)
Phenolic compound	*Nostoc muscorum*	*B. subtilis, B. cereus, E. coli, Salmonella typhi, S. aureus* (El-Sheekh et al. 2006)
Pahayokolide A	Lyngbya sp.	*Bacillus* sp. (Berry et al. 2004).

Table 13.3 Some of the cyanobacterial bioactive compounds that find application in the treatment of protozoal diseases

Products	Sources	Type of protozoa
Nostocarboline	*Nostoc* sp.	*Trypanosoma brucei, Trypanosoma cruzi, Leishmania donovani, and Plasmodium falciparum* (Singh et al. 2011)
Aerucyclamide C68	*Microcystis aeruginosa* PCC 7806	*T. brucei* (Khavari et al. 2021)
Aerucyclamide B	*Microcystis aeruginosa*	*P. falciparum* (Khavari et al. 2021)
Viridamide A	*Oscillatoria nigroviridis*	*Trypanosoma cruzi, Leishmania mexicana, Plasmodium falciparum* (Simmons et al. 2008)
Venturamides	*Oscillatoria* sp.	*P. falciparum* (Linington et al. 2008)

which are active antimycotic agents and also possess antiprotozoal activities (Torres et al. 2014). Benthic marine algae are reported to produce antiprotozoal compounds effective against infections caused by *Trypanosoma brucei*, *Leishmania* spp., and *Trypanosoma* (Khavari et al. 2021). French marine seaweed, *M. stellatus*, was reported by Vonthron-Sénécheau et al. (2011) to inhibit plasmodium. Other species like *Sargassum siliquastrum* and *S. heterophyllum* have also been studied for their anti-plasmodial activity against *Plasmodium falciparum* as well as for their cytoprotective and photoprotective properties (Afolayan et al. 2008; Heo and Jeon 2009). Some of the marine microalgal compounds that find application in protozoan treatment are listed in Table 13.3.

13.3.5 Wound Dressing Materials

Wound healing, which is highly imperative to restore normal physiological skin functions, involves complicated steps (Sen et al. 2009). Since the treatment is applied on open skin, ensuring hygienic and effective wound dressing materials is crucial. Various salient features of biopolymers, including renewability, biodegradability, and biocompatibility, enable these to be preferred to synthetic polymers (Sahana and Rekha 2018). No inflammatory reaction was observed when a unique polymer, cyanoflan (extracted from *Crocosphaera chwakensis* CCY0110), was investigated in vivo for wound healing (Costa et al. 2021).

13.4 Cosmeceuticals

As per the definition given by the European Commission Regulation No. 1223/2009, a cosmetic product is "any substance or mixture intended to be placed in contact with the external parts of the human body (epidermis, hair system, nails, lip, and external genital organs) or with the teeth and the mucous membranes of the oral cavity with a view exclusively or mainly to cleaning them, perfuming them, changing their appearance, protecting them, keeping them in good condition or correcting body odours." The cosmetic industry remains dynamic, with a constant search for innovative biomolecules beneficial for cosmetic and therapeutic purposes (Mourelle et al. 2017; Bom et al. 2019; Guillerme et al. 2017). A blend of cosmetics and pharmaceuticals constitute cosmeceuticals and serve "drug-like" beneficial purposes to the consumers (Morone et al. 2019; Rodrigues et al. 2018).

Improvement of skin properties to appear healthy is the major objective of cosmeceuticals with a primary focus on maintenance of natural moisture content of the skin, further dependent upon the filaggrin gene that in turn influences the production of NMF (natural moisturizing factor) of the skin (Ariede et al. 2017). Cyanoflan aids in sustained protein or vitamin delivery, thereby finding applications in cosmeceuticals (Costa et al. 2021). Maintaining balanced skin flora is crucial for a good skin texture, which requires the removal of various pathogenic fungi such as *Mucor ramaniannus* and *Candida albicans*. Extracellular matrix balance, improving skin tone and texture, angiogenesis induction, and management of acne and pimples are some noteworthy applications of the antimicrobial peptides released by microalgae (Alencar-Silva et al. 2018).

Skincare products are subjected to thorough quality checks to test the presence of any allergen which may induce an anaphylactic reaction. These quality checks are performed under a relatively new concept termed "cosmetovigilance" (Pereira and Pereira 2018), which includes tests for determining carcinogenicity, genetic toxicity, etc. Cosmeceuticals recovered from algal biomass should also be tested for their carcinogenic potential and allergic reactions using the Ames Salmonella test (Jain et al. 2018). Other assays, including 3T3 NRU (3T3 neutral red uptake), erythrocyte photo-hemolysis tests, etc., strongly allow the detection of phototoxicity of algal bioactive molecules (Kim et al. 2015a). Products such as soaps, skin lotions, hair care products, etc. involve diverse microalgal products (Ariede et al. 2017; Randhir et al. 2020). Microalgal pigments, including beta-carotene, chlorophyll, and astaxanthin, also find a wide range of applications in antiaging products (Khanra et al. 2018). Phycobiliproteins, such as R-phycoerythrin and C-phycocyanin, are used in the cosmetic sector owing to their enhanced antioxidant properties (Khan et al. 2018).

13.4.1 Applications in Antiaging Products

Gradual damage of the cells brought about by the influence of genetics, or environmental factors, eventually leading to cell death, is broadly defined as aging (Gkogkolou and Böhm 2012). Extrinsic and intrinsic aging are considered to be the two broad classifications of skin aging and are induced by various factors, including genetic changes, vitamin deficiencies, hormonal disorders, etc. that result in intrinsic aging and UV ray exposure, pollutants, improper skin practices, etc. that result in extrinsic aging (Rodrigues et al. 2018). The formation of advanced glycation end products (AGE), matrix metalloproteinases (MMPs), and reactive oxygen species (ROS) are some of the most significant processes leading to aging. Hence, research focus has been diverted toward studying compounds that can inhibit AGE formation. M2G (mycosporine-2-glycine) is a potential ingredient in antiaging products (Tarasuntisuk et al. 2018). Some microalgal species that actively produce M2G include *Aphanothece halophytica*, *Nostoc commune*, and *Anabaena variabilis* (Waditee-Sirisattha et al. 2014).

ROS significantly interferes with aging process and tissue degradation. They bring about several oxidative damages through signal pathway induction by attaching themselves to cell surface receptors. This can be effectively overcome with the use of natural antioxidants present on the skin surface (Wang et al. 2015a). However, with aging, these antioxidants decrease considerably, reducing the natural protective mechanism against skin deterioration by ROS (Pangestuti and Kim 2011). Some metabolites extracted from microalgae, such as carotenoids, squalene, chlorophyll, etc., can be used as strong antioxidants that can prevent skin damage caused by ROS (Ibrahim et al. 2017). *Undaria pinnatifida* is a major source of fucoxanthin used in skin products owing to its radical scavenging properties (Sachindra et al. 2007).

MMPs or matrix metalloproteinases are expressed due to skin exposure to UV radiation. MMPs degrade the extracellular matrix proteins, including elastin, collagen, proteoglycans, etc., causing skin damage (Pittayapruek et al. 2016). The association of activator protein 1 (AP-1) and nuclear k-light chain-enhancer of activated B cells (NF-kB) highly influences MMP-1 transcription. Thus, the inhibition of AP-1 and NF-kB can result in the suppression of MMP-1 and the slowing of the aging process (Favas et al. 2021). Fucosterol, a sterol metabolite obtained from microalgae, is reported to induce collagen production and exhibit an antagonistic effect on MMP-1 expression, thereby improving skin properties (Wang et al. 2015b). *Nannochloropsis salina* and *Olisthodiscus luteus* are the two major examples of microalgae that possess fucosterol in abundance (Lau et al. 2015).

One of the ways to keep skin in a healthier condition is its moisturization, which significantly improves skin elasticity and appearance. It is pivotal to wash the skin with products that have protective moisturizing agents such as HA (hydroxy acid). Various algal species, including *Anabaena variabilis*, *Chlorella pyrenoidosa*, *Phormidium foveolarum*, *Anacystis nidulans*, etc., produce alpha-hydroxy acid and beta-hydroxy acid in considerable amounts, which can be used in cosmeceuticals (Matsumoto et al. 1984). Microalgae produce EPS (exopolysaccharides) in large amounts and hence can be commercially exploited for the extraction of moisturizing compounds. Moreover, EPS is released into the medium during the microalgal cultivation making the downstream process economical. Glucuronic acid helps retain the water content in the skin and has been commercially used in cosmetics as a skin moisturizer (Liu et al. 2016). Some of the major producers of glucuronic acid are *Chlorococcum* sp. and *Scenedesmus* spp. (Liu et al. 2016).

Beta-carotene, a major microalgal pigment, helps restore skin texture by reducing photooxidative damage (Nohynek et al. 2010). *Dunaliella saline*, *Spirulina* spp., and *Scenedesmus* spp. are few beta-carotene sources (Christaki et al. 2015).

Squalene is another algal metabolite that can be applied in cosmetics as a moisturizing agent, along with its hydrogenated form, squalene (Borowitzka 2013). Skin hydration and reduction of hyperpigmentation are some of the uses of squalene (Ariede et al. 2017). Although shark liver is the primary source of squalene, microalgae can also produce the compound in abundance (Randhir et al. 2020). Among the different algae, major squalene producers include *Thraustochytrium* spp., *Botryococcus braunii* (Ariede et al. 2017), *Aurantiochytrium* spp., etc. (Borowitzka 2013).

13.4.2 Applications in Photo-protectants

Ozone layer size decrease has favored the easy penetration of UV rays to the earth's surface. UV-B rays cause DNA breakage and membrane disruptions in the organisms exposed to the sun, while UV-A rays enhance the production of ROS (Nohynek et al. 2010). Microalgae produce several metabolites, such as antioxidants, DNA repair enzymes, and compounds that filter UV rays (microbial sunscreens) to overcome this photodamage (Balskus et al. 2011). Chemical filters such as oxybenzone and octinoxate absorb UV rays and are added to several sunscreen lotions (Downs et al. 2016). However, when released in the water bodies during swimming or other activities, these compounds exert a toxicopathological effect on the marine fauna. These compounds have hence been banned in Hawaii (Raffa et al. 2019), increasing the scope of biological filters. Microalgae are exposed to sunlight and UV rays; hence, to protect themselves, they produce several biomolecules acting as UV absorbers. The accumulation of these metabolites acting as microbial screens in the algae can be commercially exploited as a cosmeceutical (Chrapusta et al. 2017). Some microalgal screens include scytonemin, MMAs (mycosporine-like amino acids), ectoine, sporopollenin, lycopene, etc. (Vijayakumar and Menakha 2015), which can protect human skin. Lutein, which is a carotenoid, is highly beneficial for the skin and is produced by *Scenedesmus almeriensis*, *Dunaliella salina*, *Galdieria sulphuraria*, *Chlorella protothecoides*, etc. (Hu et al. 2011; Levasseur et al. 2020).

Cyanobacteria produce a yellow-brown pigment, scytonemin, upon exposure to solar radiation (Pathak et al. 2019). The genes responsible for synthesizing scytonemin are expressed by UVA, and the pigment accumulates in the extracellular matrix. Some prominent scytonemin producers include *Calothrix*, *Nostoc* sp., *Phormidium* sp., *Anabaena*, etc. (Rastogi et al. 2015). Another photo-protectant synthesized by microalgae is MMAs (mycosporine-like amino acids), which protect the skin by absorbing UV radiation and preventing ROS formation (Tarasuntisuk et al. 2019). MAAs act as "broad-spectrum sunscreens," which protect against both UVA and UVB (Chrapusta et al. 2017).

13.5 Aquaculture

Microalgae are a promising source of biomass with bioactive compounds, which can be used as feed for the aquaculture industry. Microalgae are seen to improve defense mechanisms in marine species significantly. Furthermore, the advantage of natural sources over synthetically produced feeds has directed the research focus on applying microalgae in aquaculture (Robinson et al. 2005). The elevated amounts of various biomolecules, such as different pigments, fatty acids, proteins, polysaccharides, etc., in the microalgae, make them a potential candidate for aquaculture development.

Product demand is directly proportional to its color, which further influences marketing. The exoskeleton of several fishes and shellfish appear yellow, red, or orange due to the presence of carotenoids, a microalgal pigment. It is hence advised to feed carotenoids to shellfish organisms through aquafeed as they cannot produce these pigments. Microalgae hence find a major role in this scenario, where it can be used as a potential source of required carotenoids (Liñán-Cabello et al. 2002). Carotenoids play an important role in attaining the characteristic muscle color in organisms like shrimp, goldfish, salmonids, seabream, etc. Larval development is also remarkably enhanced when these organisms are fed with canthaxanthin and astaxanthin (Baker and Günther 2004). Some of the noteworthy products that are used in aquaculture include g-linolenic acid (*Arthrospira* sp.), EPA (*Nannochloropsis*), DHA (*Schizochytrium*), arachidonic acid (*Porphyridium* sp.), etc. (Paniagua-Michel 2015).

13.6 Conclusion

Micro- and macroalgae are considered promising sources of many bioactive compounds with various industrial applications, including food and health sectors. This unexplored area of algal research provides a better scope for producing several industrially important components/metabolites, raising the microalgal demand as a novel and potential candidate for the extraction of bioactive compounds. Microalgae find tremendous applications in nutraceuticals, pharmaceuticals, aquaculture, and the food industry because of their biochemical composition, including pigments, fatty acids, proteins, polysaccharides, etc.

References

Abdelaziz AE, Leite GB, Hallenbeck PC (2013) Addressing the challenges for sustainable production of algal biofuels: I. Algal strains and nutrient supply. Environ Technol 34(13–14): 1783–1805

Abe K, Hattori H, Hirano M (2007) Accumulation and antioxidant activity of secondary carotenoids in the aerial microalga Coelastrella striolata var. multistriata. Food Chem 100(2):656–661

Afolayan AF, Bolton JJ, Lategan CA, Smith PJ, Beukes DR (2008) Fucoxanthin, tetraprenylated toluquinone and toluhydroquinone metabolites from Sargassum heterophyllum inhibit the in vitro growth of the malaria parasite Plasmodium falciparum. Z Naturforsch C J Biosci 63(11–12):848–852

Agyei D, Danquah MK (2011) Industrial-scale manufacturing of pharmaceutical-grade bioactive peptides. Biotechnol Adv 29(3):272–277

Ahmed F, Fanning K, Netzel M, Turner W, Li Y, Schenk PM (2014) Profiling of carotenoids and antioxidant capacity of microalgae from subtropical coastal and brackish waters. Food Chem 165:300–306

Alencar-Silva T, Braga MC, Santana GOS, Saldanha-Araujo F, Pogue R, Dias SC, Franco OL, Carvalho JL (2018) Breaking the frontiers of cosmetology with antimicrobial peptides. Biotechnol Adv 36(8):2019–2031

Ambati RR, Phang SM, Ravi S, Aswathanarayana RG (2014) Astaxanthin: sources, extraction, stability, biological activities and its commercial applications—a review. Mar Drugs 12(1): 128–152

Ariede MB, Candido TM, Jacome ALM, Velasco MVR, de Carvalho JCM, Baby AR (2017) Cosmetic attributes of algae—a review. Algal Res 25:483–487

Ashaolu TJ, Samborska K, Lee CC, Tomas M, Capanoglu E, Tarhan Ö, Jafari SM (2021) Phycocyanin, a super functional ingredient from algae; properties, purification characterization, and applications. Int J Biol Macromol 193:2320–2331

Ayehunie S, Belay A, Baba TW, Ruprecht RM (1998) Inhibition of HIV-1 replication by an aqueous extract of Spirulina platensis (Arthrospira platensis). J Acquir Immune Defic Syndr Hum Retrovirol 18(1):7–12

Baker R, Günther C (2004) The role of carotenoids in consumer choice and the likely benefits from their inclusion into products for human consumption. Trends Food Sci Technol 15(10):484–488

Balskus EP, Case RJ, Walsh CT (2011) The biosynthesis of cyanobacterial sunscreen scytonemin in intertidal microbial mat communities. FEMS Microbiol Ecol 77(2):322–332

Bannenberg G, Mallon C, Edwards H, Yeadon D, Yan K, Johnson H, Ismail A (2017) Omega-3 long-chain polyunsaturated fatty acid content and oxidation state of fish oil supplements in New Zealand. Sci Rep 7(1):1–13

Becker EW (1994) Microalgae: biotechnology and microbiology. Cambridge University Press, Cambridge, p 230

Becker W (2004) Microalgae for aquaculture. In: Handbook of microalgal culture: biotechnology and applied phycology, vol 380. Wiley, New York

Becker EW (2007) Microalgae as a source of protein. Biotechnol Adv 25(2):207–210

Beppu F, Hosokawa M, Niwano Y, Miyashita K (2012) Effects of dietary fucoxanthin on cholesterol metabolism in diabetic/obese KK-A y mice. Lipids Health Dis 11(1):1–8

Berry JP, Gantar M, Gawley RE, Wang M, Rein KS (2004) Pharmacology and toxicology of pahayokolide A, a bioactive metabolite from a freshwater species of Lyngbya isolated from the Florida Everglades. Comp Biochem Physiol C Toxicol Pharmacol 139(4):231–238

Bhalamur GL, Valerie O, Mark L (2018) Valuable bioproducts obtained from microalgal biomass and their commercial applications: a review. Environ Eng Res 230(3):229–241

Bhateja P, Mathur T, Pandya M, Fatma T, Rattan A (2006) Activity of blue green microalgae extracts against in vitro generated Staphylococcus aureus with reduced susceptibility to vancomycin. Fitoterapia 77(3):233–235

Bhattacharya T, Rather G, Akter R, Kabir MT, Rauf A, Rahman M (2021) Nutraceuticals and bio-inspired materials from microalgae and their future perspectives. Curr Top Med Chem 21(12):1037–1051

Bilal M, Iqbal HM (2019) Marine seaweed polysaccharides-based engineered cues for the modern biomedical sector. Mar Drugs 18(1):7

Biondi N, Tredici MR, Taton A, Wilmotte A, Hodgson DA, Losi D, Marinelli F (2008) Cyanobacteria from benthic mats of Antarctic lakes as a source of new bioactivities. J Appl Microbiol 105(1):105–115

Bleakley AP (2017) Algal proteins: extraction, application, and challenges concerning production. Foods 6(5):33

Bokesch HR, O'Keefe BR, McKee TC, Pannell LK, Patterson GM, Gardella RS, Sowder RC, Turpin J, Watson K, Buckheit RW, Boyd MR (2003) A potent novel anti-HIV protein from the cultured cyanobacterium Scytonema varium. Biochemistry 42(9):2578–2584

Bom S, Jorge J, Ribeiro HM, Marto J (2019) A step forward on sustainability in the cosmetics industry: a review. J Clean Prod 225:270–290

Borowitzka MA (1995) Microalgae as sources of pharmaceuticals and other biologically active compounds. J Appl Phycol 7(1):3–15

Borowitzka MA (2013) High-value products from microalgae—their development and commercialisation. J Appl Phycol 25(3):743–756

Brasil BDSAF, de Siqueira FG, Salum TFC, Zanette CM, Spier MR (2017) Microalgae and cyanobacteria as enzyme biofactories. Algal Res 25:76–89

Carbone DA, Pellone P, Lubritto C, Ciniglia C (2021) Evaluation of microalgae antiviral activity and their bioactive compounds. Antibiotics 10(6):746

Chrapusta E, Kaminski A, Duchnik K, Bober B, Adamski M, Bialczyk J (2017) Mycosporine-like amino acids: potential health and beauty ingredients. Mar Drugs 15(10):326

Christaki E, Florou-Paneri P, Bonos E (2011) Microalgae: a novel ingredient in nutrition. Int J Food Sci Nutr 62(8):794–799

Christaki E, Bonos E, Florou-Paneri P (2015) Innovative microalgae pigments as functional ingredients in nutrition. In: Kim S-K (ed) Handbook of marine microalgae. Academic Press, New York, pp 233–243

Circuncisão AR, Catarino MD, Cardoso SM, Silva AM (2018) Minerals from macroalgae origin: health benefits and risks for consumers. Mar Drugs 16(11):400

Costa R, Costa L, Rodrigues I, Meireles C, Soares R, Tamagnini P, Mota R (2021) Biocompatibility of the biopolymer cyanoflan for applications in skin wound healing. Mar Drugs 19(3):147

Cuellar-Bermudez SP, Aguilar-Hernandez I, Cardenas-Chavez DL, Ornelas-Soto N, Romero-Ogawa MA, Parra-Saldivar R (2015) Extraction and purification of high-value metabolites from microalgae: essential lipids, astaxanthin and phycobiliproteins. Microb Biotechnol 8(2): 190–209

Cumashi A, Ushakova NA, Preobrazhenskaya ME, D'Incecco A, Piccoli A, Totani L, Tinari N, Morozevich GE, Berman AE, Bilan MI, Usov AI, Nifantiev NE (2007) A comparative study of the anti-inflammatory, anticoagulant, antiangiogenic, and antiadhesive activities of nine different fucoidans from brown seaweeds. Glycobiology 17(5):541–552

Dammak M, Haase SM, Miladi R, Ben Amor F, Barkallah M, Gosset D, Pichon C, Huchzermeyer B, Fendri I, Denis M, Abdelkafi S (2016) Enhanced lipid and biomass production by a newly isolated and identified marine microalga. Lipids Health Dis 15(1):1–13

Darwesh OM, Mahmoud RH, Abdo SM, Marrez DA (2022) Isolation of Haematococcus lacustris as source of novel anti-multi-antibiotic resistant microbes agents; fractionation and identification of bioactive compounds. Biotechnol Rep 35:e00753

Das P, Aziz SS, Obbard JP (2011) Two phase microalgae growth in the open system for enhanced lipid productivity. Renew Energy 36(9):2524–2528

Davidson BS (1995) New dimensions in natural products research: cultured marine microorganisms. Curr Opin Biotechnol 6(3):284–291

Davis TA, Volesky B, Mucci A (2003) A review of the biochemistry of heavy metal biosorption by brown algae. Water Res 37(18):4311–4330

de Amarante MCA, Braga ARC, Sala L, Kalil SJ (2020) Colour stability and antioxidant activity of C-phycocyanin-added ice creams after in vitro digestion. Food Res Int 137:109602

De Sousa APA, Torres MR, Pessoa C, De Moraes MO, Rocha Filho FD, Alves APNN, Costa-Lotufo LV (2007) In vivo growth-inhibition of Sarcoma 180 tumor by alginates from brown seaweed Sargassum vulgare. Carbohydr Polym 69(1):7–13

Del Campo JA, García-González M, Guerrero MG (2007) Outdoor cultivation of microalgae for carotenoid production: current state and perspectives. Appl Microbiol Biotechnol 74(6): 1163–1174

Dembitsky VM, Maoka T (2007) Allenic and cumulenic lipids. Prog Lipid Res 46(6):328–375

Dewapriya P, Kim SK (2014) Marine microorganisms: an emerging avenue in modern nutraceuticals and functional foods. Food Res Int 56:115–125

Downs CA, Kramarsky-Winter E, Segal R, Fauth J, Knutson S, Bronstein O, Ciner FR, Jeger R, Lichtenfeld Y, Woodley CM, Pennington P (2016) Toxicopathological effects of the sunscreen UV filter, oxybenzone (benzophenone-3), on coral planulae and cultured primary cells and its environmental contamination in Hawaii and the US Virgin Islands. Arch Environ Contam Toxicol 70(2):265–288

Durmaz Y, Monteiro M, Bandarra N, Gökpinar Ş, Işik O (2007) The effect of low temperature on fatty acid composition and tocopherols of the red microalga, Porphyridium cruentum. J Appl Phycol 19(3):223–227

El-Sheekh MM, Osman ME, Dyab MA, Amer MS (2006) Production and characterization of antimicrobial active substance from the cyanobacterium Nostoc muscorum. Environ Toxicol Pharmacol 21(1):42–50

Eskin BA, Grotkowski CE, Connolly CP, Ghent WR (1995) Different tissue responses for iodine and iodide in rat thyroid and mammary glands. Biol Trace Elem Res 49(1):9–19

Fabregas J, Herrero C (1990) Vitamin content of four marine microalgae. potential use as source of vitamins in nutrition. J Ind Microbiol Biotechnol 5(4):259–263

Favas R, Morone J, Martins R, Vasconcelos V, Lopes G (2021) Cyanobacteria and microalgae bioactive compounds in skin-ageing: potential to restore extracellular matrix filling and overcome hyperpigmentation. J Enzyme Inhib Med Chem 36(1):1829–1838

Ficner R, Huber R (1993) Refined crystal structure of phycoerythrin from Porphyridium cruentum at 0.23-nm resolution and localization of the γ subunit. Eur J Biochem 218(1):103–106

Fratelli C, Burck M, Amarante MCA, Braga ARC (2021) Antioxidant potential of nature's "something blue": something new in the marriage of biological activity and extraction methods applied to C-phycocyanin. Trends Food Sci Technol 107:309–323

Fresneda PM, Castaneda M, Blug M, Molina P (2007) Iminophosphorane-based preparation of 2,5-disubstituted oxazole derivatives: synthesis of the marine alkaloid almazole C. Synlett 2007(2):324–326

Galasso C, Gentile A, Orefice I, Ianora A, Bruno A, Noonan DM, Sansone C, Albini A, Brunet C (2019) Microalgal derivatives as potential nutraceutical and food supplements for human health: a focus on cancer prevention and interception. Nutrients 11(6):1226

Gammone MA, D'Orazio N (2015) Anti-obesity activity of the marine carotenoid fucoxanthin. Mar Drugs 13(4):2196–2214

García-Malea MC, Acién FG, Del Río E, Fernández JM, Cerón MC, Guerrero MG, Molina-Grima E (2009) Production of astaxanthin by Haematococcus pluvialis: taking the one-step system outdoors. Biotechnol Bioeng 102(2):651–657

Gellenbeck KW (2012) Utilization of algal materials for nutraceutical and cosmeceutical applications—what do manufacturers need to know? J Appl Phycol 24(3):309–313

Gkogkolou P, Böhm M (2012) Advanced glycation end products: key players in skin aging? Dermato-endocrinology 4(3):259–270

Glazer AN (1994) Phycobiliproteins—a family of valuable, widely used fluorophores. J Appl Phycol 6(2):105–112

Gong Y, Hu H, Gao Y, Xu X, Gao H (2011) Microalgae as platforms for production of recombinant proteins and valuable compounds: progress and prospects. J Ind Microbiol Biotechnol 38(12): 1879–1890

Green B, Parson WW, Parson WW (eds) (2003) Light-harvesting antennas in photosynthesis, vol 13. Springer Science and Business Media, Berlin

Gross H, Goeger DE, Hills P, Mooberry SL, Ballantine DL, Murray TF, Valeriote FA, Gerwick WH (2006) Lophocladines, bioactive alkaloids from the red alga Lophocladia sp. J Nat Prod 69(4): 640–644

Guedes AC, Amaro HM, Malcata FX (2011) Microalgae as sources of carotenoids. Mar Drugs 9(4): 625–644

Guillerme JB, Couteau C, Coiffard L (2017) Applications for marine resources in cosmetics. Cosmetics 4(3):35

Guimarães BO, de Boer K, Gremmen P, Drinkwaard A, Wieggers R, Wijffels RH, Barbosa MJ, D'Adamo S (2021) Selenium enrichment in the marine microalga Nannochloropsis oceanica. Algal Res 59:102427

Güven KC, Percot A, Sezik E (2010) Alkaloids in marine algae. Mar Drugs 8(2):269–284

Hamed I (2016) The evolution and versatility of microalgal biotechnology: a review. Compr Rev Food Sci Food Saf 15(6):1104–1123

Hammond BR, Frick JE (2007) Nutritional protection of the developing retina. Hong Kong Pract 29:128

Heo SJ, Jeon YJ (2009) Protective effect of fucoxanthin isolated from *Sargassum siliquastrum* on UV-B induced cell damage. J Photochem Photobiol B Biol 95(2):101–107

Heo SJ, Park EJ, Lee KW, Jeon YJ (2005) Antioxidant activities of enzymatic extracts from brown seaweeds. Bioresour Technol 96(14):1613–1623

Heo SJ, Ko SC, Cha SH, Kang DH, Park HS, Choi YU, Kim D, Jung WK, Jeon YJ (2009) Effect of phlorotannins isolated from Ecklonia cava on melanogenesis and their protective effect against photo-oxidative stress induced by UV-B radiation. Toxicol In Vitro 23(6):1123–1130

Horcajada JP, Montero M, Oliver A, Sorlí L, Luque S, Gómez-Zorrilla S, Benito N, Grau S (2019) Epidemiology and treatment of multidrug-resistant and extensively drug-resistant Pseudomonas aeruginosa infections. Clin Microbiol Rev 32(4):e00031-19

Hosokawa M, Miyashita T, Nishikawa S, Emi S, Tsukui T, Beppu F, Okada T, Miyashita K (2010) Fucoxanthin regulates adipocytokine mRNA expression in white adipose tissue of diabetic/obese KK-Ay mice. Arch Biochem Biophys 504(1):17–25

Hu FB, Bronner L, Willett WC, Stampfer MJ, Rexrode KM, Albert CM, Manson JE (2002) Fish and omega-3 fatty acid intake and risk of coronary heart disease in women. JAMA 287(14): 1815–1821

Hu BJ, Hu YN, Lin S, Ma WJ, Li XR (2011) Application of lutein and zeaxanthin in nonproliferative diabetic retinopathy. Int J Ophthalmol 4(3):303

Hudek K, Davis LC, Ibbini J, Erickson L (2014) Commercial products from algae. In: Bajpai R, Prokop A, Zappi M (eds) Algal biorefineries. Springer, Dordrecht, pp 275–295

Ibrahim M, Salman M, Kamal S, Rehman S, Razzaq A, Akash SH (2017) Algae-based biologically active compounds. In: Algae based polymers, blends, and composites. Elsevier, Amsterdam, pp 155–271

Ip PF, Chen F (2005) Production of astaxanthin by the green microalga Chlorella zofingiensis in the dark. Process Biochem 40(2):733–738

Jain AK, Singh D, Dubey K, Maurya R, Mittal S, Pandey AK (2018) Models and methods for in vitro toxicity. In: Dhawan A, Kwon S (eds) In vitro toxicology. Academic Press, New York, pp 45–65

Kadam SU, Tiwari BK, O'Donnell CP (2015) Extraction, structure and biofunctional activities of laminarin from brown algae. Int J Food Sci Technol 50(1):24–31

Kanekiyo K, Hayashi K, Takenaka H, Lee JB, Hayashi T (2007) Anti-herpes simplex virus target of an acidic polysaccharide, nostoflan, from the edible blue-green alga Nostoc flagelliforme. Biol Pharm Bull 30(8):1573–1575

Khan MI, Shin JH, Kim JD (2018) The promising future of microalgae: current status, challenges, and optimization of a sustainable and renewable industry for biofuels, feed, and other products. Microb Cell Factories 17(1):1–21

Khanra S, Mondal M, Halder G (2018) Downstream processing of microalgae for pigments, protein, and carbohydrate in industrial application: a review. Food Bioprod Process 110:60–84

Khavari F, Saidijam M, Taheri M, Nouri F (2021) Microalgae: therapeutic potentials and applications. Mol Biol Rep 48(5):4757–4765

Kim K, Park H, Lim KM (2015a) Phototoxicity: its mechanism and animal alternative test methods. Toxicol Res 31(2):97–104

Kim RK, Suh Y, Yoo KC, Cui YH, Hwang E, Kim HJ, Kang JS, Kim MJ, Lee YY, Lee SJ (2015b) Phloroglucinol suppresses metastatic ability of breast cancer cells by inhibition of epithelial-mesenchymal cell transition. Cancer Sci 106(1):94–101

Koller M, Muhr A, Braunegg G (2014) Microalgae as versatile cellular factories for valued products. Algal Res 6:52–63

Kuda T, Kosaka M, Hirano S, Kawahara M, Sato M, Kaneshima T, Nishizawa M, Takahashi H, Kimura B (2015) Effect of sodium-alginate and laminaran on Salmonella typhimurium infection in human enterocyte-like HT-29-Luc cells and BALB/c mice. Carbohydr Polym 125:113–119

Kumar BR, Deviram G, Mathimani T, Duc PA, Pugazhendhi A (2019) Microalgae as rich source of polyunsaturated fatty acids. Biocatal Agric Biotechnol 17:583–588

Kumari P, Kumar M, Reddy CRK, Jha B (2013) Algal lipids, fatty acids and sterols. In: Dominguez H (ed) Functional ingredients from algae for foods and nutraceuticals. Woodhead Publishing, Sawston, pp 87–134

Lanzer M, Rohrbach P (2007) Subcellular pH and Ca^{2+} in Plasmodium falciparum: implications for understanding drug resistance mechanisms. Curr Sci 92(11):1561–1570

Lau NS, Matsui M, Abdullah AAA (2015) Cyanobacteria: photoautotrophic microbial factories for the sustainable synthesis of industrial products. BioMed Res Int 2015:754934

Levasseur W, Perré P, Pozzobon V (2020) A review of high value-added molecules production by microalgae in light of the classification. Biotechnol Adv 41:107545

Li B, Lu F, Wei X, Zhao R (2008) Fucoidan: structure and bioactivity. Molecules 13:1671

Liang S, Liu X, Chen F, Chen Z (2004) Current microalgal health food R and D activities in China. In: Ang PO Jr (ed) Asian pacific phycology in the 21st century: prospects and challenges. Springer, Dordrecht, pp 45–48

Lin J-H, Lee D-J, Chang J-S (2015) Lutein production from biomass: marigold flowers versus microalgae. Bioresour Technol 184:421–428

Liñán-Cabello MA, Paniagua-Michel J, Hopkins PM (2002) Bioactive roles of carotenoids and retinoids in crustaceans. Aquac Nutr 8(4):299–309

Linington RG, Edwards DJ, Shuman CF, McPhail KL, Matainaho T, Gerwick WH (2008) Symplocamide A, a potent cytotoxin and chymotrypsin inhibitor from the marine Cyanobacterium Symploca sp. J Nat Prod 71(1):22–27

Liu L, Pohnert G, Wei D (2016) Extracellular metabolites from industrial microalgae and their biotechnological potential. Mar Drugs 14(10):191

Lordan S, Ross RP, Stanton C (2011) Marine bioactives as functional food ingredients: potential to reduce the incidence of chronic diseases. Mar Drugs 9(6):1056–1100

Lorenz RT, Cysewski GR (2000) Commercial potential for Haematococcus microalgae as a natural source of astaxanthin. Trends Biotechnol 18(4):160–167

Love RR, Leventhal H, Easterling DV, Nerenz DR (1989) Side effects and emotional distress during cancer chemotherapy. Cancer 63(3):604–612

Lunagariya J, Bhadja P, Zhong S, Vekariya R, Xu S (2019) Marine natural product bis-indole alkaloid caulerpin: chemistry and biology. Mini Rev Med Chem 19(9):751–761

Macedo NRPV, Ribeiro MS, Villaça RC, Ferreira W, Pinto AM, Teixeira VL, Cirne-Santo C, Paixao IC, Giongo V (2012) Caulerpin as a potential antiviral drug against herpes simplex virus type 1. Braz J Pharmacogn 22:861–867

Macías-Sánchez MD, Fernandez-Sevilla JM, Fernández FA, García MC, Grima EM (2010) Supercritical fluid extraction of carotenoids from *Scenedesmus almeriensis*. Food Chem 123(3): 928–935

Mahomoodally MF, Lobine D, Rengasamy KR, Gowrishankar S, Tewari D, Zengin G, Kim DH, Sivanesan I (2019) Marine algae: a potential resource of anti-HSV molecules. Processes 7(12): 887

Malla A, Rosales-Mendoza S, Phoolcharoen W, Vimolmangkang S (2021) Efficient transient expression of recombinant proteins using DNA viral vectors in freshwater microalgal species. Front Plant Sci 12:650820

Marino T, Casella P, Sangiorgio P, Verardi A, Ferraro A, Hristoforou E, Musmarra D (2020) Natural beta-carotene: a microalgae derivate for nutraceutical applications. Chem Eng Trans 79: 103–108

Marschner H (1997) Introduction, definition, and classification of mineral nutrients. In: Mineral nutrition of higher plants. Springer, Berlin, pp 3–5

Mastro TD, De Vincenzi I (1996) Probabilities of sexual HIV-1 transmission. AIDS 10:S75–S82

Matsumoto GI, Shioya M, Nagashima H (1984) Occurrence of 2-hydroxy acids in microalgae. Phytochemistry 23(7):1421–1423

Mayfield SP, Manuell AL, Chen S, Wu J, Tran M, Siefker D, Marin-Navarro J (2007) Chlamydomonas reinhardtii chloroplasts as protein factories. Curr Opin Biotechnol 18(2): 126–133

Miceli M, Cutignano A, Conte M, Ummarino R, Romanelli A, Ruvo M, Ianora A (2019) Monoacylglycerides from the diatom Skeletonema marinoi induce selective cell death in cancer cells. Mar Drugs 17(11):625

Miki W (1991) Biological functions and activities of animal carotenoids. Pure Appl Chem 63(1): 141–146

Mohamed S, Hashim SN, Rahman HA (2012) Seaweeds: a sustainable functional food for complementary and alternative therapy. Trends Food Sci Technol 23(2):83–96

Molino A, Rimauro J, Casella P, Cerbone A, Larocca V, Chianese S, Musmarra D, Mehariya S, Ferraro A, Hristoforou E (2018) Extraction of astaxanthin from microalga Haematococcus pluvialis in red phase by using generally recognized as safe solvents and accelerated extraction. J Biotechnol 283:51–61

Morançais M, Mouget JL, Dumay J (2018) Proteins and pigments. In: Levine IA, Fleurence J (eds) Microalgae in health and disease prevention. Academic Press, New York, pp 145–175

Morone J, Alfeus A, Vasconcelos V, Martins R (2019) Revealing the potential of cyanobacteria in cosmetics and cosmeceuticals—a new bioactive approach. Algal Res 41:101541

Mourelle ML, Gómez CP, Legido JL (2017) The potential use of marine microalgae and cyanobacteria in cosmetics and thalassotherapy. Cosmetics 4(4):46

Muller-Feuga A (2000) The role of microalgae in aquaculture: situation and trends. J Appl Phycol 12(3):527–534

Nagappan T, Vairappan CS (2014) Nutritional and bioactive properties of three edible species of green algae, genus Caulerpa (Caulerpaceae). J Appl Phycol 26(2):1019–1027

Neyrinck AM, Mouson A, Delzenne NM (2007) Dietary supplementation with laminarin, a fermentable marine β (1–3) glucan, protects against hepatotoxicity induced by LPS in rat by modulating immune response in the hepatic tissue. Int Immunopharmacol 7(12):1497–1506

Nohynek GJ, Antignac E, Re T, Toutain H (2010) Safety assessment of personal care products/ cosmetics and their ingredients. Toxicol Appl Pharmacol 243(2):239–259

O'Shea CJ, O'Doherty JV, Callanan JJ, Doyle D, Thornton K, Sweeney T (2016) The effect of algal polysaccharides laminarin and fucoidan on colonic pathology, cytokine gene expression and Enterobacteriaceae in a dextran sodium sulfate-challenged porcine model. J Nutr Sci 5:e15

Orjala J, Nagle D, Gerwick WH (1995) Malyngamide H, an ichthyotoxic amide possessing a new carbon skeleton from the Caribbean cyanobacterium Lyngbya majuscula. J Nat Prod 58(5): 764–768

Pádua D, Rocha E, Gargiulo D, Ramos AA (2015) Bioactive compounds from brown seaweeds: phloroglucinol, fucoxanthin and fucoidan as promising therapeutic agents against breast cancer. Phytochem Lett 14:91–98

Palthur MP, Palthur SS, Chitta SK (2010) Nutraceuticals: a conceptual definition. Int J Pharm Pharm Sci 2(3):19–27

Pangestuti R, Kim SK (2011) Biological activities and health benefit effects of natural pigments derived from marine algae. J Funct Foods 3(4):255–266

Paniagua-Michel J (2015) Microalgal nutraceuticals. In: Kim S-K (ed) Handbook of marine microalgae. Academic Press, New York, pp 255–267

Park HK, Kim IH, Kim J, Nam TJ (2013) Induction of apoptosis and the regulation of ErbB signalling by laminarin in HT-29 human colon cancer cells. Int J Mol Med 32(2):291–295

Pathak J, Ahmed H, Singh SP, Häder DP, Sinha RP (2019) Genetic regulation of scytonemin and mycosporine-like amino acids (MAAs) biosynthesis in cyanobacteria. Plant Gene 17:100172

Paul C, Reunamo A, Lindehoff E, Bergkvist J, Mausz MA, Larsson H, Pohnert G (2012) Diatom derived polyunsaturated aldehydes do not structure the planktonic microbial community in a mesocosm study. Mar Drugs 10(4):775–792

Peng J, Zheng TT, Li X, Liang Y, Wang LJ, Huang YC, Xiao HT (2019) Plant-derived alkaloids: the promising disease-modifying agents for inflammatory bowel disease. Front Pharmacol 10: 351

Pereira JX, Pereira TC (2018) Cosmetics and its health risks. Global J Med Res 18:63–70

Pittayapruek P, Meephansan J, Prapapan O, Komine M, Ohtsuki M (2016) Role of matrix metalloproteinases in photoaging and photocarcinogenesis. Int J Mol Sci 17(6):868

Plaza M, Cifuentes A, Ibáñez E (2008) In the search of new functional food ingredients from algae. Trends Food Sci Technol 19(1):31–39

Ponce NM, Pujol CA, Damonte EB, Flores ML, Stortz CA (2003) Fucoidans from the brown seaweed Adenocystis utricularis: extraction methods, antiviral activity and structural studies. Carbohydr Res 338(2):153–165

Pradhan B, Nayak R, Patra S, Jit BP, Ragusa A, Jena M (2020) Bioactive metabolites from marine algae as potent pharmacophores against oxidative stress-associated human diseases: a comprehensive review. Molecules 26(1):37

Prioto G, Kasparian S, Nqouama D, Ghorashian S, Arnold U, Ghabri S, Karunakara U (2007) Nifurtimox- eflornithine combination therapy for second-stage *Trypanosoma brucei gambiense* sleeping sickness: a randomized clinical trial in Congo. Clin Infect Dis 45:435–1444

Queiroz KCS, Medeiros VP, Queiroz LS, Abreu LRD, Rocha HAO, Ferreira CV, Juca MB, Aoyama H, Leite EL (2008) Inhibition of reverse transcriptase activity of HIV by polysaccharides of brown algae. Biomed Pharmacother 62(5):303–307

Raffa RB, Pergolizzi JV Jr, Taylor R Jr, Kitzen JM, NEMA Research Group (2019) Sunscreen bans: coral reefs and skin cancer. J Clin Pharm Ther 44(1):134–139

Rammuni MN, Ariyadasa TU, Nimarshana PHV, Attalage RA (2019) Comparative assessment on the extraction of carotenoids from microalgal sources: astaxanthin from H. pluvialis and β-carotene from D. salina. Food Chem 277:128–134

Randhir A, Laird DW, Maker G, Trengove R, Moheimani NR (2020) Microalgae: a potential sustainable commercial source of sterols. Algal Res 46:101772

Rasala BA, Muto M, Lee PA, Jager M, Cardoso RM, Behnke CA, Kirk P, Hokanson CA, Crea R, Mendez M, Mayfield SP (2010) Production of therapeutic proteins in algae, analysis of expression of seven human proteins in the chloroplast of Chlamydomonas reinhardtii. Plant Biotechnol J 8(6):719–733

Rastogi RP, Sonani RR, Madamwar D (2015) Cyanobacterial sunscreen scytonemin: role in photoprotection and biomedical research. Appl Biochem Biotechnol 176(6):1551–1563

Raveh A, Carmeli S (2007) Antimicrobial ambiguines from the cyanobacterium *Fischerella* sp. collected in Israel. J Nat Prod 70(2):196–201

Reinert RR, Low DE, Rossi F, Zhang X, Wattal C, Dowzicky MJ (2007) Antimicrobial susceptibility among organisms from the Asia/Pacific Rim, Europe and Latin and North America collected as part of TEST and the in vitro activity of tigecycline. J Antimicrob Chemother 60(5):1018–1029

Rengasamy KR, Aderogba MA, Amoo SO, Stirk WA, Van Staden J (2013) Potential antiradical and alpha-glucosidase inhibitors from Ecklonia maxima (Osbeck) Papenfuss. Food Chem 141(2):1412–1415

Rengasamy KR, Kulkarni MG, Stirk WA, Van Staden J (2014a) Advances in algal drug research with emphasis on enzyme inhibitors. Biotechnol Adv 32(8):1364–1381

Rengasamy KR, Kulkarni MG, Stirk WA, Staden JV (2014b) Bioactive metabolites and value-added products from marine macroalgae. In: Seafood processing by-products. Springer, New York, pp 423–454

Rengasamy KR, Mahomoodally MF, Aumeeruddy MZ, Zengin G, Xiao J, Kim DH (2020) Bioactive compounds in seaweeds: an overview of their biological properties and safety. Food Chem Toxicol 135:111013

Rioux LE, Turgeon SL, Beaulieu M (2009) Effect of season on the composition of bioactive polysaccharides from the brown seaweed Saccharina longicruris. Phytochemistry 70(8): 1069–1075

Robinson CB, Samocha TM, Fox JM, Gandy RL, McKee DA (2005) The use of inert artificial commercial food sources as replacements of traditional live food items in the culture of larval shrimp, Farfantepenaeus aztecus. Aquaculture 245(1–4):135–147

Rocha FD, Soares AR, Houghton PJ, Pereira RC, Kaplan MAC, Teixeira VL (2007) Potential cytotoxic activity of some Brazilian seaweeds on human melanoma cells. Phytother Res 21(2): 170–175

Rodrigues F, de la Cádiz-Gurrea ML, Nunes MA, Pinto D, Vinha AF, Linares IB, Carretero AS (2018) Cosmetics. In: Polyphenols: properties, recovery, and applications. Woodhead Publishing, Duxford

Romay CH, Gonzalez R, Ledon N, Remirez D, Rimbau V (2003) C-phycocyanin: a biliprotein with antioxidant, anti-inflammatory and neuroprotective effects. Curr Protein Pept Sci 4(3):207–216

Rupérez P, Ahrazem O, Leal JA (2002) Potential antioxidant capacity of sulfated polysaccharides from the edible marine brown seaweed Fucus vesiculosus. J Agric Food Chem 50(4):840–845

Ryu B, Kang KH, Ngo DH, Qian ZJ, Kim SK (2012) Statistical optimization of microalgae *Pavlova lutheri* cultivation conditions and its fermentation conditions by yeast, Candida rugopelliculosa. Bioresour Technol 107:307–313

Sachindra NM, Sato E, Maeda H, Hosokawa M, Niwano Y, Kohno M, Miyashita K (2007) Radical scavenging and singlet oxygen quenching activity of marine carotenoid fucoxanthin and its metabolites. J Agric Food Chem 55(21):8516–8522

Sahana TG, Rekha PD (2018) Biopolymers: applications in wound healing and skin tissue engineering. Mol Biol Rep 45(6):2857–2867

Sahin SC (2018) The potential of Arthrospira platensis extract as a tyrosinase inhibitor for pharmaceutical or cosmetic applications. S Afr J Bot 119:236–243

Sanjeewa KA, Kang N, Ahn G, Jee Y, Kim YT, Jeon YJ (2018) Bioactive potentials of sulfated polysaccharides isolated from brown seaweed *Sargassum spp* in related to human health applications: a review. Food Hydrocoll 81:200–208

Santhosh S, Dhandapani R, Hemalatha N (2016) Bioactive compounds from microalgae and its different applications—a review. Adv Appl Sci Res 7(4):153–158

Schröder HC, Badria FA, Ayyad SN, Batel R, Wiens M, Hassanein HM, Müller WE (1998) Inhibitory effects of extracts from the marine alga Caulerpa taxifolia and of toxin from Caulerpa racemosa on multixenobiotic resistance in the marine sponge *Geodia cydonium*. Environ Toxicol Pharmacol 5(2):119–126

Schubert N, García-Mendoza E, Pacheco-Ruiz I (2006) Carotenoid composition of marine red algae 1. J Phycol 42(6):1208–1216

Sen CK, Gordillo GM, Roy S, Kirsner R, Lambert L, Hunt TK, Longaker MT (2009) Human skin wounds: a major and snowballing threat to public health and the economy. Wound Repair Regen 17(6):763–771

Sheih IC, Wu TK, Fang TJ (2009) Antioxidant properties of a new antioxidative peptide from algae protein waste hydrolysate in different oxidation systems. Bioresour Technol 100(13): 3419–3425

Shim JY, Shin HS, Han JG, Park HS, Lim BL, Chung KW, Om AS (2008) Protective effects of Chlorella vulgaris on liver toxicity in cadmium-administered rats. J Med Food 11(3):479–485

Simmons TL, Engene N, Ureña LD, Romero LI, Ortega-Barría E, Gerwick L, Gerwick WH (2008) Viridamides A and B, lipodepsipeptides with antiprotozoal activity from the marine cyanobacterium Oscillatoria nigro-viridis. J Nat Prod 71(9):1544–1550

Singh IP, Bharate SB (2006) Phloroglucinol compounds of natural origin. Nat Prod Rep 23(4): 558–591

Singh S, Kate BN, Banerjee UC (2005) Bioactive compounds from cyanobacteria and microalgae: an overview. Crit Rev Biotechnol 25(3):73–95

Singh RK, Tiwari SP, Rai AK, Mohapatra TM (2011) Cyanobacteria: an emerging source for drug discovery. J Antibiot 64(6):401–412

Siqueira SF, Queiroz MI, Zepka LQ, Jacob-Lopes E (2018) Introductory chapter: microalgae biotechnology—a brief introduction. In: Microalgal biotechnology. IntechOpen, London

Soletto D, Binaghi L, Lodi A, Carvalho JCM, Converti A (2005) Batch and fed-batch cultivations of Spirulina platensis using ammonium sulphate and urea as nitrogen sources. Aquaculture 243(1–4):217–224

Sonani RR, Rastogi RP, Singh NK, Thadani J, Patel PJ, Kumar J, Tiwari AK, Devkar RV, Madamwar D (2017) Phycoerythrin averts intracellular ROS generation and physiological functional decline in eukaryotes under oxidative stress. Protoplasma 254(2):849–862

Souza CR, Bezerra WP, Souto JT (2020) Marine alkaloids with anti-inflammatory activity: current knowledge and future perspectives. Mar Drugs 18(3):147

Stapleton PD, Taylor PW (2002) Methicillin resistance in Staphylococcus aureus. Sci Prog 85:57–72

Stengel DB, Connan S, Popper ZA (2011) Algal chemodiversity and bioactivity: sources of natural variability and implications for commercial application. Biotechnol Adv 29(5):483–501

Sugiura, Y., Matsuda, K., Yamada, Y., Nishikawa, M., Shioya, K., Katsuzaki, H., , Imai, K and Amano, H. (2007). Anti-allergic phlorotannins from the edible brown alga, *Eisenia arborea*, Food Sci Technol Res, 13(1), 54-60.

Sun XM, Ren LJ, Zhao QY, Ji XJ, Huang H (2018) Microalgae for the production of lipid and carotenoids: a review with focus on stress regulation and adaptation. Biotechnol Biofuels 11(1): 1–16

Tai-fang G, Cai-yun F, Wen C (2007) Physiochemical properties of alginate in tissue engineering research and its clinical application. J Clin Rehabil Tissue Eng Res 11(18):3613

Takahashi S, Matsunaga T, Hasegawa C, Saito H, Fujita D, Kiuchi F, Tsuda Y (1998) Martefragin A, a novel indole alkaloid isolated from red alga, inhibits lipid peroxidation. Chem Pharm Bull 46(10):1527–1529

Tarasuntisuk S, Patipong T, Hibino T, Waditee-Sirisattha R, Kageyama H (2018) Inhibitory effects of mycosporine-2-glycine isolated from a halotolerant cyanobacterium on protein glycation and collagenase activity. Lett Appl Microbiol 67(3):314–320

Tarasuntisuk S, Palaga T, Kageyama H, Waditee-Sirisattha R (2019) Mycosporine-2-glycine exerts anti-inflammatory and antioxidant effects in lipopolysaccharide (LPS)-stimulated RAW 264.7 macrophages. Arch Biochem Biophys 662:33–39

Targett NM, Arnold TM (1998) Minireview—predicting the effects of brown algal phlorotannins on marine herbivores in tropical and temperate oceans. J Phycol 34(2):195–205

Tian T, Chang H, He K, Ni Y, Li C, Hou M, Chen L, Xu Z, Chen B, Ji M (2019) Fucoidan from seaweed Fucus vesiculosus inhibits 2,4-dinitrochlorobenzene-induced atopic dermatitis. Int Immunopharmacol 75:105823

Tokuşoglu Ö, Üunal MK (2003) Biomass nutrient profiles of three microalgae: Spirulina platensis, Chlorella vulgaris, and Isochrysis galbana. J Food Sci 68(4):1144–1148

Tornabene TG, Bourne TF, Raziuddin S, Ben-Amotz A (1985) Lipid and lipopolysaccharide constituents of cyanobacterium Spirulina platensis (Cyanophyceae, Nostocales). Mar Ecol Prog Ser 22(2):121–125

Torres FA, Passalacqua TG, Velásquez A, Souza RAD, Colepicolo P, Graminha MA (2014) New drugs with antiprotozoal activity from marine algae: a review. Braz J Pharmacogn 24:265–276

Vershinin A (1999) Biological functions of carotenoids-diversity and evolution. Biofactors 10(2–3):99–104

Vijayakumar S, Menakha M (2015) Pharmaceutical applications of cyanobacteria—a review. JACME 5(1):15–23

Vílchez C, Forján E, Cuaresma M, Bédmar F, Garbayo I, Vega JM (2011) Marine carotenoids: biological functions and commercial applications. Mar Drugs 9(3):319–333

Vonthron-Sénécheau C, Kaiser M, Devambez I, Vastel A, Mussio I, Rusig AM (2011) Antiprotozoal activities of organic extracts from French marine seaweeds. Mar Drugs 9(6): 922–933

Waditee-Sirisattha R, Kageyama H, Sopun W, Tanaka Y, Takabe T (2014) Identification and upregulation of biosynthetic genes required for accumulation of mycosporine-2-glycine under salt stress conditions in the halotolerant cyanobacterium Aphanothece halophytica. Appl Environ Microbiol 80(5):1763–1769

Wang L, Qu Y, Fu X, Zhao M, Wang S, Sun L (2014) Isolation, purification and properties of an R-phycocyanin from the phycobilisomes of a marine red macroalga Polysiphonia urceolata. PLoS One 9(2):e87833

Wang HMD, Chen CC, Huynh P, Chang JS (2015a) Exploring the potential of using algae in cosmetics. Bioresour Technol 184:355–362

Wang T, Gu Q, Zhao J, Mei J, Shao M, Pan Y, Zhang J, Wu H, Zhang Z, Liu F (2015b) Calcium alginate enhances wound healing by up-regulating the ratio of collagen types I/III in diabetic rats. Int J Clin Exp Pathol 8(6):6636

WHO (2022). https://www.who.int/news-room/fact-sheets/detail/herpes-simplex-virus

Wu GS, Lu JJ, Guo JJ, Huang MQ, Gan L, Chen XP, Wang YT (2013) Synergistic anticancer activity of the combination of dihydroartemisinin and doxorubicin in breast cancer cells. Pharmacol Rep 65(2):453–459

Xiao X, Xu M, Yang C, Yao Y, Liang LN, Chung PE, Long Q, Zacksenhaus E, He Z, Liu S, Ben-David Y (2018) Novel racemosin B derivatives as new therapeutic agents for aggressive breast cancer. Bioorg Med Chem 26(23–24):6096–6104

Yaakob Z, Ali E, Zainal A, Mohamad M, Takriff MS (2014) An overview: biomolecules from microalgae for animal feed and aquaculture. J Biol Res (Thessalon) 21(1):1–10

Yu P, Sun H (2014) Purification of a fucoidan from kelp polysaccharide and its inhibitory kinetics for tyrosinase. Carbohydr Polym 99:278–283

Yuan JP, Peng J, Yin K, Wang JH (2011) Potential health-promoting effects of astaxanthin: a high-value carotenoid mostly from microalgae. Mol Nutr Food Res 55(1):150–165

Zaharudin N, Salmeán AA, Dragsted LO (2018) Inhibitory effects of edible seaweeds, polyphenolics and alginates on the activities of porcine pancreatic α-amylase. Food Chem 245:1196–1203

Zhang H, Tang Y, Zhang Y, Zhang S, Qu J, Wang X, Kong R, Han C, Liu Z (2015) Fucoxanthin: a promising medicinal and nutritional ingredient. Evid Based Complement Alternat Med 2015: 723515

Zheng J, Piao MJ, Keum YS, Kim HS, Hyun JW (2013) Fucoxanthin protects cultured human keratinocytes against oxidative stress by blocking free radicals and inhibiting apoptosis. Biomol Ther 21(4):270

Zorofchian Moghadamtousi S, Karimian H, Khanabdali R, Razavi M, Firoozinia M, Zandi K, Abdul Kadir H (2014) Anticancer and antitumor potential of fucoidan and fucoxanthin, two main metabolites isolated from brown algae. ScientificWorldJournal 2014:768323

Neuroactive Peptides and Neuroprotective Molecules from Marine Sponges and Associated Bacteria: An Untapped Resource for Systemic Drug Development

14

G. B. Priyadharshini, Saqib Hassan, Ramu Meenatchi, George Seghal Kiran, C. Jayanthy, Pallaval Veera Bramhachari, and Joseph Selvin

Abstract

Neurodegenerative diseases occur when a progressive degeneration of neurons leads to the loss of neuron structure or function and even death. Common neurodegenerative disorders include Parkinson's disease, Alzheimer's disease, Huntington's disease, and other dementias. Alzheimer's is the leading neurodegenerative disease, responsible for more than 60% of dementia cases worldwide. The available treatments for such diseases are inadequate and only help to improve certain symptoms. Hence, developing new treatments and drugs for such diseases is critical. Natural marine products are rich sources of novel

G. B. Priyadharshini · C. Jayanthy
Department of Bioinformatics, School of Bio and Chemical Engineering, Sathyabama Institute of Science and Technology, Chennai, Tamil Nadu, India

S. Hassan
Department of Biotechnology, School of Bio and Chemical Engineering, Sathyabhama Institute of Science and Technology, Chennai, Tamil Nadu, India

American Society for Microbiology, Washington, DC, USA

R. Meenatchi
Department of Biotechnology, College of Science and Humanities, SRM Institute of Science and Technology, Chennai, Tamil Nadu, India

G. S. Kiran
Department of Food Science and Technology, School of Life Sciences, Pondicherry University, Puducherry, India

P. Veera Bramhachari
Department of Biotechnology, Krishna University, Machilipatnam, Andhra Pradesh, India

J. Selvin (✉)
Department of Microbiology, School of Life Sciences, Pondicherry University, Puducherry, India

P. Veera Bramhachari, C. V. Berde (eds.), *Marine Bioactive Molecules for Biomedical and Pharmacotherapeutic Applications*,
https://doi.org/10.1007/978-981-99-6770-4_14

compounds that can alter mammalian neurological activity. Though some novel compounds of medicinal importance have been reported from marine sources, this environment remains largely unexplored. Thus far, more than 75 compounds have been isolated from sponges and identified as having neuromodulatory and neuroprotective activities. While most of these compounds are still in the preclinical and clinical study phases, a few have hit the market. Marine sponges are one of the most sustainable and efficient sources to explore when bio-prospecting for neuroprotective compounds can play a role in medicine. Awareness of sponge-derived compounds' classifications and modes of action is critical to understanding their role in treating neurological disorders.

Keywords

Dementia · Marine sponge · Neuroprotective · Neurons · Neurodegeneration

14.1 Introduction

Neurodegeneration refers to the intensive loss of structure and function that finally leads to the death of neurons. Neurodegenerative disorders such as Alzheimer's disease, Parkinson's disease, Huntington's disease, and amyotrophic lateral sclerosis occur due to this neurodegenerative process (Uttara et al. 2009). In addition to aging, the factors generally responsible for neurodegeneration are genetic. They are caused by inherited genetic mutations that result in protein folding, membrane damage, and disrupted intracellular mechanisms such as mitochondrial dysfunction and degradation pathways of defective proteins and axonal transport, as well as programmed death (Perry et al. 1994; Franco-Iborra et al. 2018). There is no proper cure for these neurodegenerative disorders. Still, natural products isolated from marine organisms, especially sponges, are a promising source for producing drugs to slow the progression of such neurodegenerative disorders and aid in their symptomatic treatment (Rentas et al. 1995; Huang et al. 2019). Marine sponges hold a wealth of pharmaceutical compounds and should be further explored in the search for novel compounds with better efficacy for treating such diseases.

The oceans cover almost one-third of the Earth's surface and harbor highly diverse organisms. It has been estimated that marine species comprise 25% of the total number of species on Earth. Marine organisms have developed highly diverse mechanisms to thrive in an extremely different and challenging environment, and they are known for producing numerous secondary metabolites for their use in defense, mating, and warding off competitors. These metabolites differ from their terrestrial analogs and are unique regarding their chemical structure and biological activity (Andryukov et al. 2019). Although exploring this vast environment has only recently begun, it has already yielded more than 18,000 biologically active compounds. These marine-derived natural compounds can treat various diseases as potent pharmaceutical agents (Alves et al. 2018).

Sponges belong to the phylum Porifera and represent an important component of benthic communities worldwide (Ruocco et al. 2021; Kiran et al. 2018a, b). They mainly dwell in marine habitats but can also be found in freshwaters (Meijer et al. 2000). Sponges harbor diverse microorganisms, such as bacteria, fungi, and microalgae, which can account for more than half of the biomass of the sponge. It has been reported that these associated microbes help the sponges by serving as a nutritional source and participating in mutualistic symbioses (Hentschel and Bringmann 2010). Sponges are a collection of species with close association and complex interactions in which the host-microbiome has been less characterized when compared to host species. Based on the environment, the sponge-associated microbiome varies. Sponges produce potent chemicals to defend against predators, competitors, and infectious microorganisms (Trofimova and Robbins 2016).

Moreover, they produce highly diverse bioactive secondary metabolites that greatly interest in developing new drugs to combat various diseases. Sponges are also a superabundant source for developing bioactive compounds, which is pharmacologically vital for marine invertebrates (Trofimova and Robbins 2016; Orhan et al. 2012). Sponges produce many of the major compounds used against human diseases, compounds with cytotoxic, antitumor, antiviral, antimicrobial, and anti-inflammatory activities, as well as compounds that improve cardiovascular problems (Horikawa et al. 2013; Blunt et al. 2011; Orhan et al. 2012; Morales and Rodrigue 1991). Most of the compounds extracted and isolated from sponges have a wide range of biological activities and include terpenoids, alkaloids, macrolides, peptides, some associated polyketides, and sterols, among others (Tello-Aburto et al. 2012a; Amina and Al Musayeib 2018) (Fig. 14.1).

The establishment of numerous microbial cell factories, such as yeast cell factories, for the production of extremely useful metabolites from renewable biomass has been made possible by the discovery of genes responsible for the production of marine bioactive metabolites via metabolic pathways combined with advancements in synthetic biology tools (Kiran et al. 2018a, b).

Blunt and his coworkers explored marine organisms and reported on spongothymidine and spongouridine (bioactive nucleosides) in marine sponges (Kashman et al. 1980). Proksch and his colleagues also worked on these nucleosides, which served as template compounds for the synthesis of Ara-C, the first anticancer agent derived from an oceanic source, and the antiviral drug Ara-A (Roll et al. 1988; Sagar et al. 2010; Anjum et al. 2016). Of the 17,000 compounds reported from marine sources, more than 5300 are natural products obtained from sponges. This review aims to shed light on the neuroactive metabolites obtained from sponges or sponge-associated microorganisms.

Sponges host most marine microorganisms, such as microalgae, fungi, bacteria, cyanobacteria, and Archaea (Proksch et al. 2002). Based on the environment, the sponge-associated microbiome varies (Fathima et al. 2022). The diverse range of studies on the sponge-associated microbiome has used molecular methods such as 16S rRNA gene sequencing, fluorescence in situ hybridization, and denaturing gradient gel electrophoresis, which has unveiled that *Acidobacteria*, *Actinobacteria*, *Bacteroidetes*, *Chloroflexi*, *Cyanobacteria*, *Nitrospira*, *Poribacteria*,

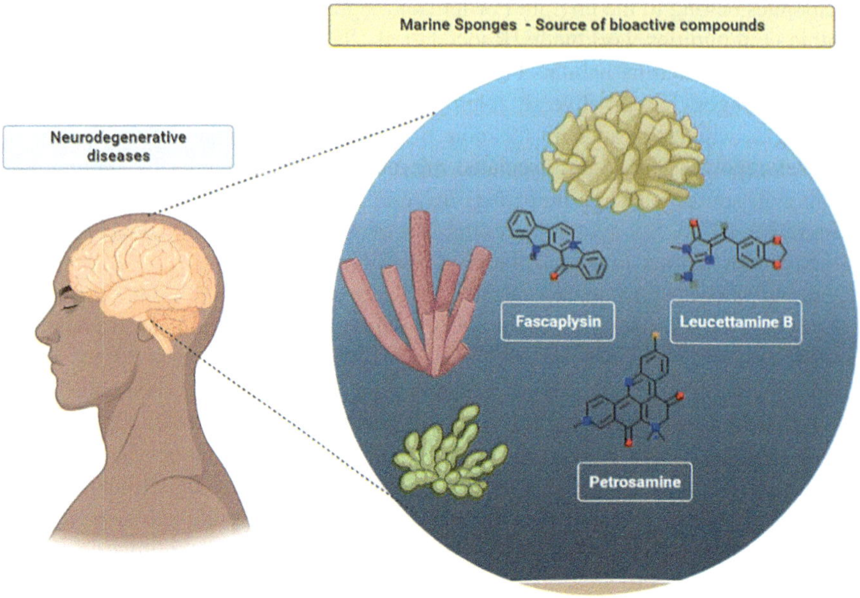

Fig. 14.1 Some bioactive/neuroactive compounds obtained from sponges

Planctomycetes, and *Proteobacteria* are generally associated with sponges (Trofimova and Robbins 2016; Zhang et al. 2012). Sponges also possess basic chemical messengers such as the diffusible amino acids glutamate and γ-aminobutyric acid (GABA) and nitric oxide (NO)—a short-lived diffusible gas found in the central nervous system (CNS) of plants, animals, and other invertebrates—which can be useful in the treatment of neurodegenerative disorders (Elliott and Leys 2010; Priyadharshini et al. 2021).

14.2 Sponge-Derived AChE Inhibitors

In patients with Alzheimer's disease, there is a deficiency of the chemical messenger acetylcholine that can be treated by inhibiting acetylcholinesterase (AChE) activity, which subsequently increases the acetylcholine concentration in nerve cells (El-Hady et al. 2014). Thus, AChE inhibitors are one promising drug target for the symptomatic treatment of Alzheimer's disease.

The first steroidal alkaloid isolated from a marine sponge in the *Corticium* genus was 4-acetoxy-plakinamine B with an IC50 value of 3.75 μM that inhibits AChE in a reversible manner, which is an AChE inhibitor (Singhal et al. 2012). Aoki and coworkers documented unique carboxylic acid derived from the sponge *Haliclona* sp., lembehyne A, which promotes neurite outgrowth (Alghazwi 2018). Lembehyne A is a neuritogenic polyacetylene that reportedly induces neuronal differentiation in neuroblastoma cells (Aoki et al. 2000, 2001; Taylor et al. 2007). Roll and his colleagues (Sakai et al. 2001) isolated the organic compound bis-indole from the sponge *Fascaplysinopsis.* A bis-indole derivative named fascaplysin is a noncompetitive AChE inhibitor with selective inhibitory activity with the IC_{50} value of 1.49 μM. Molecular docking studies have shown that fascaplysin anchors perfectly into the AChE active site (Bharate et al. 2012; Grosso 2014). In 2012, Orhan isolated oroidin from the sponge *Agelas oroides*, which is known to possess antimalarial activity, but was also found to inhibit AChE with an IC_{50} value of 55.21 μM (Forenza et al. 1971; Orhan et al. 2012). Diamine alkaloids (saraines and isosaraines) are the bioactive compounds isolated from the Mediterranean sponge *Renierasarai* (Guo et al. 1998; Cimino et al. 1986a, b; Haefner 2003; Bai et al. 2021). Saraines act as a competitive inhibitor of AChE in a reversible manner, for which the K_i value was found to be 5–10 μM (Defant et al. 2011). An Okinawan marine sponge belonging to the genus *Plakortis* produces alkaloids (platisidines A–C) (Mora et al. 2011; Kubota et al. 2010) that show inhibitory activity against AChE with inhibitory concentrations of 2.8, 2.6 and 2.1 mM, respectively. Table 14.1 summarizes the compounds derived from marine sponges having acetylcholinesterase inhibition activity.

Sponges grouped under the genus *Petrosia* are known to produce petrosamine (Moroni et al. 1995; Lee et al. 2021), which has potent AChE inhibitory activity, with an IC_{50} value of 91 nM. A crude extract of polymeric alkyl pyridinium salts from the marine sponge *Renierasarai* has also been shown to inhibit AChE (Paleari et al. 2006). It was found that micro concentrations of the extract were known to possess an IC_{50} value of 0.06 μg/mL. A novel "anti-neuroinflammatory agent" called plakortide P has been isolated from the sponge *Plakortisangulo spiculatus* (Kossuga et al. 2008). Lastly, a chloroform-methanol extract of the sponge *Halichondria panicea* exhibited increased AChE activity at higher concentrations. In contrast, a sponge extract of a *Cladiella* sp. isolated in Andaman, India, exhibited a mild inhibitory effect on AChE.

14.3 Sponge-Derived Glycogen Synthase Kinase 3 and Cyclin-Dependent Kinase Inhibitors

Glycogen synthase kinase 3 (GSK-3) is a multifunctional serine/threonine kinase highly expressed in the CNS and plays an important role in the glycogen metabolism pathway. GSK-3 has been identified as the major component responsible for most neurological disorders, including bipolar disorder, schizophrenia, Alzheimer's disease, Huntington's disease, and Parkinson's disease (Perovic et al. 1998; Bhat et al. 2004).

Gordillo and colleagues extracted palinurin, a furanosesquiterpene, from the sponge *Ircinia variabilis* (Brahmachari 2017; Alfano et al. 1979; Monks et al. 2002). This compound exhibits inhibitory activity against GSK3β with an IC_{50} value of 5.59 μM and has been found to reduce the hyper-phosphorylation of tau protein (Bidon-Chanal et al. 2013; Brahmachari 2017). The marine sponge constituent hymenialdisine is an organic compound extracted from the marine sponges

Table 14.1 List of compounds extracted from the marine sponges with acetylcholinesterase inhibition activity and their possible structures

Compound	Structure	Species	Mode of action	IC_{50} value	References
4-Acetoxy-plakinamine B	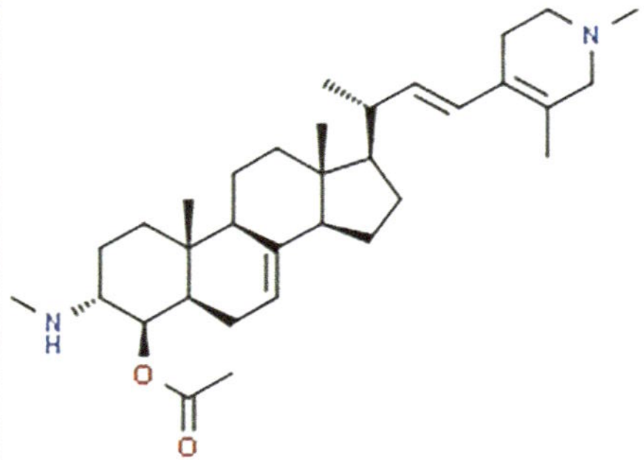	*Corticium* sp.	AChE inhibitors	3.75 μM	Singhal et al. (2012)
Lembehyne A	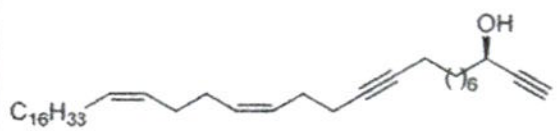 Lembehyne A	*Haliclona* sp.	AChE inhibitors	NA	Aoki et al. (2000, 2001)
Fascaplysin	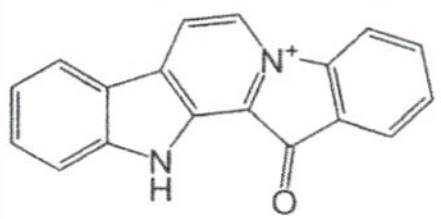	*Fascaplysinopsis* sp.	AChE inhibitors	1.49 μM	Sakai et al. (2001), Bharate et al. (2012), Grosso (2014)
Oroidin	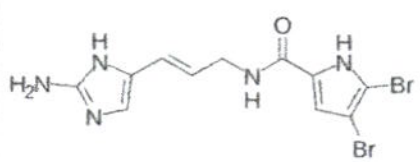	*Agelas oroides*	AChE inhibitors	55.21 μM	Gordillo et al. (2004), Pandey and Degrado (2016)

Saraines	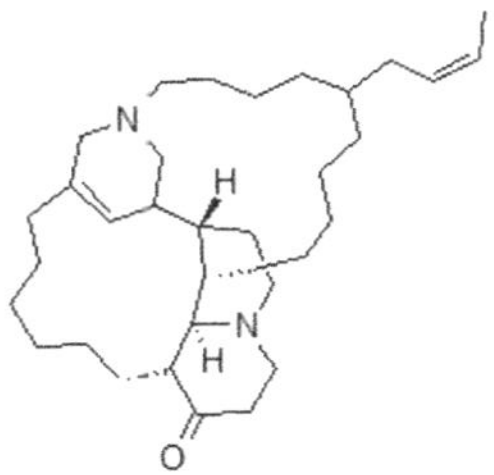	*Renierasarai*	AChE inhibitors	5–10 μM	Defant et al. (2011)
Platisidines A	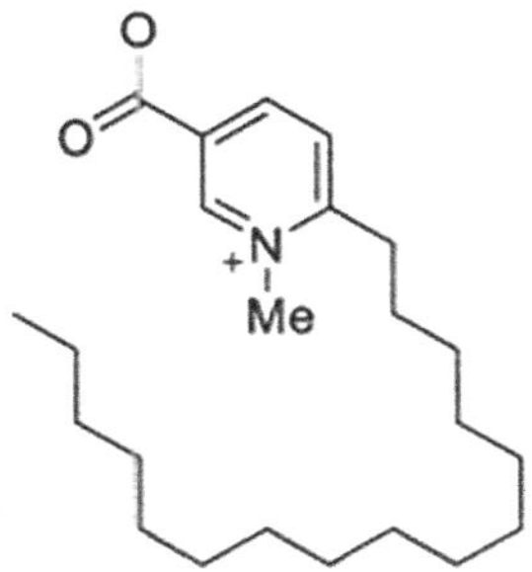	*Plakortis* sp.	AChE inhibitors	2.8, 2.6 and 2.1 mM	Mora et al. (2011)
Platisidines B	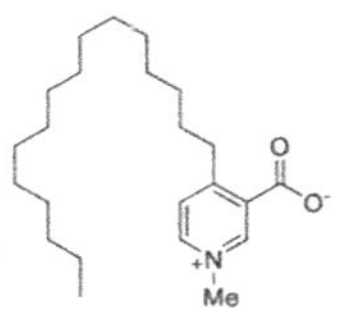				

(continued)

Table 14.1 (continued)

Compound	Structure	Species	Mode of action	IC_{50} value	References
Platisidines C	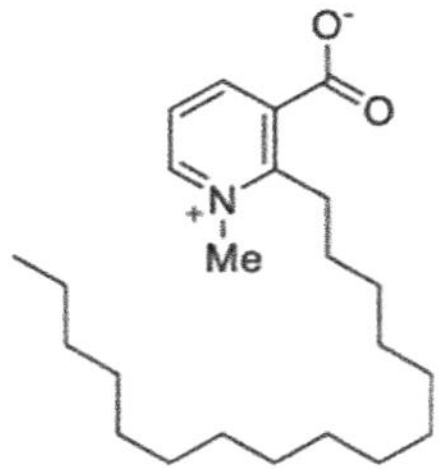				
Petrosamine	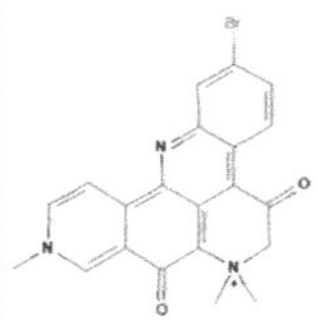	*Petrosia* sp.	AChE inhibitor	91 mM	
Polymeric alkylpyridinium salts	–	*Renierasarai*	AChE inhibitor	0.06 µg/mL	Pereira et al. (2010)

Plakortide P	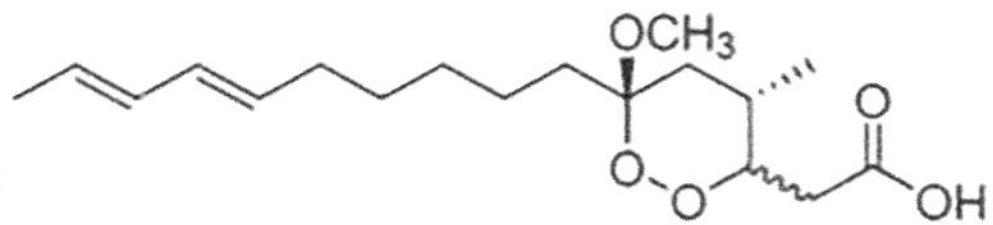	*P. angulospiculatus*	Novel anti-neuroinflammatory agent	NA	Singhal et al. (2012)
Cladidiol	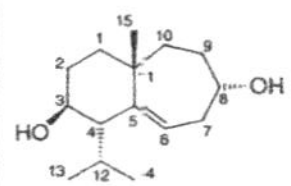	*Cladiella* sp.	Inhibitor of acetylcholinesterase	NA	Al-Sabi et al. (2006)

Acanthella aurantiaca, *Axinella verrucosa*, and *Stylissa massa*, which is effective against human neurodegenerative diseases known to inhibit tau protein phosphorylation by acting as a potent GSK 3β inhibitor with an IC50 value of 10 nM (Hamann et al. 2007; Lai and Jan 2006). Hymenialdisine is a potent inhibitor of cyclin-dependent kinases (CDKs), GSK3β, and casein kinase 1 and, as a kinase inhibitor, can be utilized in the treatment of neurodegenerative disorders (Curman et al. 2001; Lit 2006; Anjum et al. 2016). Microxine, a new purine derivative, is a CDK-1 inhibitor isolated from the Australian *Microxina* sponge genus, with an IC_{50} value of 13 μM against CDK-1.

Manzamine A, a β-carboline alkaloid acquired from the marine sponges *Haliclona* sp. and *Acanthostrongylophora* (Ang et al. 2000; Skropeta et al. 2011) first reported to have a cytotoxic effect against cancer cells (Sakai et al. 2004; Mayer et al. 2013), was later identified as having an inhibitory effect against GSK3β and the inhibitory concentration was found to be 10.2 μM (Hentschel et al. 2006). The modified analogs of (Z)-5-(4-hydroxybenzylidene)-hydantoin isolated from the red sea sponge *Hemimycale arabica* is also known to inhibit GSK3β and the IC_{50} value was calculated to be 4–20 μM (Aoki et al. 2003). A secondary metabolite (dihydropyridine) associated with sponges has also shown neuroprotective activity (Molinski et al. 1988; Mehbub et al. 2014). An Australian marine sponge *Callyspongia* sp. extract showed inhibitory effects against casein kinase 1, CDK5, and GSK3β (Aoki et al. 2003). Furthermore, variolin B, a marine compound extracted from the Antarctic sponge *Kirkpatrickia variolosa*, exhibits inhibitory activity against many CDKs and GSK3β (Murakami et al. 2001; Dayanidhi et al. 2021). The unique oroidin alkaloid agelastatin A, derived from the axinellid sponge *Agelasdendromorpha*, also exhibits inhibitory activity against GSK3β with an inhibitory concentration of 12 μM (Vippila et al. 2016; Dunbar et al. 2000; Hale et al. 2003). Table 14.2 presents a summary of the compounds derived from marine sponges with glycogen synthase kinase 3 and cyclin-dependent kinase inhibiting activities.

14.4 Role of Peptides on Voltage-Gated Ion Channels

The main subunits of the voltage-gated Na^+, Ca^{2+}, and K^+ ion channels are members of a related gene family and are functionally autonomous in their voltage-dependent activation, ion conductance, and inactivation (Catterall 2000). Pores in the plasma membrane with selective passage are embedded with highly selective membrane proteins, thus allowing only specific ions to flow through them. The gates open or close depending on the specificity of the ions, which results in conformational changes in the channel proteins that lead to alterations in the membrane potential (Feng et al. 2008). Improper channel localization in a neural network could cause extreme communication defects (Li et al. 2013). Here, we discuss some of the inhibitors obtained from marine sponges that act against voltage-gated ion channels (Table 14.3).

Table 14.2 List of sponge-derived glycogen synthase kinase 3 and cyclin-dependent kinase inhibitors

Compound	Structure	Species	Mode of action	IC_{50} value	References
Hymenialdisine	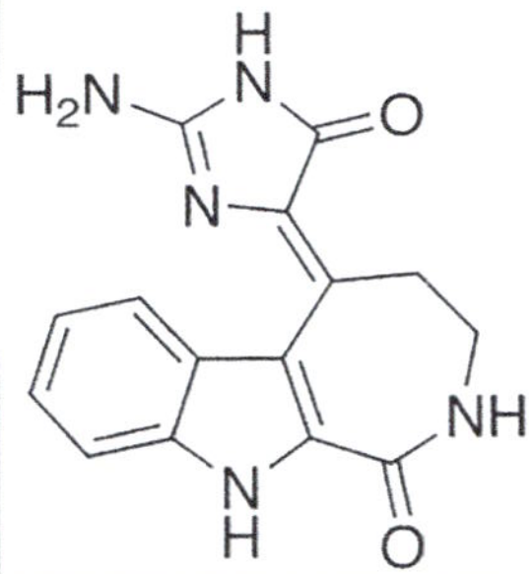	*Acanthella aurantiaca* and *Stylissa massa*	Inhibits phosphorylation of the protein tau, potent GSK3β inhibitor	10 nM	Lai and Jan (2006), Hamann et al. (2007), Choi and Choi (2015)
Palinurin	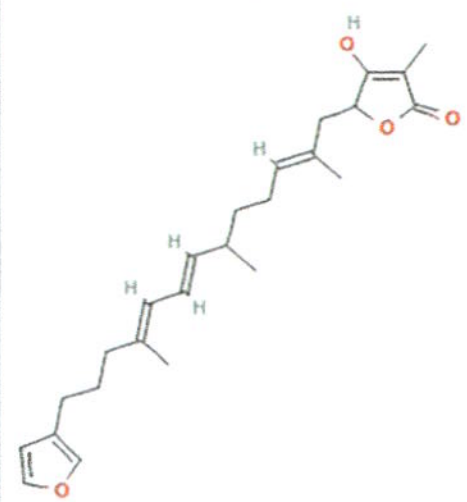	*Ircinia variabilis*	Inhibitory activity against GSK3β and reduced the hyperphosphorylation of tau protein	5.59 μM	

(continued)

Table 14.2 (continued)

Compound	Structure	Species	Mode of action	IC_{50} value	References
Debromo-hymenialdisine	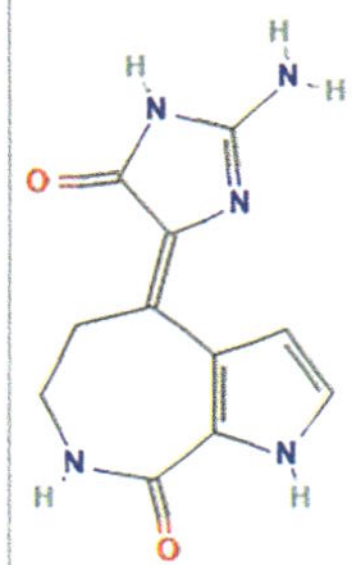		Inhibitory activity against CDK5/p25, CK1d and GSK3β	28 nM	Curman et al. (2001), Lit (2006)
Manzamine A	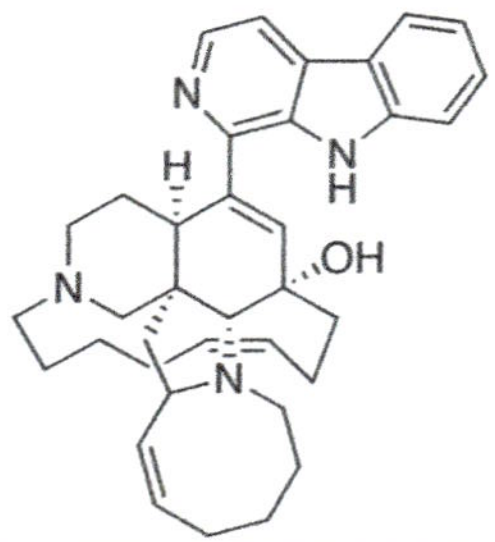	*Haliclona* sp.	Inhibitor of GSK3β	10.2 μM	Hentschel et al. (2006)

(Z)-5-(4-hydroxybenzylidene)-hydantoin	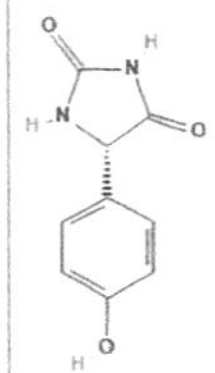	*Hemimycale arabica*	Inhibits the activity of GSK3β	20 μM	Kochanowska-Karamyan and Hamann (2010), Choi and Choi (2015), Murakami et al. (2001)
Analog of (Z)-5-(4-hydroxybenzylidene)-hydantoin	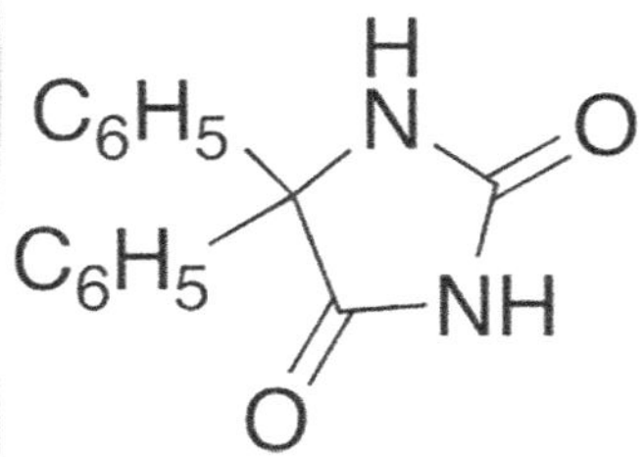	*Hemimycale arabica*	Inhibits the activity of GSK3β	NA	Kochanowska-Karamyan and Hamann (2010), Choi and Choi (2015), Murakami et al. (2001)
Dihydropyridine	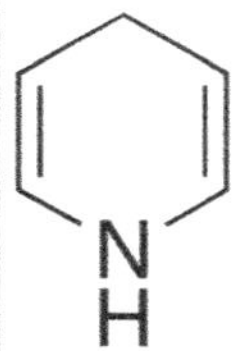		Neuroprotective activity	NA	Molinski et al. (1988)

(continued)

Table 14.2 (continued)

Compound	Structure	Species	Mode of action	IC_{50} value	References
Variolin B	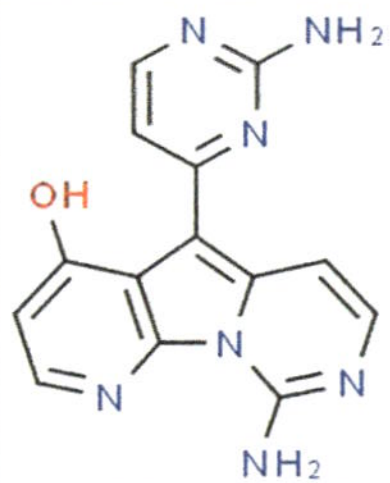	*Kirkpatrickia variolosa*	Inhibitory activity against CDKs and GSK3β	NA	Vippila et al. (2016)
Agelastatin A	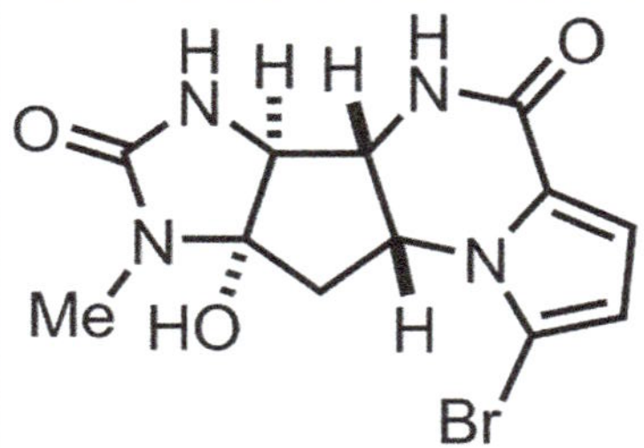	*Agelasdendromorpha*	Inhibitory activity against GSK3β	12 μM	Ang et al. (2000)

Table 14.3 List of compounds and their possible structures that exhibit inhibition activity against voltage-gated ion channels

Compound	Structure	Species	Mode of action	IC_{50} value	References
Clathrodin	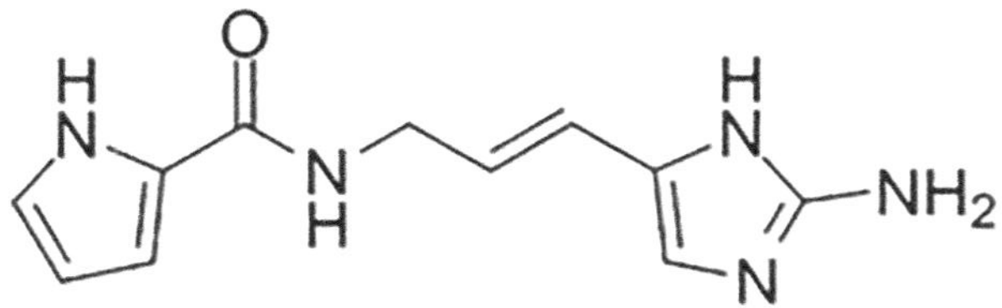	*Agelas* sp.	Na^+ channel inhibitor	~μM	Munro et al. (1999)
Dibromosceptrin	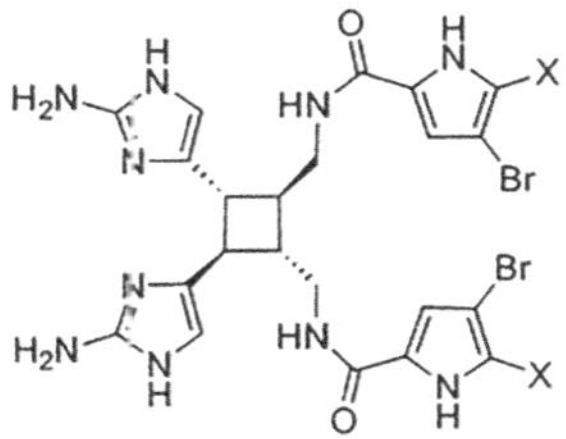	*Agelas* sp.	Na^+ channel inhibitor	~μM	Aiello et al. (2006)
Daminin	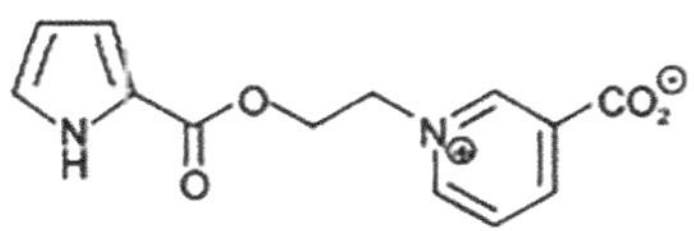	*Axinella damicornis*	Reduce Ca^{2+} levels in neuronal cells	40 μg/mL	Blunt et al. (2007)
Plakortide	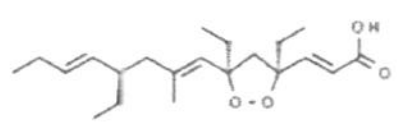	*P. angulospiculatus*	A potent inhibitor of thromboxane B2	0.93 μM	Kubota et al. (2010)

(continued)

Table 14.3 (continued)

Compound	Structure	Species	Mode of action	IC_{50} value	References
Osirisynes	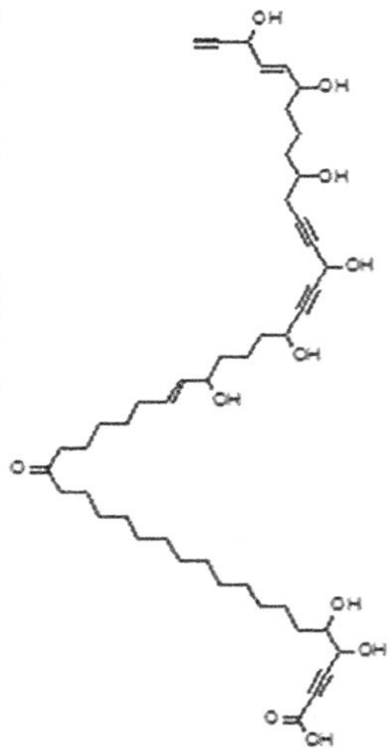	*Haliclona Osiris*	Na^+/K^+-ATPase inhibitor	~μM	
Latrunculin A	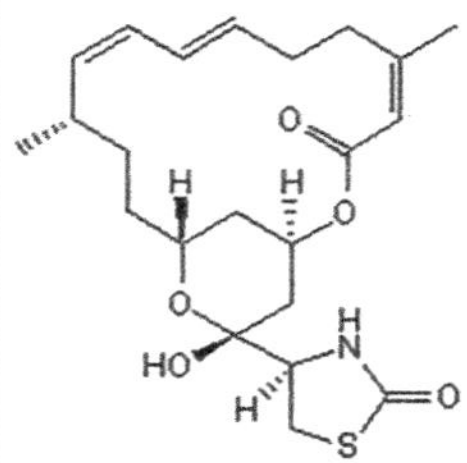	*Latrunculia magnifica*	Neurotoxin	~μM	Kawahara et al. (2012)

Petrosiacetylenes	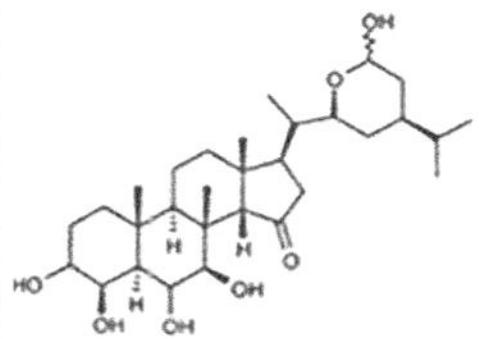	*Petrosia* sp.	Na^+/K^+-ATPase inhibitor	NA	Sipkema et al. (2005)
Biological activity of extracts		*Halichondria panacea*	Na^+-K^+ ATPase and Mg^{++} ATPase inhibitor	NA	Mioso et al. (2017)
Cyclic depsipeptidehoiamide B	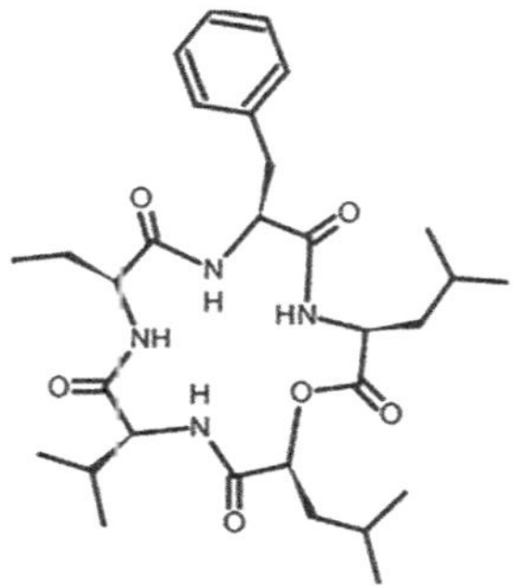	*Symploca* sp. and *Oscillatoria* cf. sp.	Stimulated sodium influx	9.3 μM	Mioso et al. (2017)

(continued)

Table 14.3 (continued)

Compound	Structure	Species	Mode of action	IC_{50} value	References
Palmyrolide A	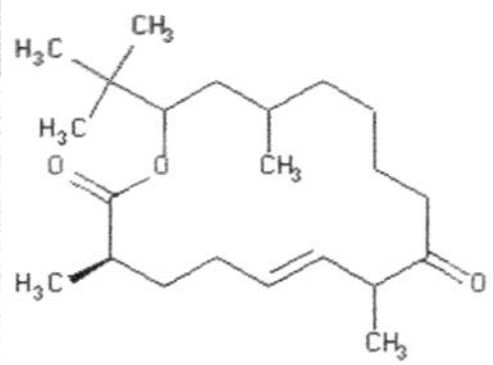	*Leptolyngbya* cf. and *Oscillatoria* sp.	Inhibited both sodium influx and Ca^{2+} oscillations	3.70 μM	Mioso et al. (2017)
Crambescidins 1–4		*Crambe crambe*	$Ca2^{+}$/channel blocker	1.5 μM	Berlinck et al. (1993), Tello-Aburto et al. (2012b)
Asteropsin A	PGluGCAFEGESCNVQFYPCCPGLGLTCIPGNPD TCYYLG	*Asteropus* sp.	Enhanced neuronal Ca^{2+} influx	14 nM	Mandelkow and Mandelkow (1998), Blunt et al. (2015)
Gombamide A	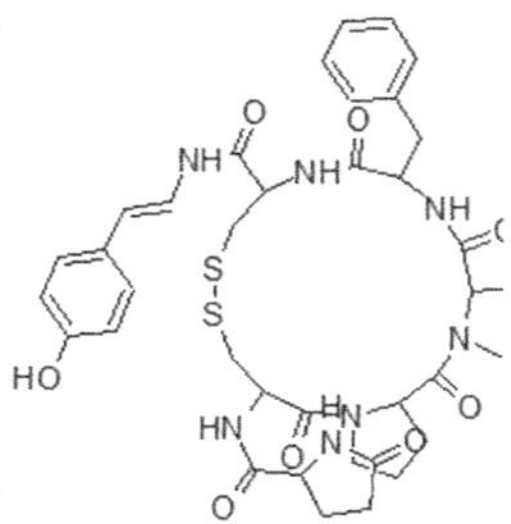	*C. gombawuiensis*	Na^{+}/K^{+}-ATPase inhibitor	17.8 μM	Agrawal et al. (2016), Barrantes and Lindsley (2016), Perry et al. (1994)

Agelas tropical sponges produce alkaloid compounds that have shown promising results by inhibiting the membrane receptors of nerves, muscles, and channels involving Na^+ ions. Morales and Rodriguez described the alkaloids clathrodin and dibromosceptrin (Munro et al. 1999; Sakai et al. 1986; Peigneur et al. 2014), which were isolated from Caribbean sponges in the *Agelas* genus. When these compounds were tested on sympathetic ganglia cells from chick embryos, there was a 27–40% decrease in the amplitude of the inward Na^+ current (Aiello et al. 2006). Also, the inhibitory concentration of the compounds against voltage-gated sodium channels was calculated as very low micromolar concentrations (Schumacher et al. 2011)—similarly, daminin—an alkaloid extracted from the sponge *Axinella damicornis* reduces Ca^{2+} levels in neuronal cells. Daminin has also been shown to reduce glutamic acid and *N*-methyl-D-aspartate, which is responsible for increasing Ca^{2+} concentrations, and the inhibitory concentration was calculated to be more than 40 μg/mL (Aiello et al. 2005; Blunt et al. 2007). Kossuga et al. (2008) isolated a novel anti-neuro-inflammatory compound called plakortide, which is a polyketide from a marine sponge with excellent inhibition activity against thromboxane B2 release with an IC_{50} value of 0.93 μM (Kubota et al. 2010; Swanson and Sakai 2009). Shin and coworkers extracted osirisynes from the sponge species *Haliclona osiris*, which proved to be polyacetylenes with high oxygen content. These compounds exhibit moderate cytotoxicity and inhibitory activities against Na^+/K^+-ATPase and reverse transcriptase with an inhibitory concentration in micromolar values (Trimmer and Rhodes 2004). The neurotoxin latrunculin A is produced by the sponge *Latrunculia magnifica* (toxic finger sponge), which inhibits the activity of Na^+/K^+-ATPase with a minimal inhibitory concentration (Kawahara et al. 2012). The bioactive metabolite petrosiacetylenes was isolated from a *Petrosia* sp. sponge which also exhibits inhibitory activity against Na^+/K^+-ATPase (Sipkema et al. 2005). Similarly, extracts from the marine sponge *Halichondria panicea* increased the activities of Na^+/K^+-ATPase and Mg^{2+}-ATPase (Mioso et al. 2017). Two other marine compounds, hoiamide B and palmyrolide A, have been shown to inhibit neuronal Ca^{2+} oscillations, a process that is dependent on voltage-gated sodium channel activation with IC_{50} values of 9.3 and 3.70 μM, respectively (Mioso et al. 2017). Pentacyclic guanidine derivatives (crambescidins 1–4) were isolated from *Crambe crambe* (Poecilosclerida) and found to act as a potent Ca^{2+} channel blocker with an inhibitory concentration of 1.5 μM (Berlinck et al. 1993; Erijman et al. 1991).

Asteropsin A, a peptide containing a cysteine knot and unusual N-terminal pyro-glutamate residues, is isolated from the sponge *Asteropus* sp. It has been found to enhance neuronal calcium influx in cerebrocortical neurons, and the calculated inhibitory concentration was 14 nM (Mandelkow and Mandelkow 1998; Li et al. 2013). The cyclic thiopeptide gombamide A with a molecular formula $C_{38}H_{45}N_7O_8S_2$ obtained from *Clathria gombawuiensis, a* sponge species in Korea, has shown moderate inhibitory activity against Na^+/K^+-ATPase (Agrawal et al. 2016; Gautam et al. 2017). Woo with his coworkers stated that gombamide A consists of a *para*-hydroxystyrylamide with a pyro-glutamic acid subunit (Long et al. 1998) which was responsible for the inhibitory activity of sodium-potassium (Na^+/K^+)-ATPase channel with an LC_{50} value of 17.8 μM.

14.5 Beta-Amyloid Cleaving Enzyme (BACE)-1 Inhibitors Obtained from Marine Sponges

Beta-amyloid cleaving enzyme-1 (BACE1) is the transmembrane aspartyl protease that cleaves amyloid precursor protein (APP) at the β site. Table 14.4 lists the compounds derived from sponges having BACE inhibition activity and their structures. Studies have reported that BACE1 inhibitors can potentially decrease amyloid-beta peptide concentrations in the brain, thus preventing the progression of Alzheimer's disease (Citron 2002; Coimbra et al. 2018). Xestosaprols D and F, extracted from a marine sponge species in the genus *Xestospongia*, are known for their biological activity against BACE1 with inhibitory concentrations of 93.2 and 135.0 μM, respectively (Millán-Aguiñaga et al. 2010; Molinski et al. 1988). Dictyodendrins (1–5) are rare marine alkaloids extracted from *Ianthella* sp. and *Dictyodendrilla* sp. that have been demonstrated to be good BACE1 inhibitors with a calculated IC_{50} value of 1.5, 1.0, 2.0, 2.0, and 2.0 μM, respectively (Trimurtulu et al. 1994). The dictyodendrins (1–5) obtained from *Dictyodendrilla* sp. and topsentinol K trisulfate from the sponge *Topsentia* sp. have both been found to act as potent BACE1 inhibitors with an IC_{50} value of 1.2 μM (Dai et al. 2010; Guo et al. 1996). Ianthellidone F and lamellarins O and O2 are in a new class of pyrrolidones derived from an *Ianthella* marine sponge. Bastadin, a dipeptide obtained from the sponge *Ianthella basta* (Pallas), is a tyrosine-derived product that inhibits BACE1 by reducing the effect of APP (Essa et al. 2012; Brunner et al. 2009). All three compounds isolated from *Ianthella* marine sponge exhibited inhibitory activity against BACE1 with an inhibitory concentration calculated to be 40% (at lower concentrations of 10 μM).

14.6 Serotonin, Glutamate, Nitric Oxide, and Gamma-Aminobutyric Acid (GABA)nergic Inhibitors Isolated from Marine Sponges

Neurotransmitters are chemical messengers that transmit signals from one neuron to another. They also help in regulating the release of other neurotransmitters. Neurotransmitter disorders can lead to insomnia, depression, Parkinson's disease, memory loss, and attention deficit hyperactive disorder, among other problems (Wadsworth et al. 2012; Webster and Taylor 2012). Aiello et al. (2005) performed experiments on *Axinella verrucose*, a Mediterranean sponge, and found that it produced a complex mixture of bromopyrrole alkaloid compounds (1–4). These compounds have shown potent neuroprotective activity against serotonin and glutamate agonists in vitro by inhibiting neuronal Ca^{2+} entry (Mehbub et al. 2014; Aiello et al. 2005). Dysiherbaine, the first member of a new structural class of marine-derived ionotropic

Table 14.4 Compounds that possess BACE inhibition activity and their structures

Compound	Structure	Species	Mode of action	IC_{50} value (μM)	References
Xestosaprols D	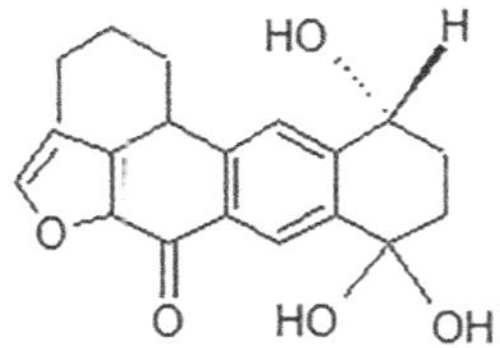	*Xestospongia* sp.	BACE1 inhibitor	93.2	Molinski et al. (1988)
Xestosaprols E	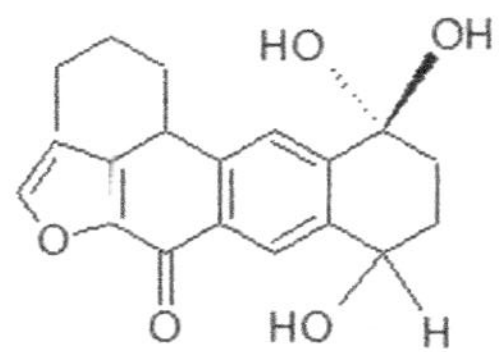	*Xestospongia* sp.	BACE1 inhibitor	135.0	Molinski et al. (1988)
Dictyodendrin 1	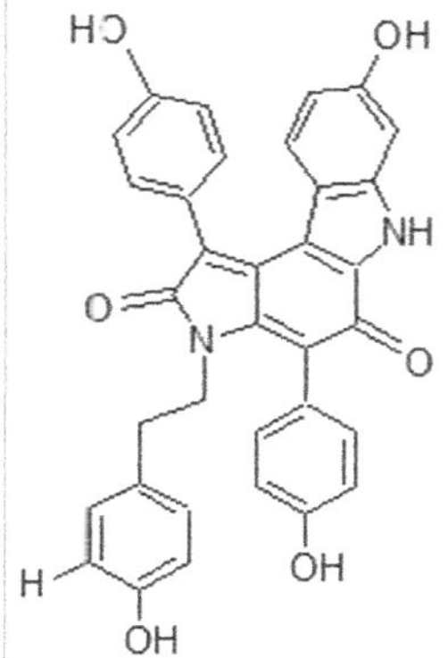	*Ianthella* sp., and *Dictyodendrilla* sp.	BACE1 inhibitor	1.5, 1.0, 2.0, 2.0 and 2.0	Guo et al. (1996), Aoki et al. (2003)

(continued)

Table 14.4 (continued)

Compound	Structure	Species	Mode of action	IC_{50} value (μM)	References
Dictyodendrin 2	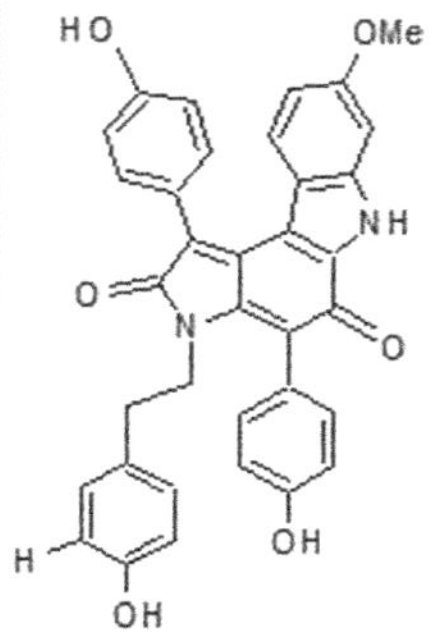				
Dictyodendrin 3	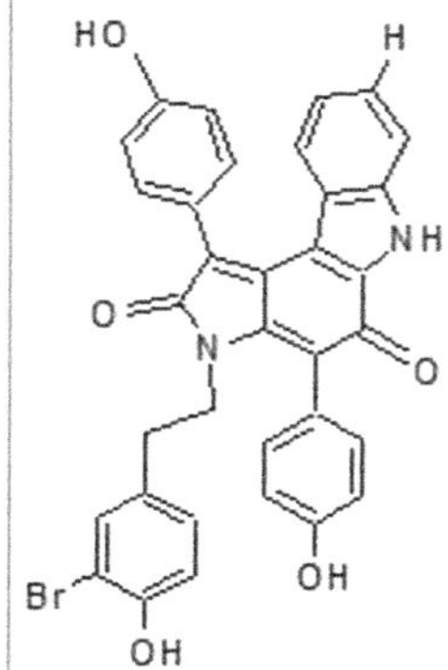				

Dictyodendrin 4	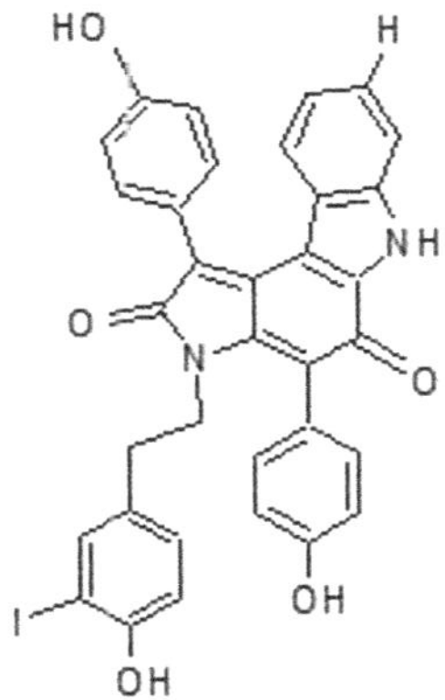				
Dictyodendrin 5	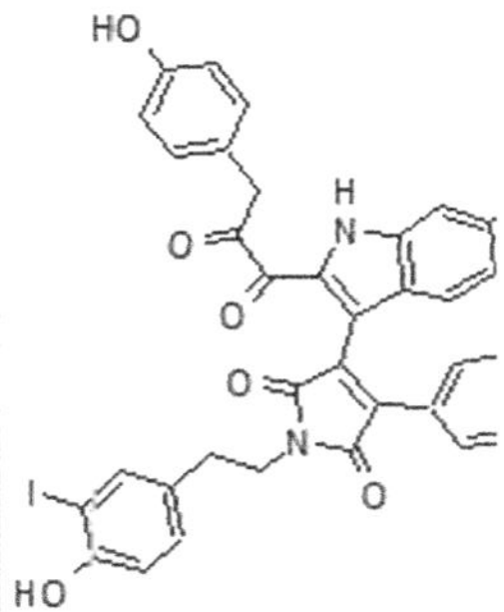				

(continued)

Table 14.4 (continued)

Compound	Structure	Species	Mode of action	IC_{50} value (μM)	References
Topsentinol K trisulfate	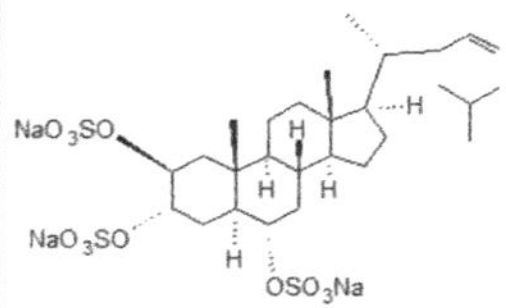	*Topsentia* sp.	BACE1 inhibitor	1.2	Guo et al. (1996), Aoki et al. (2003)
Ianthellidone F	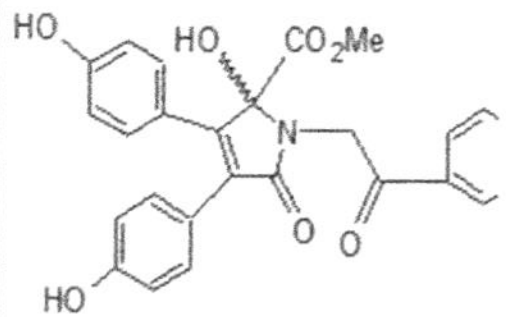	*Ianthella* sp.	BACE1 inhibitor	10	Aoki et al. (2003)
Lamellarin O2	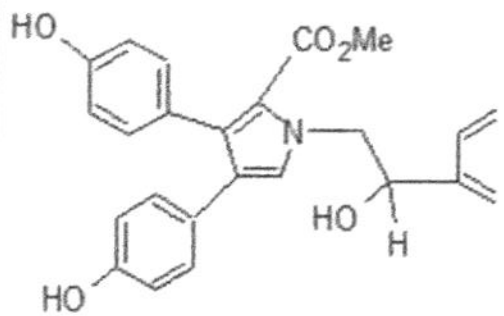				
Lamellarin O	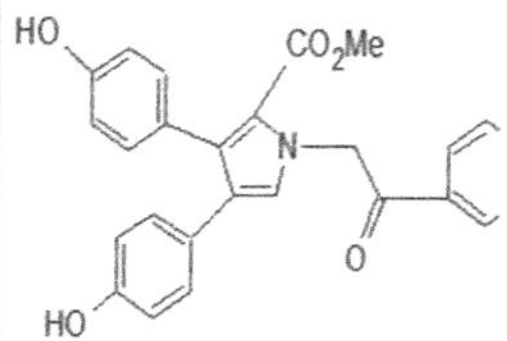				

glutamate receptor agonists and a neurotoxic amino acid with an inhibitory concentration of 59 ± 7.8 nM, is highly specific for kinate receptors. Sakai and coworkers found that when an aqueous extract of the marine sponge *Lendenfeldia chodrodes* was injected into mice, they exhibited a potent muscle contraction.

Naamine D, an imidazole alkaloid extracted from *Leucetta* cf. *chagosensis* (Calcinea), has been identified as a NO synthetase inhibitor with a calculated inhibitory concentration of 1.0 mM (Tello-Aburto et al. 2012a; Dunbar et al. 2000; Sakurada et al. 2010). A *Plakortis* species of sponge from the Canary Islands contains the compound plakolide A, which is a c-lactone and helps in the inhibition of inducible NO synthase activity with a calculated IC_{50} value of 0.2 μg/mL (Sakai et al. 2004; Tello-Aburto et al. 2012b). Sakurada (Sakurada et al. 2010) reported a novel purine from the Palauan sponge *Cribrocha linaolemda* that showed inhibition activity against GABAergic transmission in hippocampal neurons. Van-Lear et al. (1973) reported brominated antibacterial compounds isolated from the marine sponges *Polyfibrospongiamaynardii* and *Hyrtios erecta* that inhibited the neuronal isoform of NO synthase (Woo et al. 2013; Kossuga et al. 2008). Purines 1–4, which are neurologically active, were isolated from Haplosclerida sponges in the Republic of Palau, and with a CD_{50} value of 2–4 nmol/mouse, compound 1 caused convulsions in mice when injected intracerebroventricularly. At higher doses, purines 2–4 were active in mice bioassays. In electrophysiological recordings from hippocampal neurons, compound 1's seizurogenic activity was linked to inhibiting neuronal GABAergic transmission with little impact on excitatory signaling (Senthilkumar and Kim 2013). Compounds produced by sponges having inhibitory activity against neurotransmitters are tabulated in Table 14.5.

14.7 *N*-Methyl-D-Aspartate Receptors and Other Secondary Metabolites Inhibiting Neural Outgrowth

Sakai and colleagues extracted structurally novel betaines, such as dysibetaines CPa and CPb (neodysiherbaine A) from the sponge *Dysidea herbacea*, that showed a weak affinity against glutamate receptors (Schneemann et al. 2010; Nukoolkarn et al. 2008). Comparing the two compounds, dysibetaines CPa displaces the kainate and *N*-methyl-D-aspartate (NMDA) types of receptor ligands, whereas dysibetaine CPb (neodysiherbaine A) inhibits kainate binding (Trimmer and Rhodes 2004; Sakai et al. 2001). Based on this study, two pipecolic acid analogs—cribronic acid ((2*S*, 4*R*, 5*R*)-5-hydroxy-4-sulfooxy-piperidine2-carboxylic acid) isolated from the Palauan sponge *Cribrochalinaolemda* and (2*S*, 4*S*) 4-sulfooxy-piperidine-2-carboxylic acid (*trans*-4-hydoroxypipecolic acid sulfate, *t*-HPIS) from the Micronesian sponges *Axynellacarteri* and *Stylotella aurantium*—showed activity on NMDA receptors with IC_{50} values of 63 and 83 nM, respectively (Evans et al. 1985). The *Petrosia* sp. sponges act as a novel source of polyacetylenes. *Petrosiastrongylata* (Ishigaki Island, Okinawa) produces petrosiols A–E, which induce nerve growth factor-like neuronal differentiation in PC12 cells (Blunt et al. 2015). Motuporamine C, an alkaloid from the marine sponge *Xestospongia exigua*, with

Table 14.5 Structure of some inhibitors of neurotransmitters

Compound	Structure	Species	Mode of action	IC_{50} value	References
Bromopyrrole alkaloid compounds 1–4	1 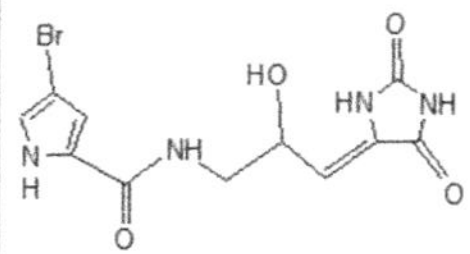	*Axinella verrucosa*	Serotonin and glutamate agonist	NA	Aiello et al. (2006), Mehbub et al. (2014)
	2 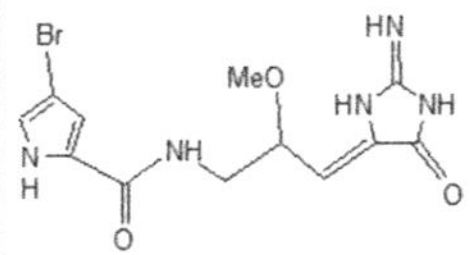				
	3 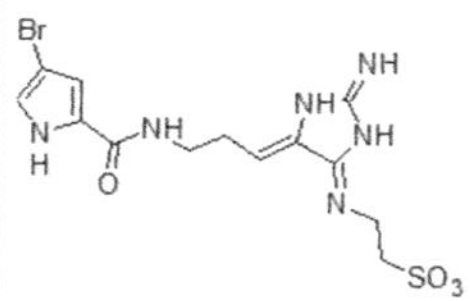				
	4 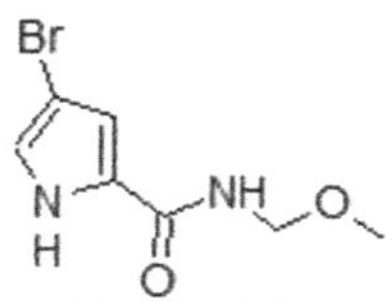				

Dysiherbaine	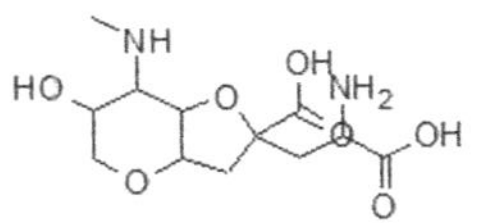	*Lendenfeldia chodrodes*	Marine-derived iGluR agonists	59 ± 7.8 μM	Sakai et al. (2010), Thakur and Müller (2004)
Naamine D	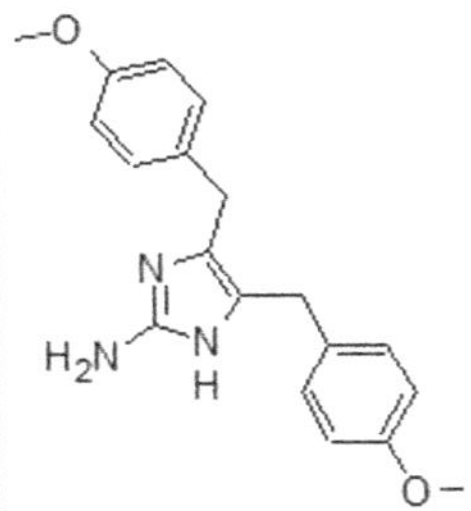	*Leucetta* cf. *chagosensis*	Nitric oxide synthetase inhibitor	1.0 mM	Tello-Aburto et al. (2012b), Dunbar et al. (2000)
Plakolide A	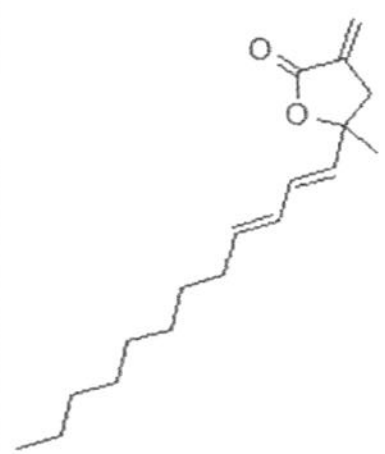	*Plakortis* sp.	Inhibited GABAergic transmission	0.2 μg/mL	Tello-Aburto et al. (2012b), Dunbar et al. (2000)

(continued)

Table 14.5 (continued)

Compound	Structure	Species	Mode of action	IC_{50} value	References
5,6-Dibromo-3-(2-methylaminoethyl) indole	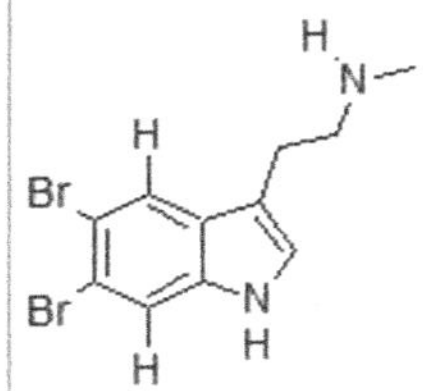	*Polyfibrospongiamaynardii*	Inhibits nitric oxide synthase	NA	Yak and Qian (2003)
5,6-Dibromo-3-(aminoethyl)indole	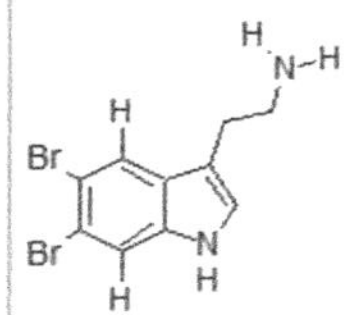	*Hyrtios erecta*	Inhibits nitric oxide synthase	NA	Aoki et al. (2001), Kossuga et al. (2008)

an inhibitory concentration of 3 μM has been noted for its significant upregulation of the Rho-Rho-kinase collapse pathway of neuronal outgrowth (Mehbub et al. 2014; Van-Lear et al. 1973; Gogineni and Hamann 2018). Lembehynes A, B, and C, obtained from the Indonesian sponge *Haliclona* sp., exhibit excellent neuritogenic activity against neuroblastoma cells with an IC50 value of 3 μg/mL (Joseph et al. 2012). Polydiscamides B–D, isolated from the *Ircinia* sp., are strong human sensory neuron-specific G protein-coupled receptor (SNSR) agonists with EC_{50} values of 1.26, 3.57, and 2.80 μM, respectively (Feng et al. 2008; Malloy 2011). Table 14.6 presents the structures of sponge-derived compounds effective against NMDA receptors.

14.8 Neuroactive Peptides Isolated from Marine Sponge-Associated Microorganisms

JBIR-59 and JBIR-124 extracted from the mycelium of the fungal strain *Penicillium citrinum* SpI080624G1f01 that is associated with marine sponges exhibit scavenging activity for the 2,2-diphenyl-1-picrylhydrazylradicals, with IC_{50} values of 25 and 30 μM, respectively. Both compounds belong to a class of hexaketides called sorbicillinoids (Khanfar et al. 2009). *Aspergillus sydowii* was isolated from the sponges *Amphimedon viridis* and *Agelas* sp. and it has shown potent inhibitory activity against AChE (El-Hady et al. 2014). *Streptomyces coelescens* ICSSB 1021, isolated from the sponge *Halichondria panicea*, produces germicidin, an inhibitor of porcine Na^+/K^+-activated ATPase (Plisson et al. 2014; Shin et al. 1998). Neuroactive compounds have been found in other bacterial species (*Psychroserpens burtonensis* and *Antarcticum vesiculatum*) associated with this sponge that's how activity against the NMDA receptor that upregulates intercellular calcium levels (Pietra 1997). Crude extracts of bacterial isolates obtained from the sponge *H. panacea* also show activity against neuronal ionotropic glutamate receptors (Pietra 1997). Choi and colleagues (Choi et al. 2010) reported that hoiamide B—a cyclic depsipeptide obtained from the cyanobacteria *Symposia* sp. and *Oscillatoria* sp. in an assemblage from Papua New Guinea—stimulates Na^{2+} influx ($IC_{50} = 3.9$) in a murine model. Pereira and colleagues (Thomas et al. 2010; To et al. 2006) isolated an effective neuroactive macrolide, palmyrolide A, from marine sponge-associated cyanobacteria assemblages of *Leptolyngbya* sp. and *Oscillatoria* sp. collected from the Palmyra Atoll in the Northern Pacific Ocean (Zhang et al. 2012; Pereira et al. 2010). The compound exhibited good inhibition against Na^{2+} influx ($IC_{50} = 5.2$ μM) in mouse neuroblastoma cells. The compound also suppressed impulsive Ca^{2+} oscillations with an IC_{50} of 3.7 μM in primary cultures of murine cerebrocortical neurons, making it a potent candidate for further exploration for pharmacological applications (Petersen et al. 1993).

Table 14.6 Structure of compounds that inhibits NMDA receptors and other secondary metabolites

Compound	Structure	Species	Mode of action	IC_{50} value	References
Dysibetaine CPa	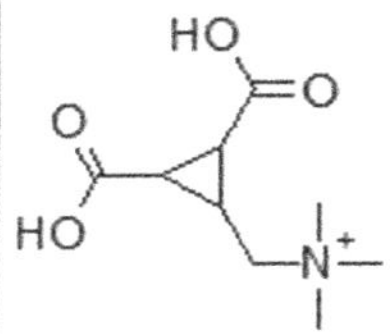	*Dysidea herbacea*	Displaced the kainate and NMDA type receptor—ligands	NA	Schneemann et al. (2010)
Dysibetaine CPb	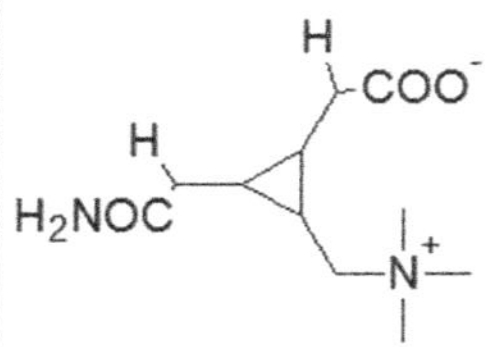	*Dysidea herbacea*	Inhibited the binding of kainate	NA	Schneemann et al. (2010)
Neodysiherbaine A	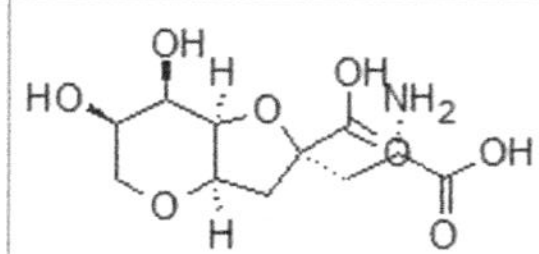	*Dysidea herbacea*	Displaced the kainate and NMDA type receptor—ligands	66 5.2 and 227 40 nM	Sakai et al. (2001)
(2*S*, 4*R*, 5*R*)-5-Hydroxy-4-sulfooxy-Piperidine2-carboxylic acid	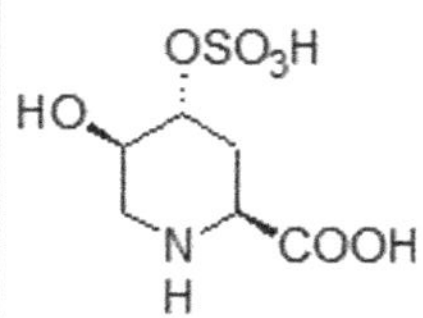	*Cribrochalinaolemda*	Active on NMDA receptors	63 nM	Evans et al. (1985), Nukoolkarn et al. (2008), Thakur and Müller (2004)

(2*S*, 4*S*) 4-Sulfooxy-piperidine-2-carboxylic acid	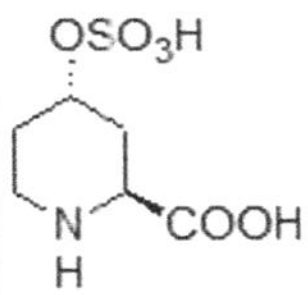	*Axynellacarteri* and *Stylotella aurantium*	Active on NMDA receptors	83 nM	Evans et al. (1985), Nukoolkarn et al. (2008), Thakur and Müller (2004)
Petrosiol A	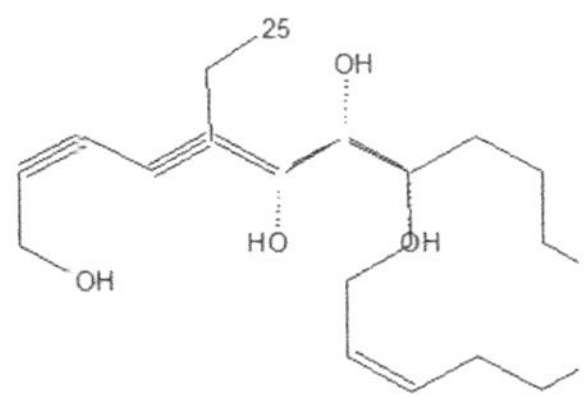	*Petrosiastrongylata*	Inducer of nerve growth factor	NA	Blunt et al. (2015), Jha and Zi-Rong (2004)
Petrosiol B	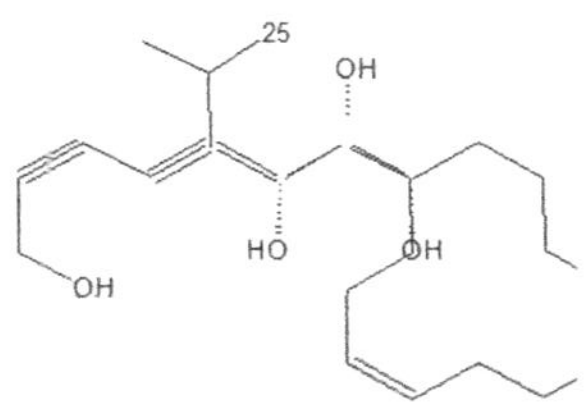				

(continued)

Table 14.6 (continued)

Compound	Structure	Species	Mode of action	IC_{50} value	References
Petrosiol C	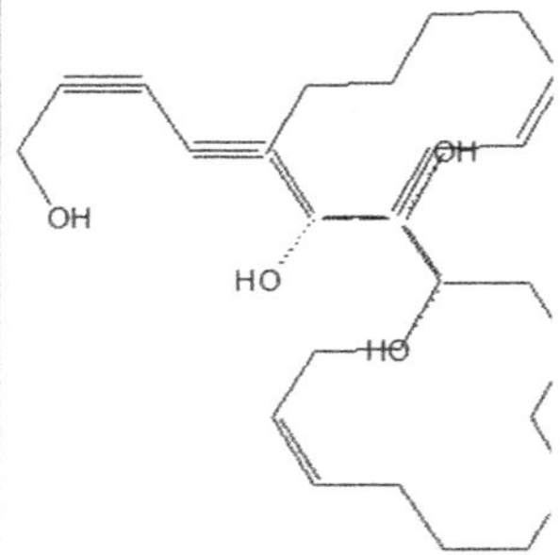				
Petrosiol D	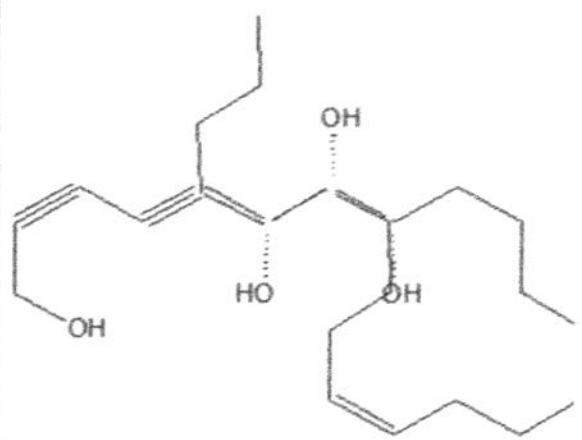				

Petrosiol E	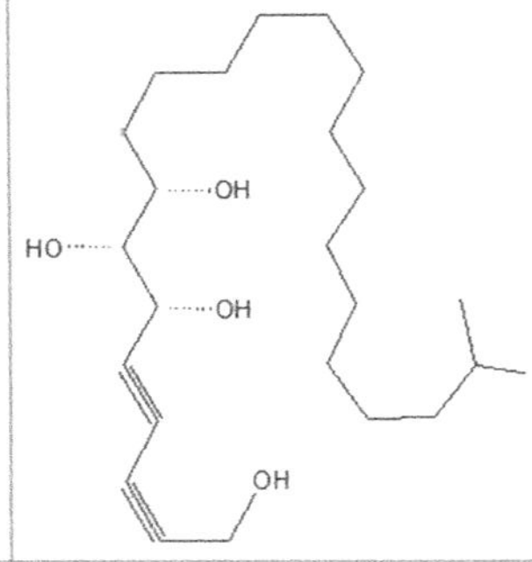				
Motuporamine C	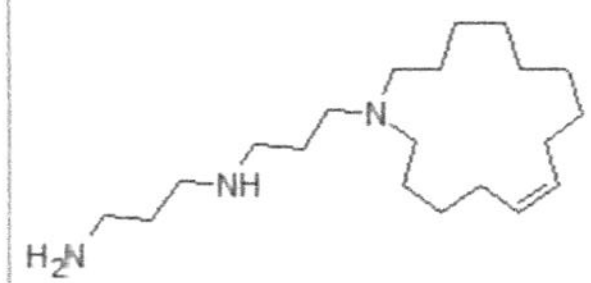	*Xestospongia exigua*	Upregulated the Rho-Rho-kinase collapse pathway	3 μM	Mehbub et al. (2014), Van-Lear et al. (1973)
Lembehyne B	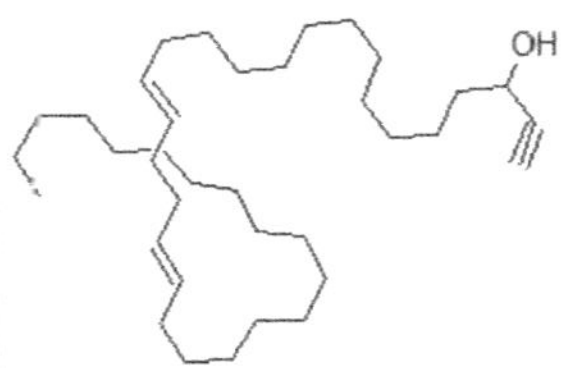	*Haliclona* sp.	Neuritogenic activity against neuroblastoma cells	3 μg/mL	Joseph et al. (2012)
Lembehyne C					

(continued)

Table 14.6 (continued)

Compound	Structure	Species	Mode of action	IC_{50} value	References
	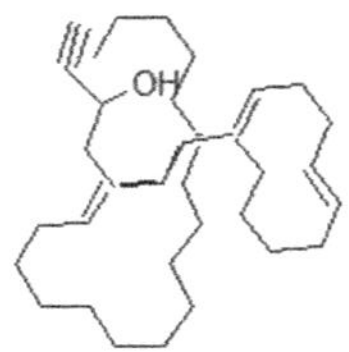				
Polydiscamides B–D	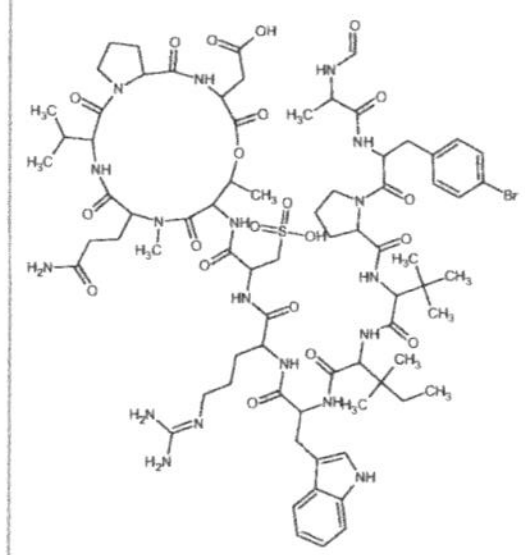	*Ircinia* sp.	Sensory neuron-specific G protein-coupled receptor (SNSR) agonists	1.26 μM	Huang et al. (2019)

14.9 Conclusion

The largely unexplored marine environment harbors many novel compounds that need to be further investigated. Natural compounds derived from the marine environment, especially those from marine sponges, are biologically active and have a high demand for pharmaceutical development and medicinal chemistry. Apart from the efficient production of sponge-derived products, engineered microorganisms provide a promising source for developing these bioactive molecules that deserve our attention, and they can be used as a sustainable source for more effective drug discoveries.

References

Agrawal S, Adholeya A, Deshmukh SK (2016) The pharmacological potential of non-ribosomal peptides from marine sponge and tunicates. Front Pharmacol 7:333

Aiello A, D'Esposito M, Fattorusso E, Menna M, Müller WE, Perović-Ottstadt S, Tsuruta H, Gulder TA, Bringmann G (2005) Daminin, a bioactive pyrrole alkaloid from the Mediterranean sponge Axinella damicornis. Tetrahedron 61:7266–7270

Aiello A, D'Esposito M, Fattorusso E, Menna M, Müller WE, Perović-Ottstadt S, Schröder HC (2006) Novel bioactive bromopyrrole alkaloids from the Mediterranean sponge *Axinella verrucosa*. Bioorganic Med Chem 14:17–24

Alfano G, Cimino G, De Stefano S (1979) Palinurin, a new linear sesterterpene from a marine sponge. Experientia 35:1136–1137

Alghazwi M (2018) Discovery of potential neuroprotective compounds from marine sponges and algae (Doctoral dissertation, Flinders University, College of Medicine and Public Health)

Al-Sabi A, McArthur J, Ostroumov V, French RJ (2006) Marine toxins that target voltage-gated sodium channels. Mar Drugs 4:157–192

Alves C, Silva J, Pinteus S, Gaspar H, Alpoim MC, Botana LM, Pedrosa R (2018) From marine origin to therapeutics: the antitumor potential of marine algae-derived compounds. Front Pharmacol 9:777

Amina M, Al Musayeib NM (2018) Biological and medicinal importance of sponge. In: Biological resource of waters, vol 25. IntechOpen, London, pp 201–230

Andryukov B, Mikhailov V, Besednova N (2019) The biotechnological potential of secondary metabolites from marine bacteria. J Mar Sci Eng 7:176

Ang KK, Holmes MJ, Higa T, Hamann MT, Kara UA (2000) In vivo antimalarial activity of the beta-carboline alkaloid Manzamine A. Antimicrob Agents Chemother 44:1645–1649

Anjum K, Abbas SQ, Shah SA, Akhter N, Batool S, Ul Hassan SS (2016) Marine sponges as a drug treasure. Biomol Ther 24:347

Aoki S, Matsui K, Tanaka K, Satari R, Kobayashi M (2000) Lembehyne A, a novel neuritogenic polyacetylene, from a marine sponge of Haliclona sp. Tetrahedron 56:9945–9948

Aoki S, Matsui K, Takata T, Hong W, Kobayashi M (2001) Lembehyne A, a spongean polyacetylene, induces neuronal differentiation in neuroblastoma cell. Biochem Biophys Res Commun 289:558–563

Aoki S, Matsui K, Takata T, Kobayashi M (2003) In situ photo affinity labeling of the target protein for Lembehyne A, a neuronal differentiation inducer. FEBS Lett 544:223–227

Bai X, Liu Y, Wang H, Zhang H (2021) Natural products from the marine sponge subgenus Reniera. Molecules 26:1097

Barrantes PMG, Lindsley CW (2016) Total synthesis of Gombamide A. Org Lett 18:3810–3813

Berlinck RG, Braekman JC, Daloze D, Bruno I, Riccio R, Ferri S, Spampinato S, Speroni E (1993) Polycyclic guanidine alkaloids from the marine sponge Crambe crambe and Ca++ channel blocker activity of crambescidin 816. J Nat Prod 56:1007–1015

Bharate SB, Manda S, Joshi P, Singh B, Vishwakarma RA (2012) Total synthesis and anti-cholinesterase activity of marine derived bis-indole alkaloid fascaplysin. Med Chem Commun 3:1098–1103

Bhat RV, Haeberlein SLB, Avila J (2004) Glycogen synthase kinase 3: a drug target for CNS therapies. J Neurochem 89:1313–1317

Bidon-Chanal A, Fuertes A, Alonso D, Pérez DI, Martínez A, Luque FJ, Medina M (2013) Evidence for a new binding mode to GSK-3: allosteric regulation by the marine compound palinurin. Eur J Med Chem 60:479–489

Blunt JW, Copp BR, Hu WP, Munro MH, Northcote PT, Prinsep MR (2007) Marine natural products. Nat Prod Rep 24:31–86

Blunt JW, Copp BR, Munro MH, Northcote PT, Prinsep MR (2011) Marine natural products. Nat Prod Rep 28:196–268

Blunt JW, Copp BR, Keyzers RA, Munro MH, Prinsep MR (2015) Marine natural products. Nat Prod Rep 32:160–258

Brahmachari G (2017) Neuroprotective natural products: clinical aspects and mode of action. Wiley, New York

Brunner E, Ehrlich H, Schupp P, Hedrich R, Hunoldt S, Kammer M, Machill S, Paasch S, Bazhenov VV, Kurek DV, Arnold T (2009) Chitin-based scaffolds are an integral part of the skeleton of the marine demosponge Ianthella basta. J Struct Biol 168:539–547

Catterall WA (2000) From ionic currents to molecular review mechanisms: the structure and function of voltage-gated sodium channels. Neuron 26:13–25

Choi DY, Choi H (2015) Natural products from marine organisms with neuroprotective activity in the experimental models of Alzheimer's disease, Parkinson's disease and ischemic brain stroke: their molecular targets and action mechanisms. Arch Pharm Res 38:139–170

Choi H, Pereira AR, Cao Z, Shuman CF, Engene N, Byrum T, Matainaho T, Murray TF, Mangoni A, Gerwick WH (2010) The Hoiamides, structurally intriguing neurotoxic lipopeptides from Papua New Guinea marine cyanobacteria. J Nat Prod 73:1411–1421

Cimino G, De Rosa S, De Stefano S, Sodano G (1986a) Marine natural products: new results from Mediterranean invertebrates. Pure Appl Chem 58:375–386

Cimino G, Stefano SD, Scognamiglio G, Sodano G, Trivellone E (1986b) Sarains: a new class of alkaloids from the marine sponge Reniera sarai. Bull Soc Chim Belg 95:783–800

Citron M (2002) β-Secretase as a target for the treatment of Alzheimer's disease. J Neurosci Res 70:373–379

Coimbra JR, Marques DF, Baptista SJ, Pereira CM, Moreira PI, Dinis TC, Santos AE, Salvador JA (2018) Highlights in BACE1 inhibitors for Alzheimer's disease treatment. Front Chem 6:178

Curman D, Cinel B, Williams DE, Rundle N, Block WD, Goodarzi AA, Hutchins JR, Clarke PR, Zhou BB, Lees-Miller SP, Andersen RJ (2001) Inhibition of the G2 DNA damage checkpoint and of protein kinases Chk1 and Chk2 by the marine sponge alkaloid debromohymenialdisine. J Biol Chem 276:17914–17919

Dai J, Sorribas A, Yoshida WY, Kelly M, Williams PG (2010) Topsentinols, 24-isopropyl steroids from the marine sponge Topsentia sp. J Nat Prod 73:1597–1600

Dayanidhi DL, Thomas BC, Osterberg JS, Vuong M, Vargas G, Kwartler SK, Schmaltz E, Dunphy-Daly MM, Schultz TF, Rittschof D, Eward WC (2021) Exploring the diversity of the marine environment for new anticancer compounds. Front Mar Sci 7:614766

Defant A, Mancini I, Raspor L, Guella G, Turk T, Sepčić K (2011) New structural insights into saraines A, B, and C, macrocyclic alkaloids from the Mediterranean sponge Reniera (Haliclona) sarai. Eur J Org Chem 20–21:3761–3767

Dunbar DC, Rimoldi JM, Clark AM, Kelly M, Hamann MT (2000) Anti-cryptococcal and nitric oxide synthase inhibitory imidazole alkaloids from the calcareous sponge Leucetta cf chagosensis. Tetrahedron 56:8795–8798

El-Hady FK, Abdel-Aziz MS, Shaker KH, El-Shahid ZA (2014) Tyrosinase, acetylcholinesterase inhibitory potential, antioxidant and antimicrobial activities of sponge derived fungi with correlation to their GC/MS analysis. Int J Pharm Sci Rev Res 58:338–345

Elliott GRD, Leys SP (2010) Evidence for glutamate, GABA and NO in coordinating behaviour in the sponge, Ephydatia muelleri (Demospongiae, Spongillidae). J Exp Biol 213:2310–2321

Erijman EAJ, Sakai R, Rinehart KL (1991) Crambescidins: new antiviral and cytotoxic compounds from the sponge Crambe crambe. J Org Chem 56:5712–5715

Essa MM, Vijayan RK, Castellano-Gonzalez G (2012) Neuroprotective effect of natural products against Alzheimer's disease. Neurochem Res 37:1829–1842

Evans SV, Shing TK, Aplin RT, Fellows LE, Fleet GW (1985) Sulphate ester of trans-4-hydroxypipecolic acid in seeds of peltophor. Phytochem Rev 24:2593–2596

Fathima A, Arafath Y, Hassan S, Bramhachari PV, Kiran GS, Selvin J (2022) Novel insights into environmental niche adaptations and secondary metabolite biosynthesis potential of marine sponge microbiome. In: Veera Bramhachari P (ed) Understanding the microbiome interactions in agriculture and the environment. Springer, Singapore. https://doi.org/10.1007/978-981-19-3696-8_8

Feng Y, Carroll AR, Pass DM, Archbold JK, Avery VM, Quinn RJ (2008) Polydiscamides B–D from a marine sponge Ircinia sp. as potent human sensory neuron-specific G protein coupled receptor agonists. J Nat Prod 71:8–11

Forenza S, Minale L, Riccio R, Fattorusso EJ (1971) New bromo-pyrrole derivatives from the sponge Agelas oroides. J Chem Soc D18:1129–1130

Franco-Iborra S, Vila M, Perier C (2018) Mitochondrial quality control in neurodegenerative diseases: focus on Parkinson's disease and Huntington's disease. Front Neurosci 12:342

Gautam M, Acharya D, Ali Bhat Z, Kumar D (2017) Future leads: natural products as anti-infective agent. J Nat Prod 7:84–96

Gogineni V, Hamann MT (2018) Marine natural product peptides with therapeutic potential: chemistry, biosynthesis, and pharmacology. Biochim Biophys Acta Gen Subj 1862:81–196

Gordillo D, Diaz I, Martinez-Gil A, del Pliego G, Huerta A, Puerto MJ, Aparicio E, Navarro D, Padilla M, Noscira SA (2004) GSK-3 inhibitors isolated from marine organisms. U.S. Patent Application 10/596

Grosso C (2014) Bioactive marine drugs and marine biomaterials for brain diseases. Mar Drugs 12: 2539–2589

Guo Y, Madaio A, Trivellone E, Scognamiglio G, Cimino G (1996) Further studies of alkaloids from Reniera sarai: structures of saraine-3 and isosaraine-3, absolute stereochemistry of saraine-1 and saraine-2. Tetrahedron 52:14961–14974

Guo Y, Trivellone E, Scognamiglio G, Cimino G (1998) Absolute stereochemistry of isosaraine-1 and isosaraine-2. Tetrahedron Lett 39:463–466

Haefner B (2003) Drugs from the deep: marine natural products as drug candidates. Drug Discov Today 8:1359–6446

Hale KJ, Domostoj MM, Tocher DA, Irving E, Scheinmann F (2003) Enantiospecific formal total synthesis of the tumor and GSK-3β inhibiting alkaloid, (−)-Agelastatin A. Org Lett 5:2927–2930

Hamann M, Alonso D, Martín-Aparicio E, Fuertes A, Pérez-Puerto MJ, Castro A, Morales S, Navarro ML, del Monte-Millán M, Medina M, Pennaka H (2007) Glycogen synthase kinase-3 (GSK-3) inhibitory activity and structure activity relationship (SAR) studies of the manzamine alkaloids. Potential for Alzheimer's disease. J Nat Prod 70:1397–1405

Hentschel U, Bringmann G (2010) Marine Schwämme: Füllhorne der Meere. Pharm Unserer Zeit 39:62–66

Hentschel U, Usher KM, Taylor MW (2006) Marine sponges as microbial fermenters. FEMS Microbiol Ecol 55:167–177

Horikawa K, Yagyu T, Yoshioka Y, Fujiwara T, Kanamoto A, Okamoto T, Ojika M (2013) Petrosiols A-E, neurotrophic diyne tetraols isolated from the Okinawan sponge Petrosia strongylata. Tetrahedron 69:101–106

Huang C, Zhang Z, Cui W (2019) Marine derived natural compounds for the treatment of Parkinson's disease. Mar Drugs 17:221

Jha RK, Zi-Rong X (2004) Biomedical compounds from marine organisms. Mar Drugs 2:123–146

Joseph B, Nair VM, Sujatha S (2012) Pharmacological perception of peptides from marine sponge: a review. IJPSR 3:4689–4696

Kashman Y, Groweiss A, Shmeuli U (1980) Latrunculin, a new 2-thiazolidinone macrolide from the marine sponge Latrunculia magnifica. Tetrahedron Lett 21:3629–3632

Kawahara T, Takagi M, Shin-ya K (2012) JBIR-124: a novel antioxidative agent from a marine sponge-derived fungus Penicillium citrinum SpI080624G1f01. J Antibiot 65:45–47

Khanfar MA, Asal BA, Mudit M, Kaddoumi A, El Sayed KA (2009) The marine natural-derived inhibitors of glycogen synthase kinase-3b phenylmethylene hydantoins: in vitro and in vivo activities and pharmacophore modelling. Bioorg Med Chem 17:6032–6039

Kiran GS, Ramasamy P, Sekar S, Ramu M, Hassan S, Ninawe AS, Selvin J (2018a) Synthetic biology approaches: towards sustainable exploitation of marine bioactive molecules. Int J Biol Macromol 112:1278–1288

Kiran GS, Sekar S, Ramasamy P, Thinesh T, Hassan S, Lipton AN et al (2018b) Marine sponge microbial association: towards disclosing unique symbiotic interactions. Mar Environ Res 140: 169–179

Kochanowska-Karamyan AJ, Hamann MT (2010) Marine indole alkaloids: potential new drug leads for the control of depression and anxiety. Chem Rev 110:4489–4497

Kossuga MH, Nascimento AM, Reimão JQ, Tempone AG, Taniwaki NN, Veloso K, Ferreira AG, Cavalcanti BC, Pessoa C, Moraes MO, Mayer AM (2008) Antiparasitic, antineuroinflammatory, and cytotoxic polyketides from the marine sponge Plakortis angulospiculatus collected in Brazil. J Nat Prod 71:334–339

Kubota T, Ishiguro Y, Yamamoto S (2010) Platisidines A-C, N-methylpyridinium alkaloids from an Okinawan marine sponge of Plakortis species. Heterocycles 80:1407–1412

Lai HC, Jan LY (2006) The distribution and targeting of neuronal voltage-gated ion channels. Nat Rev Neurosci 7:548–562

Lee YJ, Cho Y, Tran HN (2021) Secondary metabolites from the marine sponges of the genus Petrosia: a literature review of 43 years of research. Mar Drugs 19:122

Li H, Bowling JJ, Fronczek FR, Hong J, Jabba SV, Murray TF, Ha NC, Hamann MT, Jung JH (2013) Asteropsin A: an unusual cystine-crosslinked peptide from porifera enhances neuronal Ca2+ influx. Biochim Biophys Acta 1830:2591–2599

Lit M (2006) A marine literature database produced and maintained by the Department of Chemistry. University of Canterbury, Canterbury

Long BH, Carboni JM, Wasserman AJ, Cornell LA, Casazza AM, Jensen PR, Lindel T, Fenical W, Fairchild CR (1998) Eleutherobin, a novel cytotoxic agent that induces tubulin polymerization, is similar to paclitaxel. Cancer Res 58:1111–1115

Malloy KL (2011) Structure elucidation of biomedically relevant marine cyanobacterial natural products. University of California, San Diego

Mandelkow EM, Mandelkow E (1998) Tau in Alzheimer's disease. Trends Cell Biol 8:425–427

Mayer A, Rodriguez A, Taglialatela-Scafati O, Fuselani N (2013) Marine pharmacology in 2009–2011: marine compounds with antibacterial, antidiabetic, antifungal, anti-inflammatory, antiprotozoal, antituberculosis, and antiviral activities; affecting the immune and nervous systems, and other miscellaneous mechanisms of action. Mar Drugs 11:2510–2573

Mehbub M, Lei J, Franco C, Zhang W (2014) Marine sponge derived natural products between 2001 and 2010: trends and opportunities for discovery of bioactives. Mar Drugs 12:4539–4577

Meijer L, Thunnissen AM, White AW, Garnier M, Nikolic M, Tsai LH, Walter J, Cleverley KE, Salinas PC, Wu YZ, Biernat J (2000) Inhibition of cyclin-dependent kinases, GSK-3β and CK1 by Hymenialdisine, a marine sponge constituent. Chem Biol 7:51–63

Millán-Aguiñaga N, Soria-Mercado IE, Williams P (2010) Xestosaprol D and E from the Indonesian marine sponge Xestospongia sp. Tetrahedron Lett 51:751–753

Mioso R, Marante F, Bezerra R, Borges F, Santos B, Laguna I (2017) Cytotoxic compounds derived from marine sponges. A review (2010–2012). Molecules 22:208
Molinski TF, Fahy E, Faulkner DJ, Van Duyne GD, Clardy J (1988) Petrosamine, a novel pigment from the marine sponge Petrosia sp. J Org Chem 53:1340–1341
Monks NR, Lerner C, Henriques AT, Farias FM, Schapoval EE, Suyenaga ES, da Rocha AB, Schwartsmann G, Mothes B (2002) Anticancer, antichemotactic and antimicrobial activities of marine sponges collected off the coast of Santa Catarina, southern Brazil. J Exp Mar Biol Ecol 28(1):1–12
Mora C, Tittensor DP, Adl S, Simpson AG, Worm B (2011) How many species are there on earth and in the ocean? PLoS Biol 9:e1001127
Morales JJ, Rodrigue AD (1991) The structure of Clathrodin, a novel alkaloid isolated from the Caribbean Sea sponge Agelas Clathrodes. J Nat Prod 54:629–631
Moroni F, Galli A, Mannaioni G, Carla V, Cozzi A, Mori F, Marinozzi M, Pellicciari R (1995) NMDA receptor heterogeneity in mammalian tissues: focus on two agonists, (2S, 3R, 4S) cyclopropylglutamate and the sulfate ester of 4-hydroxy-(S)-pipecolic acid. Naunyn Schmiedebergs Arch Pharmacol 351:371–376
Munro MH, Blunt JW, Dumdei EJ, Hickford SJ, Lill RE, Li S, Battershill CN, Duckworth AR (1999) The discovery and development of marine compounds with pharmaceutical potential. J Biotechnol 70:15–25
Murakami N, Nakajima T, Kobayashi M (2001) Total synthesis of Lembehyne A, a neuritogenic spongean polyacetylene. Tetrahedron Lett 42:1941–1943
Nukoolkarn VS, Saen-oon S, Rungrotmongkol T, Hannongbua S, Ingkaninan K, Suwanborirux K (2008) Petrosamine, a potent anticholinesterase pyridoacridine alkaloid from a Thai marine sponge Petrosian. sp. Bioorg Med Chem 16:6560–6567
Orhan IE, Ozcelik B, Konuklugil B, Putz A, Kaban UG, Proksch P (2012) Bioactivity screening of the selected Turkish marine sponges and three compounds from Agelas oroides. Rec Nat Prod 6: 356–367
Paleari L, Trombino S, Falugi C, Gallus L, Carlone S, Angelini C, Sepcic K, Turk T, Faimali M, Noonan DM, Albini A (2006) Marine sponge derived polymeric alkylpyridinium as a novel tumor chemotherapeutic targeting the cholinergic system lung tumors. Int J Oncol 29:1381–1388
Pandey MK, Degrado TR (2016) Glycogen synthesis kinase-3 (GSK-3)-targeted therapy and imaging. Theranostics 6:571–593
Peigneur S, Žula A, Zidar N, Chan-Porter F, Kirby R, Madge D, Ilaš J, Kikelj D, Tytgat J (2014) Action of clathrodin and analogues on voltage-gated sodium channels. Mar Drugs 12:2132–2143
Pereira AR, Cao Z, Engene N, Soria-Mercado IE, Murray TF, Gerwick WH (2010) Palmyrolide A, an unusually stabilized neuroactive macrolide from Palmyra atoll cyanobacteria. Org Lett 12: 4490–4493
Perovic S, Wichels A, Schütt C, Gerdts G, Pahler S, Steffen R, Müller WE (1998) Neuroactive compounds produced by bacteria from the marine sponge Halichondria panicea: activation of the neuronal NMDA receptor. Environ Toxicol Pharmacol 6:125–133
Perry NB, Ettouati L, Litaudon M, Blunt JW, Munro MH, Parkin S, Hope H (1994) Alkaloids from the antarctic sponge Kirkpatrickia varialosa: part 1: variolin B, a new antitumor and antiviral compound. Tetrahedron 50:3987–3992
Petersen F, Zähner H, Metzger JW, Freund S, Hummel RP (1993) Germicidin, an autoregulative germination inhibitor of Streptomyces viridochromogenes, NRRLB-1551. J Antibiot 46:1126–1138
Pietra F (1997) Secondary metabolites from marine microorganisms: bacteria, protozoa, algae and fungi. Nat Prod Rep 14:453–464
Plisson F, Prasad P, Xiao X, Piggott AM, Huang XC, Khalil Z, Capon RJ (2014) Callyspongisines A–D: bromopyrrole alkaloids from an Australian marine sponge, Callyspongia sp. Org Biomol Chem 12:1579–1584

Priyadharshini GB, Jaynthy C, Kiran GS, Selvin J (2021) In vitro analysis of neuroprotective and antioxidant property of sponge extract. Ann Rom Soc Cell Biol 7760–7771
Proksch P, Edrada RA, Ebel R (2002) Drugs from the seas: current status and microbiological implications. Appl Microbiol Biotechnol 59:125–134
Rentas AL, Rosa R, Rodríguez AD, De Motta GE (1995) Effect of alkaloid toxins from tropical marine sponges on membrane sodium currents. Tooxwon 33:491–497
Roll DM, Ireland CM, Lu HS, Clardy J (1988) Fascaplysin, an unusual antimicrobial pigment from the marine sponge Fascaplysinopsis sp. J Org Chem 53:3276–3278
Ruocco N, Esposito R, Zagami G, Bertolino M, De Matteo S, Sonnessa M, Andreani F, Crispi S, Zupo V, Costantini M (2021) Microbial diversity in Mediterranean sponges as revealed by metataxonomic analysis. Sci Rep 11:1–12
Sagar S, Kaur M, Minneman KP (2010) Antiviral lead compounds from marine sponges. Mar Drugs 10:2619–2638
Sakai R, Higa T, Jefford CW, Bernardinelli G (1986) Manzamine A, a novel antitumor alkaloid from a sponge. J Am Chem Soc 108:6404–6405
Sakai R, Koike T, Sasaki M, Shimamoto K, Oiwa C, Yano A, Suzuki K, Tachibana K, Kamiya H (2001) Isolation, structure determination, and synthesis of neodysiherbaine A, a new excitatory amino acid from a marine sponge. Org Lett 3:1479–1482
Sakai R, Suzuki K, Shimamoto K, Kamiya H (2004) Novel betaines from a micronesian sponge Dysidea herbacea. J Org Chem 69:1180–1185
Sakai R, Suzuki K, Shimamoto K, Kamiya H (2010) Novel N-methylated 8-oxoisoguanines from pacific sponges with diverse neuroactivities. J Med Chem 53:6089–6099
Sakurada T, Gill MB, Frausto S, Copits B, Noguchi K, Shimamoto K, Swanson GT, Sakai R (2010) Novel N-methylated 8-oxoisoguanines from Pacific sponges with diverse neuroactivities. J Med Chem 53:6089–6099
Schneemann I, Nagel K, Kajahn I, Labes A, Wiese J, Imhoff JF (2010) Comprehensive investigation of marine actinobacteria associated with the sponge Halichondria panacea. Appl Environ Microbiol 76:3702–3714
Schumacher M, Kelkel M, Dicato M, Diederich M (2011) Gold from the sea: marine compounds as inhibitors of the hallmarks of cancer. Biotechnol Adv 29:531–547
Senthilkumar K, Kim SK (2013) Marine invertebrate natural products for anti-inflammatory and chronic diseases. Evid Based Complement Alternat Med 2013:572859
Shin J, Seo Y, Cho KW, Rho JR, Paul VJ (1998) Osirisynes A-F, highly oxygenated polyacetylenes from the sponge Haliclona Osiris. Tetrahedron 54:8711–8720
Singhal AK, Naithani V, Bangar OP (2012) Medicinal plants with a potential to treat Alzheimer and associated symptoms. J Nutr Pharmacol Neurol Dis 2:84–91
Sipkema D, Franssen MC, Osinga R, Tramper J, Wijffels RH (2005) Marine sponges as pharmacy. Mar Biotechnol 7:142–162
Skropeta D, Pastro N, Zivanovic A (2011) Kinase inhibitors from marine sponges. Mar Drugs 9: 2131–2154
Swanson GT, Sakai R (2009) Ligands for ionotropic glutamate receptors. Prog Mol Subcell Biol 46: 123–157
Taylor MW, Radax R, Steger D, Wagner M (2007) Sponge-associated microorganisms: evolution, ecology, and biotechnological potential. Microbiol Mol Biol Rev 71:295–347
Tello-Aburto R, Johnson EM, Valdez CK, Maio WA (2012a) Asymmetric total synthesis and absolute stereochemistry of the neuroactive marine macrolide palmyrolide A. Org Lett 14:2150–2153
Tello-Aburto R, Newar TD, Maio WA (2012b) Evolution of a protecting-group-free total synthesis: studies en route to the neuroactive marine macrolide (−)-palmyrolide A. J Org Chem 77:6271–6289
Thakur NL, Müller WE (2004) Biotechnological potential of marine sponges. Curr Sci 86:1506–1512

Thomas TRA, Kavlekar DP, Lokabharathi PA (2010) Marine drugs from sponge-microbe association—a review. Mar Drugs 8:1417–1468
To KC, Loh KT, Roskelley CD, Andersen RJ, O'Connor TP (2006) The anti-invasive compound motuporamine C is a robust stimulator of neuronal growth cone collapse. J Neurosci 139:1263–1274
Trimmer JS, Rhodes KJ (2004) Localization of voltage-gated ion channels in mammalian brain. Annu Rev Physiol 66:477–519
Trimurtulu G, Faulkner DJ, Perry NB, Ettouati L, Litaudon M, Blunt JW, Munro MH, Jameson GB (1994) Alkaloids from the antarctic sponge Kirkpatrickia varialosa. Part 2: variolin A and N (3′)-methyl tetrahydro-variolin B. Tetrahedron 50:3993–4000
Trofimova I, Robbins TW (2016) Temperament and arousal systems: a new synthesis of differential psychology and functional neurochemistry. Neurosci Biobehav Rev 64:382–402
Uttara B, Singh AV, Zamboni P, Mahajan RT (2009) Oxidative stress and neurodegenerative diseases: a review of upstream and downstream antioxidant therapeutic options. Curr Neuropharmacol 7:65–74
Van-Lear GE, Morton GO, Fulmor W (1973) New bacterial bromo indole metabolites from the marine sponge Polyfibrospongia maynardii. Tetrahedron Lett 14:299–300
Vippila MR, Nikhar S, Gracia AP, Cuny GD (2016) Divergent approach for the synthesis of Gombamide A and derivatives. Org Lett 18:4726–4729
Wadsworth AD, Furkert DP, Sperry J, Brimble MA (2012) Total synthesis of the initially reported and revised structures of the neuroprotective agent palmyrolide A. Org Lett 14:5374–5377
Webster NS, Taylor MW (2012) Marine sponges and their microbial symbiont: love and other relationships. Environ Microbiol 14:335–346
Woo JK, Jeon JE, Kim CK, Sim CJ, Oh DC, Oh KB, Shin J (2013) Gombamide A, acyclic thiopeptide from the sponge Clathriagom bawuiens. J Nat Prod 76:1380–1383
Yak E, Qian ZM (2003) Iron mis-regulation in the brain: a primary cause of neurodegenerative disorders. Lancet Neurol 2:246–253
Zhang H, Conte MM, Khalil Z, Huang XC, Capon RJ (2012) New dictyodendrins as BACE inhibitors from a southern Australian marine sponge, Ianthella sp. RSC Adv 2:4209–4214

SPRINGER NATURE

GPSR Compliance

The European Union's (EU) General Product Safety Regulation (GPSR) is a set of rules that requires consumer products to be safe and our obligations to ensure this.

If you have any concerns about our products, you can contact us on ProductSafety@springernature.com

In case Publisher is established outside the EU, the EU authorized representative is:

Springer Nature Customer Service Center GmbH
Europaplatz 3
69115 Heidelberg, Germany

The manufacturer's authorised representative in the EU is Springer Nature Customer Service Centre GmbH, Europaplatz 3, 69115 Heidelberg, Germany. If you have any concerns regarding our products, please contact ProductSafety@springernature.com

Printed and bound by CPI Group (UK) Ltd, Croydon, CR0 4YY
30/09/2026
02225324-0011